Lecture Notes in Computer Science 6083

Ching-Hsien Hsu Victor Malyshkin (Eds.)

Methods and Tools of Parallel Programming Multicomputers

Second Russia-Taiwan Symposium, MTPP 2010
Vladivostok, Russia, May 16-19, 2010
Revised Selected Papers

Volume Editors

Ching-Hsien Hsu
Chung Hua University
Department of Computer Science
and Information Engineering
Hsinchu, 300 Taiwan, China
E-mail: chh@chu.edu.tw

Victor Malyshkin
Supercomputer Software Department
Institute of Computational Mathematics
and Mathematical Geophysics
630090 Novosibirsk, Russia
E-mail: malysh@ssd.sscc.ru

Library of Congress Control Number: 2010931448

CR Subject Classification (1998): C.2.4, H.4, D.2, F.2, D.4

LNCS Sublibrary: SL 2 – Programming and Software Engineering

ISSN 0302-9743
ISBN-10 3-642-14821-2 Springer Berlin Heidelberg New York
ISBN-13 978-3-642-14821-7 Springer Berlin Heidelberg New York

springer.com

Printed in Germany

Typesetting: Camera-ready by author, data conversion by Scientific Publishing Services, Chennai, India
Printed on acid-free paper 06/3180

Preface

It is our great pleasure to present the proceedings of the second Russia–Taiwan Symposium on Methods and Tools of Parallel Programming (MTPP 2010).

MTPP is the main regular event of the Russia–Taiwan scientific forum that covers the many dimensions of methods and tools of parallel programming, algorithms and architectures, encompassing fundamental theoretical approaches, practical experimental projects, and commercial components and systems. As applications of computing systems have permeated every aspect of daily life, the power of computing systems has become increasingly critical. Therefore, MTPP is intended to play an important role allowing researchers to exchange information regarding advancements in the state of the art and practice of IT-driven services and applications, as well as to identify emerging research topics and define the future directions of parallel computing.

We received a large number of high-quality submissions this year. In the first stage, all papers submitted were screened for their relevance and general submission requirements. These manuscripts then underwent a rigorous peer-review process with at least three reviewers per paper. At the end, 33 papers were accepted for presentation and included in the main proceedings. To encourage and promote the work presented at MTPP 2010, we are delighted to inform the authors that some of the papers will be accepted in special issues of the *Journal of Supercomputing*, which has played a prominent role in promoting the development and use of parallel and distributed processing.

The success of MTPP 2010 required the support of many people. First of all, we would like to thank all advisory committees, for nourishing the symposium and guiding its course. We appreciate the participation of all delegates whose speeches greatly benefited the audience. We are also indebted to the members of the Program Committee, who put in hard work and long hours to review each paper in a professional way. Thanks to them all for their valuable time and effort in reviewing the papers. Without their help, this program would not have been possible. Special thanks go to Maxim Gorodnichev and Andrey Kovtanyuk for their help with the symposium website, administrative matters and much detailed work, which facilitated the overall process. Thanks also go to the entire local Arrangements Committee for their help in making the symposium a wonderful success. We take this opportunity to thank all the authors, participants and Session Chairs for their valuable efforts, many of whom traveled long distances to attend this symposium and make their valuable contributions. Last but not least, we would like express our gratitude to all of the organizations that supported our efforts to bring the symposium to fruition. We are grateful to Springer for publishing the proceedings.

The symposium was held in the beautiful city of Vladivostok, which provided our guests with ample natural and cultural beauty to enjoy apart from the events of the symposium. This symposium owes its success to the support of many academic and industrial organizations, and most importantly, and to all the symposium participants. We are proud to share these proceedings with you and hope you enjoy them.

May 2010

Victor Malyshkin
Ching-Hsien (Robert) Hsu
Sergey M. Abramov
Chu-Sing Yang

Organization

Symposium Committees

General Chairs

Victor Malyshkin	RAS, Russia
Robert C. H. Hsu	Chung Hua University, Taiwan

Program Chairs

Sergey M. Abramov	Institute of Program Systems, RAS, Russia
Chu-Sing Yang	National Cheng Kung University, Taiwan

Advisory Committee

Yuri Kulchin	Far East Branch of RAS, Russia
Vladimir Kurilov	Far East State University, Russia
Viktor K. Prasanna	University of Southern California, USA
Sartaj Sahni	University of Florida, USA
Yeh-Ching Chung	National Tsing Hua University, Taiwan
Sergei Smagin	Far East Branch of RAS, Russia
Ivan Stojmenovic	University of Ottawa, Canada
Satoshi Matsuoka	Tokyo Institute of Technology, Japan
Peter Sloot	University of Amsterdam, The Netherlands
Hai Jin	Huazhong University of Science and Technology, China
Pen-Chung Yew	Academia Sinica, Taiwan
Marcin Paprzycki	SWPS and IBS PAN, Warsaw, Poland
Laurence T. Yang	St.Francis Xavier University Canada

Publication Chairs

Ol'ga L. Bandman	RAS, Russia

Publicity Chairs

Hsi-Ya Chang	National Center for High-Performance Computing, Taiwan
Ren-Hung Hwang	National Chung Cheng University, Taiwan
Alekcander P. Vazhenin	University of Aizu, Japan

Registration Chairs

Maxim A. Gorodnichev	Novosibirsk State Technical University, Russia
Margarita Knyazeva	Far East State University, Russia
Georgi Tarasov	Far East Branch of RAS, Russia
Eugenia Nikitina	Far East State University, Russia
Valeria Gribova	Far East Branch of RAS, Russia

Local Arrangements Chairs

Boris Reznik	Far East State University, Russia
Igor Soppa	Far East State University, Russia
Mikhail Guzev	Far East Branch of RAS, Russia
Andrey Kovtanyuk	Far East Branch of RAS, Russia
Alexander Kolobov	Far East State University, Russia
Alexander Abramov	Far East State University, Russia
Marina Aleksanina	Far East Branch of RAS, Russia
Valeriy Dikarev	Far East State University, Russia
Vladimir Korochencev	Far East State University, Russia
Andrey Velichko	Far East Branch of RAS, Russia
Natalia Shamray	Far East Branch of RAS, Russia
Alexander Oleinikov	Far East Branch of RAS, Russia
Igor Prokhorov	Far East Branch of RAS, Russia
Anton Zhuplev	Far East State University, Russia
Alexander Shaturin	Far East State University, Russia
Polina Zamorova	Far East State University, Russia
Ekaterina Gerasimenko	Far East State University, Russia

International Program Committee

Raj Buyya	Melbourne University / Manjrasoft, Australia
Hsi-Ya Chang	National Center for High Performance Computing, Taiwan
Barbara Chapman	University of Houston, USA
Jinjun Chen	Swinburne University of Technology, Australia
Wenguang Chen	Tsinghua University, China
Boris N. Chetverushkin	Institute of Applied Mathematics RAS, Russia
Geoffrey Fox	Indiana University, USA
Maria Ganzha	System Research Institute Polish Academy of Sciences (SRI PAS) and University of Gdansk, Poland
Yuri G. Karpov	Technical University of Saint Petersburg RAS, Russia
Chung-Ta King	National Tsing Hua University, Taiwan
Jenq-Kuen Lee	National Tsing Hua University, Taiwan
Alexander I. Legalov	Federal State University of Krasnoyarsk, Russia
Kuan-Ching Li	Providence University, Taiwan

Pangfeng Liu	National Taiwan University, Taiwan
T. Ludwig	German Climate Computing Centre, University Hamburg, German
Nikolay N. Mirenkov	University of Aizu, Japan
Vitalii Novoseltsev	Institute of Program Systems, RAS, Russia
Eugeni Nurminskii	Far East Branch of RAS, Russia
Omer F. Rana	Cardiff University, UK
Yu-Chee Tseng	National Chiao Tung University, Taiwan
Cho-Li Wang	Hong Kong University, Hong Kong
Roman Wyrzykowski	Czestochowa University of Technology, Poland
Jingling Xue	University of New South Wales, Australia
Chao-Tung Yang	Tung-Hai University, Taiwan
Wuu Yang	National Chiao-Tung University, Taiwan
Kun-Ming Yu	Chung Hua University, Taiwan
Albert Zomaya	University of Sydney, Australia

Table of Contents

Languages, Systems and Optimizations

Multi-core Technologies, Embedded Systems and Compilers

Cluster Computing

Cloud and Grid Computing

Simulations and Modeling

Algorithms and Applications

Performance Measurement and Analysis

Optimization of Parallel Execution of Numerical Programs in LuNA Fragmented Programming System

Victor Malyshkin and Vladislav Perepelkin

Institute of Computational Mathematics and Mathematical Geophysics
Russian Academy of Sciences
Prospekt Akademika Lavrentjeva 6, Novosibirsk, Russia
{malysh,perepelkin}@ssd.sscc.ru

Abstract. Organization of high performance execution of fragmented programs met the problem of choice of acceptable way of their execution. The possibilities of execution optimization on the stages of fragmented program development, compilation and execution are considered. The methods and algorithms of optimizations are suggested to be included both in fragmented programming language and in run-time system.

Keywords: parallel programming, fragmented programming, high performance computing, program execution optimization.

1 Introduction

The idea of data and algorithms fragmentation is exploited in programming at least from the early 1970th [1–8]. Generally, the model of a program in this approach looks as follows. The data are fragmented and values of the simple (atomic) variables can be the data aggregates (*data fragment* (DF)) that usually reflect the essence of an object domain. For example, a cell of a 3D-mesh in Particle-In-Cell method can be considered as atomic part of the description of the minimal part of a simulated phenomenon. The variables' description can reflect the structure of its value, i.e., a DF structure. In particular, in numerical algorithms a sub-matrix of a matrix can be defined as a DF and the whole matrix is represented as an array of its sub-matrices. An operation computes the values of output variables from the input variables. An operation plus input and output variables is called a *fragment of computation* (FC) (Fig. 1). DFs have unique names that predefine single assignment mode of programming. Therefore, if two or more FCs compute the value of a certain variable, then the same DF is yielded by each FC as the value of this variable (fig. 1.b and 1.c). This is the restriction on the set of permitted interpretations. Any FC has also unique name. A *fragmented program* (FP) is represented as a computable set of FC. A FC can be executed once if certain values are assigned to all of its input variables. Formal definitions and more details can be found in [3].

C.H. Hsu and V. Malyshkin (Eds.): MTPP 2010, LNCS 6083, pp. 1–10, 2010.

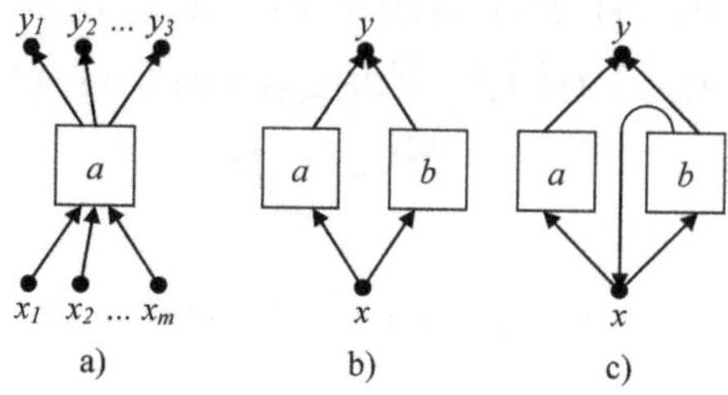

Fig. 1. Fragments of computation and data fragments

Different modifications of this general model were embodied in programming systems as commercial [2, 4] or academic [5] product. In [2] instead of usually used run-time system for FCs execution a special operating system was developed. Many programming systems use run-time systems for computation organization [6–12]. Problems of the set of the FCs execution are well known and shortly can be formulated as:

a. dynamic resources allocation,
b. dynamic DFs and FCs distribution and their migration among the processor elements (PE) of a multicomputer,
c. dynamic choice of a certain FC for execution.

Obviously, the algorithms, used to overcome the problems above, cardinally influence on providing such important properties of a program as dynamic tunability to all the available resources, dynamic load balancing, data transfer in parallel with computations and so on.

Taking into account the necessity to solve the above listed problems, basing on the experience of the other related developments, we started the development of our fragmented programming system LuNA, based on the theory of parallel program synthesis [3]. The system is oriented to the development of FPs, implementing large scale numerical models. First planned application of LuNA is the creation of parallel numerical subroutine library. Every subroutine should be automatically provided with all the necessary dynamic properties. Aside from the different problems of the LuNA creation we are concentrated here on the problem of high performance execution of a FP.

2 LuNA Model of a Program

The general model is modified in order to meet LuNA needs. LuNA is the programming system, not the system of program synthesis. Therefore, there is no necessity to include in the LuNA model the single assignment. In order to facilitate the resources allocation and the data distribution the variables are defined similar to variables definition in programming languages, i.e., the name of a variable denotes the memory extent where different values are kept at different moments.

Contrary to variable's names an operation name denotes a certain execution of an operation. In particular, if an operation (procedure) b should be applied to every entry of the array x (fig. 2.a), then i-th execution of b is denoted as b^i . Two sets of the FCs are defined in fig. 2.a: $\{a^i, i = 1, 2, 3, \ldots\} = \{i = 1, 2, 3, \ldots | x[i] \rightarrow a^i \rightarrow x[i+1]\}$, $\{b^i, i = 1, 2, 3, \ldots\} = \{i = 1, 2, 3, \ldots | x[i] \rightarrow b^i \rightarrow y[i]\}$. Each FC is named by the indexed name of its operation.

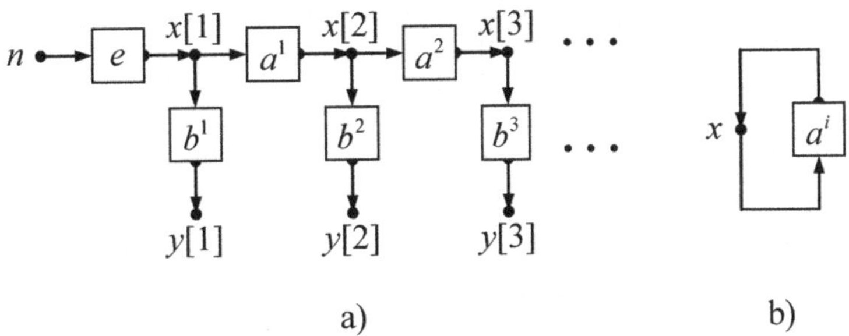

Fig. 2. Sample FP (a) and loop computations (b)

Loopwise computations are shown in the fig. 2.b. The set of the FCs $\{i = 1, 2, 3, \ldots | x \rightarrow a^i \rightarrow x\}$ is defined. The order of the FCs execution is defined by the binary partial order relation $\rho = \{i = 1, 2, 3, \ldots | < a^i, a^{i+1} >\}$ (*relation* ρ). This means, that FC a^i should be executed before a^{i+1}. For computation in fig. 2.a the order of the FCs execution is defined by the information dependencies between operations whereas for computations in fig. 2.b the order should be defined explicitly. The resources allocation for the array x implementation is done in fig. 2.b by a user.

A certain FC can be chosen for execution, if its execution does not contradict to the relation ρ. The relation ρ should guarantee the correct execution of the FP.

As result, LuNA language contains at least the facilities for definition of the DFs, the FCs and the relation ρ on the set of FCs. Therefore, the LuNA user has the possibility to define at least partially the resources allocation (see fig. 2.b). This might substantially improve the results of automatic resources allocation and the FCs distribution done by LuNA compiler and run-time system.

3 Fragmented Algorithms and Their Execution

3.1 Matrices Multiplication

Fragmented version of the algorithm of two square matrices A and B multiplication, $C = A \times B$, is considered. Matrices are represented by the square $K \times K$ matrices of square sub-matrices $A_{i,k}$, $B_{k,j}$, $C_{i,j}$ (see fig. 3). Sub-matrices $A_{i,k}$, $B_{k,j}$, $C_{i,j}$ are the DFs here.

Intermediate values are kept in DFs $C_{i,j,k}$. The FCs $\mathcal{F}_{i,j,k}$ and $\mathcal{S}_{i,j}$ perform matrices multiplication $A_{i,k} \times B_{k,j} = C_{i,j,k}$ and summation $C_{i,j} = \sum_{k=1}^{K} C_{i,j,k}$ respectively. Necessary order to compute the product C correctly is $\mathcal{F}_{i,j,k} < \mathcal{S}_{i,j} \forall i, j$.

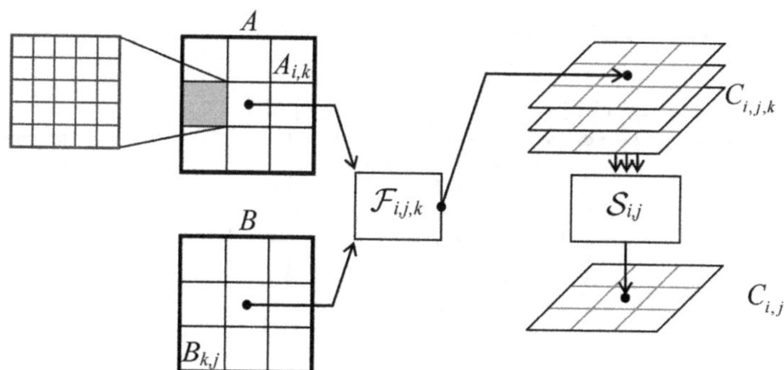

Fig. 3. Schema of fragmented algorithm of matrices multiplication

Run-time system chooses a certain FC for execution in any order, which does not contradict to the relation ρ . In this case the correct result of the FP execution will be produced, but the FP execution performance might be poor. For example, execution of any FC $\mathcal{F}_{i,j,k}$ produces a DF $C_{i,j,k}$, therefore some memory extent should be allocated to keep its value. On the other hand, after $\mathcal{S}_{i,j}$ execution the memory, allocated for the DFs $C_{i,j,k}$, is released. Run-time system should take this into account when FC is chosen for execution, otherwise the computer memory might be exhausted not productively. Good (recommended) order would be the one with FCs $\mathcal{S}_{i,j}$ executed as soon as possible (but only after all the $\mathcal{F}_{i,j,k}$ with the same i and j are finished).

Another problem here is data distribution. To what PE a certain DF should be allocated? Random distribution results in huge communications and some PEs can be idle due to load imbalance. In LuNA there are opportunities, which are considered in section 4, to control the DFs distribution and migration to provide good performance.

3.2 LU-Factorization

Another example is the fragmentation of LU-factorization algorithm. Square $n \times n$ matrix A is factorized into lower triangular matrix L and upper triangular matrix U, $A = L \times U$.

$$\begin{array}{l} u_{1,j} = a_{1,j}, j = 1, \ldots, n \\ l_{j,1} = \frac{a_{j,1}}{u_{1,1}}, j = 2, \ldots, n \\ u_{i,j} = a_{i,j} - \sum_{k=1}^{i-1} l_{i,k} u_{k,j}, i = 2, \ldots, n; j = i, \ldots, n \\ l_{j,i} = \frac{1}{u_{i,j}} (a_{j,i} - \sum_{k=1}^{i-1} l_{j,k} u_{k,j}) i = 2, \ldots, n; j = i+1, \ldots, n \end{array} \tag{1}$$

Matrix A is represented by the $K \times K$ matrix of $A_{i,j}$ sub-matrices. Matrices L and U are both represented by the same $K \times K$ matrix of $L_{i,j}$ and $U_{i,j}$ sub-matrices (fig. 4.a). They are the DFs. Each DF $A_{i,j}$ is processed in accordance to formulas (1) by the FCs $\mathcal{L}_{i,j}$, $\mathcal{D}_i$ or $\mathcal{U}_{i,j}$, for lower diagonal, diagonal and upper diagonal DFs respectively, as shown in fig. 4.a. The relation ρ is defined by the information dependencies in (1) as follows: The FCs $\mathcal{D}_i$ should be executed after $\mathcal{L}_{i,j-1}$ and $\mathcal{U}_{i-1,j}$, the FCs $\mathcal{U}_{i,j}$ should be executed after $\mathcal{U}_{i-1,j}$ and $\mathcal{D}_i$, and the FCs $\mathcal{L}_{i,j}$ should be executed after $\mathcal{L}_{i,j-1}$ and $\mathcal{D}_j$ (fig 4.b).

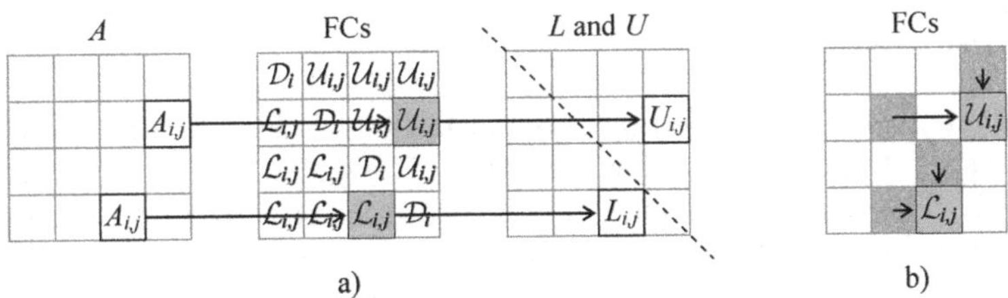

Fig. 4. LU factorization. General schema (a) and order schema (b).

It is clear from 4.b, that execution of the FC $\mathcal{D}_i$ increases the number of the FCs ready to be executed (*ready FC*), whereas the execution of the other FCs reduces it. To exploit the algorithm's parallelism the set of the ready FCs must be large enough to load all the PEs over time. Therefore, run-time system should provide the execution of the FCs $\mathcal{D}_i$ before the other FCs' execution. The fig. 5 illustrates two FC choice algorithms. The first is to choose the $\mathcal{D}_i$ FCs last. It leads to situations, when only one FC is ready (fig. 5.a), and all the PEs are idle waiting for only one FC to be completed. The second algorithm chooses the FCs according to the diagonal front, as shown in fig. 5.b. It provides more parallelism and enables more effective program execution. The opportunities to control the FC choice are considered in section 4.

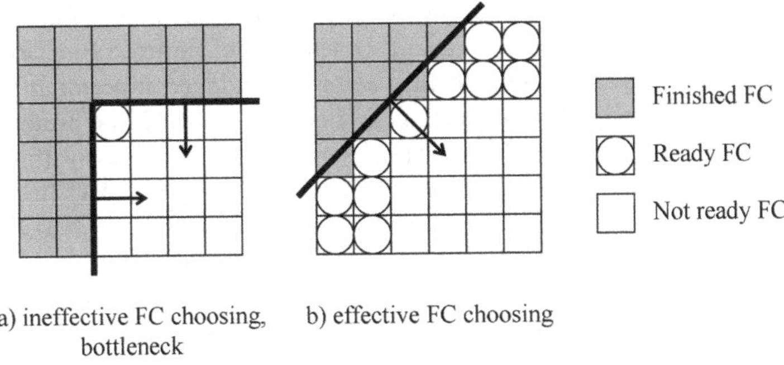

Fig. 5. LU factorization, FC choosing

4 The other Opportunity to Optimize the Execution of a Set of FC

There are several technological opportunities, used in LuNA system, to improve a set of FC execution. Those are the means to express supplementary information about a FP and recommended ways of its execution. Note, that different hardware requires different recommendations; therefore both general and hardware-dependent recommendations can be provided by user. Run-time system selects the most suitable recommendation set.

4.1 Priority

A real number called *priority* is assigned to each FC. At any moment, run-time system tries to choose for execution a fragment with the highest priority available. This allows controlling the FC execution flow to reach better resources usage. The LuNA run time system not only chooses the highest priority FC for execution, but also schedules FCs execution in such a way, that high priority FCs become ready sooner. In the LU-factorization FP (example in section 3.2 above) the DFs $\mathcal{D}_i$ should have higher priorities, than the rest of the FCs.

4.2 Groups and Derivation Algorithm Use

LuNA language has the facilities in order a user could describe a *group* of FCs. Usually, the information depended FCs are included into the group. The FCs, belonging to such a group, are executed in accordance with the MGF strategy (Member of Group First). With the MGF strategy if a certain FC, included into a group, was chosen for execution, then the higher priority is assigned to all the other FCs, belonging to the same group. This strategy leads to consumption of the intermediate DFs soon after they were computed.

The groups can be formed by LuNA compiler using the derivation algorithm [3]. This algorithm processes the countable set of the FCs, that constitute the FP, re-constructs the set of functional terms, implemented by the FP, and then folds them into the finite sets of indexed functional terms (see fig. 2.b). A range of different optimizing transformations of the sets of indexed functional terms is also provided. Any FCs, included into a certain indexed functional term, is also included into a group. Construction of these sets of indexed functional terms permits the use automatically the MGF strategy in run-time system.

In the matrices multiplication algorithm above (section 3.1) in order to optimize the resources usage the groups' definition can be exploited. All the FCs $\mathcal{F}_{i,j,k}$ with the same values of their indices i and j are included into the same group. If a certain FC is chosen for execution, the priorities of all the other FCs from its group are increased. Thus, mostly this group's FCs will execute. As result, all the intermediate resources keeping the DFs $D_{i,j,k}$ will be released.

4.3 Weight of FC

Weight of FC is a real-valued function, defined on the set of FCs. It represents an estimation of the FC's execution time. In the LU-factorization (section 3.2) the time of computation of the FCs increases in the direction to the right bottom corner of the matrix. The value of Weight gives the run-time system this info.

4.4 Neighborhood Relation

A binary neighborhood relation ν is defined on the set of the DFs. Two DFs are defined to be neighbor-related if it is recommended to keep them close to each other, for example, in the memory of the same PE. Usually this is done for the DFs, which are the input values of a certain FC, and location of them in the same PE leads to reduction of the total communication overhead. This relation can be constructed automatically, basing on the structure of information dependences of the FP, but in general case, neighborhood relation ν is better defined by the user.

Neighborhood relation ν is considered when performing initial data distribution with low communication overhead, or dynamic load balancing, that also should keep the neighborhood relation ν .

The numerical algorithms employ limited number of spatial data structures, like vectors, matrices, arrays, 3D meshes. LuNA supports explicit declaration of such data structures and implements a number of algorithms to perform initial distribution and structure-keeping dynamic load balancing on commonly used hardware network topologies, like 3D torus, cluster or complete graph.

5 Performance Tests

The ideas presented were implemented in experimental LuNA functional programming system. It comprises the language of FPs description, the translator to an executable representation and the run-time system. A number of tests was performed. Priority and group testings were performed on a 8-core SMP multiprocessor. Weights and neighborhood relation testings were performed on a cluster.

5.1 Priority Testing

The test should demonstrate the advantages of the priority use. Three tests were accomplished for LU-factorization (example in section 3.2):

1. *Ineffective.* The relation ρ is defined in such a way, that the ineffective order of the FCs execution would be implemented (fig. 5.a.).
2. *Priority-based.* The relation ρ reflects only information dependences between the FCs. Two different priorities were assigned to the FCs. The higher priority was assigned to $\mathcal{D}_i$ FCs and the lower for the rest of the FCs. Certain order was chosen dynamically by the run-time system.
3. *Effective.* The relation ρ is defined in such a way, that the effective order of the FCs execution would be implemented (fig. 5.b).

The results of testing are shown in fig. 6.a.

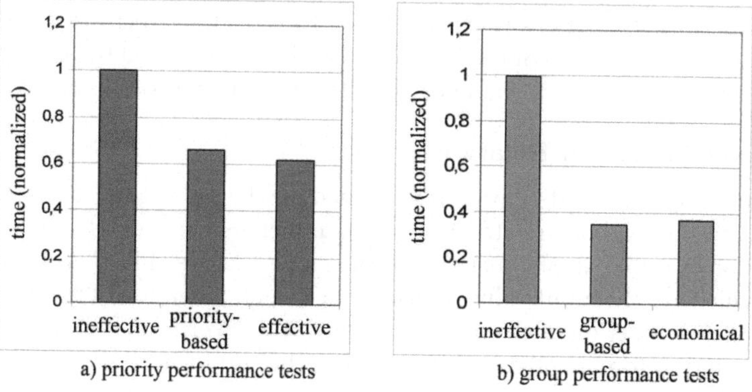

Fig. 6. Priority and group performance tests

5.2 Group Testing

Groups' use influences to the execution time of the matrices multiplication program (example in section 3.1) was tested in 3 tests:

1. *Ineffective.* The relation ρ is such defined that all the FCs $\mathcal{S}_{i,j}$ are executed the last. As a result all the DFs $C_{i,j,k}$ are kept in the memory long time. This is the most time- and memory-consuming FP execution.
2. *Group-based.* The priorities of the FCs of the same group are dynamically increased.
3. *Economical.* The FCs execution from another group is never started before the execution of all the FCs from a currently executed group are completed.

The results are shown in fig. 6.b.

5.3 Neighborhood Relation and FC Weight Testing

The model of fragmented algorithm of Particle-In-Cell method (PIC) [15] implementation was used for testing. This is explicit finite differences 3D scheme. A 3D mesh is represented by 3D grid of DFs. Processing of each DF requires the values from its 26 neighbors. The 3D grid of the DFs is processed iteratively by the FC $\mathcal{F}^t_{i,j,k}$, where i, j and k are the indices of the FCs name, and t is the iteration number. The execution time of the FC $\mathcal{F}^t_{i,j,k}$ is defined by the function $f_{i,j,k}(t)$. Different definitions of $f_{i,j,k}(t)$ lead to different model behaviors. The $f_{i,j,k}(t)$ was chosen in such a way that PIC model imitated the soliton orbiting a massive center. Correspondingly the time of the FCs $\mathcal{F}^t_{i,j,k}$ execution was changing.

On the graphics four balancing versions are shown. The abscissa axis is the iteration number, the ordinate axis is the iteration execution time.

The *None* version has no dynamic load balancing, the DFs don't migrate. The execution time remains about the same, but a lot of PEs' time is wasted, since the workload is not uniformly distributed.

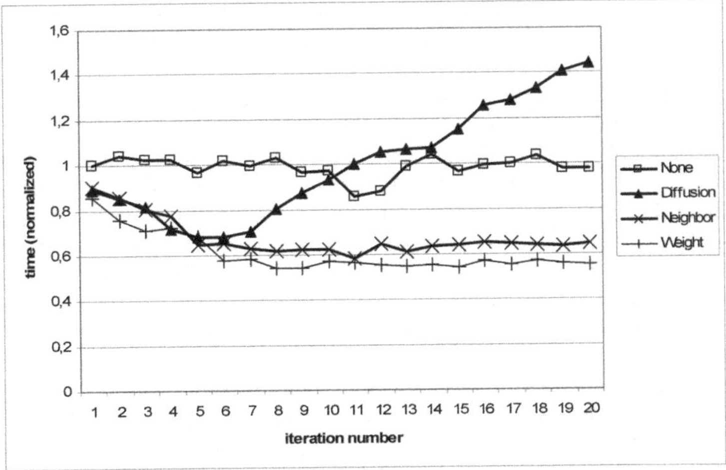

Fig. 7. Neighborhood relation influence on dynamic load balancing

The *Diffusion* version is a certain diffusion dynamic load balancing algorithm. The load is being balanced, but the communication overhead grows, since the DFs are mixing up, and the total execution time even exceeds the unbalanced version.

The *Neighbor* variant is the diffusion load balancing with neighborhood relation taken into account. Those DFs migrate, which have more neighbors in the target PE. Such load balancing keeps communication overhead at certain level, and it doesn't grow with the lapse of time.

The *Weight* variant is the same as the *Neighbor*, but the FCs' weights are taken into account by run-time system. The function $f_{i,j,k}(t)$ is used as the FCs' weight. The load balancing algorithm works more accurate, compared to Neighbor variant, and the execution time is a bit less.

6 Conclusion

The LuNA system of fragmented programming is yet under development and improvement. Our next step is the development of the parallel numerical subroutine library on the basis of LuNA system.

References

1. Glushkov, V.M., Ignatiev, M.V., Myasnikov, V.A., Torgashev, V.A.: Recursive machines and computing technologies. In: IFIP Congress, vol. 1, pp. 65–70. North-Holland Publish. Co., Amsterdam (1974)
2. Torgashev, V.A., Tsarev, I.V.: Programming facilities for organization of parallel computation in multicomputers of dynamic architecture. Programmirovanie (4), 53–67 (2001) (in Russian) (Sredstva organizatsii parallelnykh vychislenii i programmirovaniya v multiprocessorakh s dynamicheskoi architechturoi)

3. Valkovskii, V.A., Malyshkin, V.E.: Parallel Program Synthesis on the Basis of Computational Models. Novosibirsk, Nauka (in Russian) (Sintez parallel'nykh program i system na vychislitel'nykh modelyakh) (1988)
4. Cell Superscalar, `http://www.bsc.es/cellsuperscalar`
5. Charm++, `http://charm.cs.uiuc.edu`
6. Shu, W., Kale, L.V.: Chare Kernel – a Runtime Support System for Parallel Computations. Journal of Parallel and Distributed Computing 11(3), 198–211 (1991)
7. Kalgin, K.V., Malyskin, V.E., Nechaev, S.P., Tschukin, G.A.: Runtime System for Parallel Execution of Fragmented Subroutines. In: Malyshkin, V.E. (ed.) PaCT 2007. LNCS, vol. 4671, pp. 544–552. Springer, Heidelberg (2007)
8. Blumofe, R.D., Joerg, C.F., Kuszmaul, B.C., Leiserson, C.E., Randall, K.H., Zhou, Y.: Cilk: An Efficient Multithreaded Runtime System. ACM SIGPLAN Notices 30(8), 207–216 (1995)
9. Foster, I., Kesselman, C., Tuecke, S.: Nexus: Runtime Support for Task-Parallel Programming Languages. Cluster Computing, Issue 1(1), 95–107 (1998)
10. Chien, A.A., Karamcheti, V., Plevyak, J.: The Concert System – Compiler and Runtime Support for Efficient, Fine-Grained Concurrent Object-Oriented Programs. UIUC DCS Tech. Report R-93-1815 (1993)
11. Grimshaw, A.S., Weissman, J.B., Strayer, W.T.: Portable Run-Time Support for Dynamic Object-Oriented Parallel Processing. ACM Transactions on Computer Systems (TOCS) 14(2), 139–170 (1996)
12. Benson, G.D., Olsson, R.A.: A Portable Run-Time System for the SR Concurrent Programming Language. In: Workshop on Run-Time Systems for Parallel Programming, RTSPP (1997)
13. Malyshkin, V.E., Sorokin, S.B., Chajuk, K.G.: Fragmentation of Numerical Algorithms for the Parallel Subroutines Library. In: Malyshkin, V. (ed.) PaCT 2009. LNCS, vol. 5698, pp. 331–343. Springer, Heidelberg (2009)
14. Kraeva, M.A., Malyshkin, V.E.: Implementation of PIC Method on MIMD Multicomputers with Assembly Technology. In: Malyshkin, V.E. (ed.) PaCT 1997. LNCS, vol. 1277, pp. 541–549. Springer, Heidelberg (1997)
15. Kraeva, M.A., Malyshkin, V.E.: Assembly Technology for Parallel Realization of Numerical Models on MIMD-Multicomputers. Int. Journal on Future Generation Computer Systems 17(6), 755–765 (2001)

On Numerical Solution of Integral Equations for Three-Dimensional Diffraction Problems*

A.A. Kashirin and S.I. Smagin

Computing Center FEB RAS (65, Kim Yu Chen St., Khabarovsk, 680000)
elomer@mail.ru, smagin@as.khb.ru

Abstract. Questions of solution of three-dimensional diffraction problems are considered. The problems are formulated as weakly singular integral equations of 1 kind with alone unknown density. Discretization of these equations is realized by means of special smoothing method of fit integral operators. Numerical solutions of systems of linear algebraic equations, approximating integral equations of diffraction problems, were found by using of the variational iterative method and parallel computing technology. We gave the numerical experiment results.

Keywords: diffraction, numerical method, boundary integral equations, variational iterative method, parallel computing.

1 Introduction

Mathematical modeling of stationary acoustic waves processes propagation in media with three-dimensional inclusions plays an important role in various areas of science and engineering and leads to statement of problems of mathematical physics which are customarily called diffraction or scattering problems. Exact analytical solutions of such problems can be constructed only in exceptional cases; therefore, the basic direction of their research is direct computer modeling.

Using computer assumes preliminary construction of a discrete analogue of the considered problem which can be carried out in various ways. While choosing the method of discretization it is necessary to consider the following properties of diffraction problems: solutions of the problems are searched in unlimited domains, depend on three space variables, slowly decrease at infinity, can be fast oscillating functions and must satisfy radiation conditions at infinity.

The enumerated peculiarities lead to ineffectiveness of algorithms of numerical solution of diffraction problems based on differential statements, leading to finite difference or projective-grid schemes.

This fact stimulated the development of integral methods of diffraction problems solution. N.M. Gyunter, V.D. Kupradze, S.G. Mikhlin, A.G. Sveshnikov,

* The work was financially supported by the Russian Foundation for Basic Research (project no. 08-01-00947) and by Far Eastern Branch of the Russian Academy of Sciences (projects no. 09-I-P2-01, 09-III-B-01-012).

C.H. Hsu and V. Malyshkin (Eds.): MTPP 2010, LNCS 6083, pp. 11–19, 2010.

V.I. Dmitriev, S.M. Belonosov, V.A. Tsetsokho, V.V. Voronin, C. Muller, D. Colton, R. Kress, J.-C. Nedelec, W. McLean, etc. were engaged in investigations in this area.

Methods applied in these works allow us to reduce diffraction problems to various systems of two integral equations with two unknown functions (density). From the computing point of view the approach when such problems are formulated in the form of 1 weakly singular Fredholm integral equation of the 1st or 2nd kind with alone unknown density, which has been developing since late 80-es - early 90-es (S.I. Smagin, R.E. Kleinman, P.A. Martin, P. Ola) is more perspective. At the same time within the limits of the specified approach each problem allows various equivalent statements. In the given work integral equations of the 1st kind were used for numerical solution of scalar problems of diffraction.

A special method of averaging integral operators with weak singularities in kernels, co-ordinated with a grid pitch, which had been developed and used earlier for solution of integral equations of boundary value problems of acoustics, was applied to numerical solution of the received equations.

Solutions of discretized equations were found by means of iterative method of variation type. The check of correctness and accuracy of numerical method was carried out with application of three-dimensional problems of diffractions having known analytical solutions.

2 Integral Equations of Scalar Diffraction Problem

Problem 1 (Generalized statement of scalar problem of diffraction). In bounded domain Ω_i of three-dimensional Euclidean space $\mathbb{R}^3$ and in unlimited domain $\Omega_e = \mathbb{R}^3 \backslash \bar{\Omega}_i$, divided by closed surface Γ of Holder class $C^{r+\beta}$, $r + \beta > 1$, find complex-valued functions $\varphi_{i(e)} \in H^1(\Omega_{i(e)})$, satisfying integral identities

$$\int\limits_{\Omega_{i(e)}} \nabla\varphi_{i(e)} \nabla\psi_{i(e)}\, dx - k_{i(e)}^2 \int\limits_{\Omega_{i(e)}} \varphi_{i(e)}\psi_{i(e)}\, dx = 0 \qquad \forall\psi_{i(e)} \in H_0^1\left(\Omega_{i(e)}\right) \ , \quad (1)$$

interface conditions on the boundary of media contact from Ω_i and Ω_e

$$\langle \varphi_i - \varphi_e, \mu\rangle_\Gamma = \langle \varphi_0, \mu\rangle_\Gamma \quad \forall\mu \in H^{-1/2}(\Gamma) \ , \quad (2)$$

$$\langle \eta, p_i N\varphi_i - p_e N\varphi_e\rangle_\Gamma = \langle \eta, p_e\varphi_1\rangle_\Gamma \quad \forall\eta \in H^{1/2}(\Gamma) \ ,$$

and also to the radiation condition at infinity for φ_e

$$\partial\varphi_e/\partial\,|x| - ik_e\varphi_e = o\left(|x|^{-1}\right), \quad |x| \to \infty \ , \quad (3)$$

if on Γ functions $\varphi_0 \in H^{1/2}(\Gamma)$ and $\varphi_1 \in H^{-1/2}(\Gamma)$ are set.

Here and below $\langle\cdot,\cdot\rangle_\Gamma$ is a duality ratio on $H^{1/2}(\Gamma) \times H^{-1/2}(\Gamma)$, generalizing scalar product in $H^0(\Gamma)$,

$$\langle f, \bar{g} \rangle_\Gamma = \int_\Gamma f \bar{g} d\Gamma \quad \forall f, g \in H^0(\Gamma) \ ,$$

$\bar{g}$ is a complex conjugate function to g, $N = \partial/\partial n$, $n(x)$ is the unit exterior normal to the surface Γ at a point x,

$$p_{i(e)} = \left(\rho_{i(e)}\omega\left(\omega + i\gamma_{i(e)}\right)\right)^{-1}, \quad k_{i(e)}^2 = \omega\left(\omega + i\gamma_{i(e)}\right)/c_{i(e)}^2, \quad \mathrm{Im}(k_{i(e)}) \geq 0 \ ,$$

ω is a circular frequency of oscillations, $c_{i(e)}$, $\rho_{i(e)}$, $\gamma_{i(e)}$ are velocities of sound, denseness and absorption factors of media filling domains $\Omega_{i(e)}$, $c_{i(e)} > 0$, $\rho_{i(e)} > 0$, $\gamma_{i(e)} \geq 0$.

Theorem 1. *Problem 1 has no more than one solution.*

The proof of the theorem is avalable in work [1].

Let's introduce designations

$$\left(A_{i(e)}q\right)(x) \equiv \left\langle G_{i(e)}(x,\cdot), q\right\rangle_\Gamma, \ \left(B_{i(e)}q\right)(x) \equiv \left\langle N_x G_{i(e)}(x,\cdot), q\right\rangle_\Gamma \ ,$$

$$\left(B'_{i(e)}\varphi\right)(x) \equiv \left\langle \varphi, N_{(\cdot)} G_{i(e)}(x,\cdot)\right\rangle_\Gamma, \ G_{i(e)}(x,y) = \exp\left(ik_{i(e)}|x-y|\right)/(4\pi|x-y|)$$

and consider potentials

$$\varphi_e(x) = (A_e q)(x), \quad x \in \Omega_e \ , \tag{4}$$

$$\varphi_i(x) = p_{ei}\left(A_i\left(N\varphi_e + \varphi_1\right)^+\right)(x) - \left(B'_i\left(\varphi_e + \varphi_0\right)^+\right)(x), \quad x \in \Omega_i \ .$$

Here $q \in H^{-1/2}(\Gamma)$ is an unknown density, $p_{ei} = p_e/p_i$, signs "+" mark limiting values of corresponding expressions on Γ, when $x \to \Gamma$ from domain Ω_e.

Kernels of the potentials (4) are fundamental solutions of Helmholtz equations and their normal derivatives. Therefore they satisfy identities (1) and radiation condition at infinity (3). Besides, for them meeting the first condition of interface (2) automatically involves meeting the second condition of interface. Substituting potentials (4) into the first interface condition, we obtain 1 kind weakly singular integral equation of Fredholm type, equivalent to diffraction problem

$$\langle (Cq), \mu \rangle_\Gamma = \langle \psi_0, \mu \rangle_\Gamma \ \forall \mu \in H^{-1/2}(\Gamma) \ , \tag{5}$$

where

$$(Cq)(x) \equiv \left(\left(0.5\left(A_e + p_{ei}A_i\right) + \left(B'_i A_e - p_{ei}A_i B_e\right)\right)q\right)(x) \ ,$$

$$\psi_0(x) = -0.5\varphi_0(x) + p_{ei}\left(A_i\varphi_1\right)(x) - \left(B'_i\varphi_0\right)(x) \ .$$

The next theorem is true [1].

Theorem 2. *Let $\varphi_0 \in H^{1/2}(\Gamma)$, $\varphi_1 \in H^{-1/2}(\Gamma)$, $\gamma_e > 0$ or ω is not an eigenfrequency problem*

$$\Delta\varphi + k_e^2\varphi = 0, \quad x \in \Omega_i, \qquad \varphi = 0, \quad x \in \Gamma \ . \tag{6}$$

Then the equation (5) is well posed solvable in the class of density $q \in H^{-1/2}(\Gamma)$ and formulas (4) define a solution of problem 1.

Problem 1 allows one more equivalent statement in the form of Fredholm the 1 kind integral equation with a weak singularity in the kernel. We will search for its solution in the form of

$$\varphi_i(x) = (A_i q)(x), \quad x \in \Omega_i \ , \tag{7}$$

$$\varphi_e(x) = \left(A_e(\varphi_1 - p_{ie} N\varphi_i)^-\right)(x) - \left(B'_e(\varphi_0 - \varphi_i)^-\right)(x), \quad x \in \Omega_e \ ,$$

where $p_{ie} = p_i/p_e$, signs "–" mark limiting values of corresponding expressions on Γ, when $x \to \Gamma$ from domain Ω_i. In this case problem 1 is reduced to equation

$$\langle (Dq), \mu \rangle_\Gamma = \langle \varphi_0, \mu \rangle_\Gamma \ \forall \mu \in H^{-1/2}(\Gamma) \ , \tag{8}$$

$$(Dq)(x) \equiv ((0.5(A_i + p_{ie}A_e) + (p_{ie}A_eB_i - B'_eA_i))\, q)(x)$$

and takes place [1]

Theorem 3. *Let $\varphi_0 \in H^{1/2}(\Gamma)$, $\varphi_1 \in H^{-1/2}(\Gamma)$, $\gamma_e > 0$ or ω is not an eigenfrequency problem (6). Then equation (8) is correctly solvable in class of density $q \in H^{-1/2}(\Gamma)$ and formulas (7) define a solution of problem 1.*

When it is necessary to calculate the wave field in domain Ω_e, it is preferable to reduce problem 1 to equation (5), which allows to calculate the reflected field according to a simpler formula. For a similar reason, in cases when it is necessary to calculate the passing wave field in domain Ω_i, it is preferable to use equation (8).

3 Method of Numerical Solution

The stated method of solution of integral equations with weak singularities in kernels was first offered in work [2]. Its idea is that an unknown density is in the form of a linear combination of smooth finite functions generating partition of a unit on the boundary of inclusion. Such approach does not demand preliminary triangulation of a surface and is equally simply realized both on regular, and on irregular grids. During discretization of equations superficial integrals are approximately substituted by expressions containing integrals on $\mathbb{R}^3$, which are calculated then analytically that allows us to find factors of systems of linear algebraic equations approximating integral equations simply enough.

Let's shortly describe the general scheme of realization of the method. Let's construct a covering of surface Γ by system $\{\Gamma_m\}_{m=1}^M$ neighborhoods of knot points $x'_m \in \Gamma$, lying within spheres of radiuses h_m with centers in x'_m, and designate through $\{\varphi_m\}$ a partition of unit subordinate to it. As φ_m we will use functions

$$\varphi_m(x) = \varphi'_m(x)\left(\sum_{k=1}^{M} \varphi'_k(x)\right)^{-1}, \qquad \varphi'_m(x) = \begin{cases} \left(1 - r_m^2/h_m^2\right)^3, & r_m < h_m, \\ 0, & r_m \geq h_m, \end{cases}$$

where $x \in \Gamma$, $r_m = |x - x'_m|$, $\varphi_m \in C^1(\Gamma)$ at $\Gamma \in C^{r+\beta}$, $r + \beta > 1$.

We will search for approximate solutions of integral equations on grid $\{x_m\}$,

$$x_m = \frac{1}{\bar{\varphi}_m} \int\limits_{\Gamma} x\varphi_m d\Gamma, \quad \bar{\varphi}_m = \int\limits_{\Gamma} \varphi_m d\Gamma \ ,$$

which knots are the centers of weights of function φ_m. We will assume that for all $m = 1, 2, \ldots, M$ inequalities

$$0 < h' \leq |x_m - x_n|, \quad m \neq n, \quad n = 1, 2, \ldots, M \ ,$$

$$h' \leq (2\pi)^{1/2}\sigma_m \leq h_m \leq h, \quad h/h' \leq q_0 < \infty \ ,$$

are fulfilled. Where h, h' are positive numbers depending on M, q_0 does not depend on M,

$$\sigma_m^2 = 0.5\bar{\varphi}_m \ .$$

Instead of unknown function q set on Γ we will search for generalized function $q\delta_\Gamma$, operating according to rule $(q\delta_\Gamma, \varphi)_{\mathbb{R}^3} = \langle q, \varphi \rangle_\Gamma$. We will approximate this function by expression

$$q\,(x)\, \delta_\Gamma\,(x) \approx \sum_{n=1}^{M} q_n \psi_n(x) \ ,$$

where q_n are unknown factors, $\psi_n(x) = (\pi\sigma_n^2)^{-3/2} \exp\left(-(x - x_n)^2/\sigma_n^2\right)$.

Let's multiply both parts of integral equation by φ_m and integrate on Γ. Then, if M is big enough, integral operator of the 1st kind is approximated according to formula [3]

$$\int\limits_{\Gamma} \left(A_{i(e)}q\right) \varphi_m d\Gamma \approx \bar{\varphi}_m \sum_{n=1}^{M} A_{i(e)}^{mn} q_n, \quad m = 1, 2, ..., M \ , \tag{9}$$

and integral operator of the 2nd kind – according to formula [4]

$$\int\limits_{\Gamma} \left(aq + \left(B_{i(e)}q\right)\right) \varphi_m d\Gamma \approx \bar{\varphi}_m \sum_{n=1}^{M} B_{i(e)}^{mn} q_n, \quad m = 1, 2, ..., M \ . \tag{10}$$

Here

$$A_{i(e)}^{mn} = 0.5\beta_{i(e)}^{mn} \exp\left(-\gamma_{mn}^2\right) \left(w\,(z_1) - w\,(z_2)\right)/r_{mn}, \quad m \neq n \ ,$$

$$\beta_{i(e)}^{mn} = \bar{\varphi}_n\,(4\pi)^{-1} \left(1 - \left(\mu_{i(e)}^{mn}\right)^2 + 0.5\left(\mu_{i(e)}^{mn}\right)^4\right)^{-1}, \quad \mu_{i(e)}^{mn} = 0.5k_{i(e)}\sigma_{mn} \ ,$$

$$r_{mn} = |x_m - x_n|, \quad \sigma_{mn} = \left(\sigma_m^2 + \sigma_n^2\right)^{1/2}, \quad \gamma_{mn} = r_{mn}/\sigma_{mn} \ ,$$

$$z_1 = \mu_{i(e)}^{mn} - i\gamma_{mn}, \quad z_2 = \mu_{i(e)}^{mn} + i\gamma_{mn} \ ,$$

$$w\,(z) = \exp\left(-z^2\right) \left(1 + 2\pi^{-1/2} i \int\limits_0^z \exp\left(t^2\right) dt\right) \ ,$$

$$A_{i(e)}^{mm} = \beta_{i(e)}^{mm} \left[ik_{i(e)} w \left(\mu_{i(e)}^{mm} \right) + 2 \left(\sigma_{mm} \pi^{1/2} \right)^{-1} + \kappa_{i(e)}^{m} \right] ,$$

$$\kappa_{i(e)}^{m} = 2\pi^{1/2} \sigma_{mm} \left(1 - 2 \left(\mu_{i(e)}^{mm} \right)^2 \Big/ 3 \right) \left(\bar{\varphi}_m \right)^{-1} ,$$

$$B_{i(e)}^{mn} = \left(4\pi r_{mn}^3 \right)^{-1} n_{mn}^* \left(\exp\left(ik_{i(e)} r_{mn} \right) \left(ik_{i(e)} r_{mn} - 1 \right) + 1 + \lambda_{mn} \right) \bar{\varphi}_n ,$$

$$\lambda_{mn} = 2\pi^{-1/2} \left(\int_0^{\gamma_{mn}} \exp\left(-t^2 \right) dt - \gamma_{mn} \exp\left(-\gamma_{mn}^2 \right) \right) ,$$

$$n_{mn}^* = \sum_{l=1}^{3} n_{lm} \left(x_{lm} - x_{ln} \right) ,$$

$$B_{i(e)}^{mm} = -\mathrm{Gs}(x_m) \quad \text{at } a = 0.5, \quad B_{i(e)}^{mm} = -1 - \mathrm{Gs}(x_m) \quad \text{at } a = -0.5 ,$$

$$\mathrm{Gs}\left(x_m \right) = \sum_{n \neq m}^{M} \frac{n_{mn}^* \bar{\varphi}_n}{4\pi r_{mn}^3} \approx \frac{1}{4\pi} \int_{\Gamma} \frac{\partial}{\partial n_y} \frac{1}{|x_m - y|} d\Gamma_y = -\frac{1}{2} .$$

Integral operators in the left parts of equations (5) and (8) are a composition of integral operators (9) and (10), let's approximate them according to formulas

$$\int_{\Gamma} (Cq) \varphi_m d\Gamma \approx \sum_{n=1}^{M} \left(A_e^{mn} B_i^{mn} - p_{ei} A_i^{mn} B_e^{mn} \right) q_n, \quad m = 1, 2, ..., M , \tag{11}$$

$$\int_{\Gamma} (Dq) \varphi_m d\Gamma \approx \sum_{n=1}^{M} \left(p_{ie} A_e^{mn} B_i^{mn} - A_i^{mn} B_e^{mn} \right) q_n, \quad m = 1, 2, ..., M , \tag{12}$$

And right parts of equations (5) and (8) according to formulas

$$\int_{\Gamma} \psi_0 \varphi_m d\Gamma \approx \bar{\varphi}_m \left(\mathrm{Gs}(x_m) \varphi_0(x_m) + \sum_{n=1}^{M} \bar{\varphi}_n \left[p_{ei} A_i^{mn} \varphi_1(x_n) - \varphi_0(x_n) B_i^{nm} \right] \right) ,$$

$$\int_{\Gamma} \varphi_0 \varphi_m d\Gamma \approx \bar{\varphi}_m \varphi_0(x_m), \quad m = 1, 2, ..., M .$$

Solving corresponding system of linear algebraic equations, we will find approximate values of density in points of discretization for each of integral equations. After that, required solutions of problems of acoustics can be calculated equally simply and precisely both in distant, and in near zones.

4 Algorithm Decomposition

The algorithm of numerical solution of diffraction problems consists of the following modules, carried out sequentially:

- Discretization of surface: deriving of three-dimensional co-ordinates of points of discretization;
- Calculation of normal vectors at each point of discretization, calculation of matrixes of transition from local co-ordinates to global ones and back to local;
- Calculation of integrals $\bar{\varphi}_m$;
- Calculation of the vector of the right part of system of linear algebraic equations;
- Approximation of integral operator according to formula (9) or (10) and solution of systems of linear algebraic equations for determining factors q_n;
- Calculation of solution of diffraction problem in an arbitrary point of space, calculation of solution inaccuracy.

The first 4 modules were carried out in a uniprocessor condition. The last two most exacting modules to computing and temporary resource – a subprogram for solving systems of linear algebraic equations and a subprogram for calculating solution inaccuracy of diffraction problem in exterior area were exposed to deparallelizing. For numerical solution of systems of linear algebraic equations, approximating integral equations of diffraction problems, the iterative method of variation type – a generalized method of minimal residual (GMRES) [5] – was used. Its most labour-consuming part is matrix-vector multiplication which complexity consists of $O(M^2)$ operations on each iteration. This procedure deparallelized by matrix decomposition into blocks consisting of l lines. Each block was transmitted to one of p processors ($l = M/p$), where matrix-vector multiplication was carried out. Then the resultant vector was gathered from p parts and the rest of iteration was fulfilled in a uniprocessor condition. The calculation of solution inaccuracy in exterior domain was carried out according to the same scheme.

5 Outcomes of Numerical Experiments

Correctness and accuracy of numerical method were checked with the help of test problems having exact analytical solutions. As the latter, interior and exterior boundary value problems of Dirichlet and Neumann in the form of integral equations for Helmholtz equation [6] in the domains bounded to an unit sphere and an ellipsoid with semi axes $(0.4, 1, 0.2)$, and a problem on dispersion of a plain acoustic wave on an unit ball [7] were chosen. Integral operators in these problems look like (9)–(12), wave numbers $k_{i(e)}$ were chosen out of the range from 0 to 30, number of points of discretization $M = 500 \div 128000$.

Calculations were carried out on a cluster of Computing Center Far Eastern Branch of the Russian Academy of Sciences, consisting of eight HP Proliant DL360G5 knots and having peak productivity of 204 GFlops. The cluster was

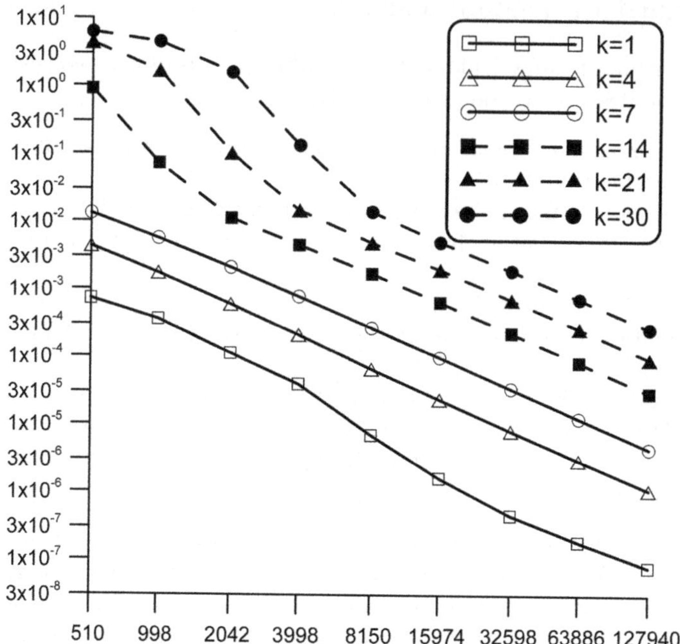

Fig. 1. The dependence of the relative errors of solutions of the exterior Dirichlet problem for Helmholtz equation on the number of discretization points and on the wave numbers. A boundary of the domain is the unit sphere.

driven by Linux CentOS 4.4 operational system, Intel FORTRAN 10.1.012 was used as a compiler. Total number of processors p, depending on the load of the cluster, varied from 8 to 20.

Numerical experiments have shown that GMRES allows to calculate required solutions for a small, in comparison with the order of systems of linear algebraic equations, number of iterations which slightly depends on M, but increases with the growth of $k_{i(e)}$. At the same time magnitude of inaccuracy in considered problems for big M has an order not worse than $h^2 \sim M^{-1}$, and a relative error of calculations in the norm of grid space $H_h^0(\Omega_{i(e)})$ for all problems at $M = 128000$ belongs to the interval from 10^{-7} to $5 \cdot 10^{-4}$ depending on input data. As an example the dependence of the relative errors of solutions of the exterior Dirichlet problem for Helmholtz equation on the number of discretization points and on the wave numbers shown in Fig. 1. A boundary of the domain is the unit sphere.

References

1. Kashirin, A.A., Smagin, S.I.: Generalized Solutions of the Integral Equations of a Scalar Diffraction Problem. Differential Equations 42(1), 79–90 (2006) (in Russian)
2. Smagin, S.I.: Numerical Solution of Integral Equation of the 1st Kind with a Weak Singularity on a Closed Surface. FAS USSR 303(5), 1048–1051 (1988) (in Russian)

3. Kashirin, A.A.: Research and Numerical Solution of Integral Equations of Three-dimensional Stationary Problems of Diffraction of Acoustic Waves: Thesis... of the candidate of physical and mathematical sciences. Khabarovsk (2006) (in Russian)
4. Ershov, N.E., Smagin, S.I.: Numerical Solution of a Three-dimensional Stationary Problem of Diffraction of Acoustic Waves on Three-dimensional Elastic Inclusion. Vladivostok (1989) (in Russian)
5. Saad, Y.: Iterative Methods for Sparse Linear Systems. PWS Publ. Co., Boston (2000)
6. Colton, D., Kress, R.: Integral Equation Method in Scattering Theory: Translated from English. Mir, Moscow (1987) (in Russian)
7. Selezov, I.T., Krivonos, Y.G., Jakovlev, V.V.: Scattering of Waves by Local Heterogeneities in Continuous Media. Naukova dumka, Kiev (1985) (in Russian)

Discrete Models of Physicochemical Processes and Their Parallel Implementation*

Olga Bandman

Supercomputer Software Department
ICM&MG, Siberian Branch, Russian Academy of Sciences
Pr. Lavrentieva, 6, Novosibirsk, 630090, Russia
bandman@ssd.sscc.ru

Abstract. Discrete simulation method of physicochemical kinetic processes is proposed and investigated. The method is based on formal representation of classical Von-Neumann's Cellular Automaton (CA) extension, which allow all kind of discrete alphabets, probabilistic transition functions, and asynchronous mode of operation. Some techniques for simple CA composition are given for simulating complex processes. Transformation of asynchronous CA into block-synchronous type is used to provide high efficiency of parallel implementation.

1 Introduction

Discrete models of spatial dynamics and particularly of physicochemical kinetics on micro and nano levels are of great interest nowadays. They are based on Von-Neumann's Cellular Automaton, which is considered as a discrete expression of spatial-time function [1,2]. A CA is regarded as a discrete space (a cellular array) in each point of which (in each cell) the function values (cell states) are updated by application discrete function of the cell states values in a certain proximity. All cells may be updated simultaneously or in a random order changing in such a way the global state of the cellular array. The system is autonomous: being set in an initial state it starts an iterative process called the CA evolution, which imitates a spatially distributed process. A proper choice of cell transition functions allows to construct a CA, directly expressing the events representing the process under simulation: particles movements, phase transition, chemical interactions between abstract, or, sometimes, real, particles. Such point of view at CA is motivated, mainly, by two reasons. The first is the availability of powerful computer systems capable to simulate CA with huge number (billions) of cells. The second is weakness of traditional mathematical models based on partial differential equations, which are incapable to simulate such phenomena as chemical reactions, biological evolution, living beings behavior, and many others that may be essentially nonlinear, dissipative, self organizing or having other exotic properties.

* Supported by 1) Presidium of Russian Academy of Sciences, Basic Research Program N 2 (2009), 2) Siberian Branch of Russian Academy of Sciences, Interdisciplinary Project 32 (2009).

C.H. Hsu and V. Malyshkin (Eds.): MTPP 2010, LNCS 6083, pp. 20–29, 2010.

It is clear, that simulating capabilities of a classical Von-Neumann's CA is not strong enough to satisfy new requirements. So, many modifications were proposed allowing integer or symbolic alphabet, arbitrary transition functions and modes of operation, but preserving two basic properties.

1) The transition function for all cells is identical and local, i.e. next state value depends only on states of cells allocated in a certain locality.

2) Computation of cell transition function are independent, i.e. they may be performed in any order, which conditions simplicity and efficiency of coarse grain parallel implementation.

The above two properties determine a wide class of discrete computations, combined under the concept of *fine grained parallelism* [3].

A large part of diffusion-reaction CA are of probabilistic asynchronous type, referred to in chemical literature as *Kinetic Monte Carlo Method.* They are mostly used in scientific investigations for simulation physicochemical processes, such as reactions on catalysts [4,5], epitaxial growth of crystals [6], formation of coating, erosion on metals [7], etc.. Most of them have limited capabilities and have problems in parallel implementation. Moreover, since the results of these investigations have particular aims and are scattered about specialized journals, there is no possibility to use each other experience. Meanwhile, the concept of CA covers all above and many other similar discrete models simulation capability and may be used as a base for obtaining a generalized model. Particularly, some attempts have been made to [8] create a unique formalism [9] and develop a methodology of efficient parallelization [10]. The aim of this paper is to improve the available formalism by including some controlling functions for allocating computations in time and space, in particular, for obtaining high efficiency of parallel implementation.

The paper contains five sections. The Introduction is followed by Second section where the main concepts and formal representation of CA–models are given. In Third section correctness conditions and modes of functioning are described. Fourth section is devoted to kinetic CA-models, and the Fifth one reports on its implementation on multiprocessor systems.

2 Formal Definitions of CA-Model Concepts

Formal definitions of CA concepts follow the ideas of "Parallel Substitution Algorithm "(PSA) [9]. The PSA formal model, although worked out for parallel VLSI architecture design has a strict and general background, which allows to use its main principles for constructing formalisms for specific types of fine-grained algorithms. In particular, PSA principles turned out to be very useful for CA-models formal representation due to two following reasons.

1) In PSA *context configurations* are defined that are needed to represent probabilistic and timing conditions,

2) In PSA multi cellular transition rules are allowed, that are necessary for representing diffusion.

A CA-model is further defined by four sets $\aleph = \langle A, M, T, \Theta\rangle$ together with indication of operation mode (synchronous, asynchronous or their combination), where A is an alphabet, M – a set of points in discrete space, $T \subset M$ is a translational template, Θ – a set of local operators. A main concept of a CA - model is a *cell* which is a pair (v, m), where $v \in A$ is a cell state from an alphabet A, and $m \in M$ is a cell name from a finite set of discrete space points coordinates. The set of cells $\Omega = \{(v, m) : v \in A, m \in M$ form a *cellular array.* Locality of cells interaction is given by a concept of a *main template*

$$T(m) = \{(m, \dots, \phi_k(m), \dots, \phi_q(m))\}. \tag{1}$$

In 2D case when $M = \{(i,j) : i,j = 0, \dots, N\}$, the most used main template contains $\phi_k(m) = (i+a, j+b), a, b \in \{-r, \dots, 0, \dots, r\}$, called *naming functions,* defined in M_B. The set of cells

$$S(m) = \{(v_0, m), \dots, (v_k, \phi_k(m), \dots, (v_q, \phi_q(m))\}, \tag{2}$$

is called a *main local configuration.* The neighboring cells in $S(m)$ have fixed numbers $0, 1, \dots, q$, determined by the underlying template $T(m)$, which are preserved in all expressions of the model. So, in any $S_h(m) \subseteq S(m)$, $S_h(m) = \{(v_{h_k}, n_{h_k}(m)) : h_k \in \{0, \dots, r\}\}$, as well as in its underlying template $T_h(m) = \{(m, \dots, \phi_{h_1}(m), \dots, \phi_{h_r}(m))\}$ the neighboring cells are indexed according to its main template location.

A local *state configuration* $V(m)$ as well as its subsets

$$V_h(m) = \{v_{h_k} : h_k \in \{0, \dots, q\}\}, \quad V_h(m) \subseteq V(m), \tag{3}$$

may be separated from it. Three local configurations

$$\begin{aligned} S_h(m) &= \{(v_0, m), (v_{h_1}, \phi_{h_1}(m)), \dots, (v_{h_r}, \phi_{h_r}(m))\}, \\ S'_h(m) &= \{(u_0, m), u_{h_1}, \phi_{h_1}(m)), \dots, (u_{h_p}, \phi_{h_p}(m))\}, \\ S''_h &= \{(w_{h_1}, \mu_{h_1}), \dots, (w_{h_s}, \mu_{h_s}\}. \end{aligned} \tag{4}$$

expressed in the form of a substitution

$$\theta_h(m) : \; S_h(m) \star S''_h \to S'_h(m), \tag{5}$$

represent *a local operator.* Local configuration $S_h(m)$ in (4) is a *basic local configuration* or simply *a base.* $S'_h(m)$ called *next state configuration* has the underlying template $T'_h(m) \subseteq T_h(m)$. Its state configuration components are values of *transition functions*

$$u_{h_k} = F_{h_k}(v_0, v_{h_1}, \dots, v_{h_r}, w_{h_1}, \dots, w_{h_s}\}, \quad k = 0, \dots, n. \tag{6}$$

A subset $T''_h(m) = T_h(m) \setminus T'_h$ is the underlying template of a main context $S''_h(m)$ whose cell states serve as arguments in (5). The set of cells S''_h comprise the *controlling context,*

A local operator $\theta_h(m)$ (4) *is applicable*, if two following condition hold

$$S_h(m) \subseteq \Omega, \qquad \text{and} \;\; \forall l \in \{1, \dots, s\} \;\; w_{h_l} = 1, \tag{7}$$

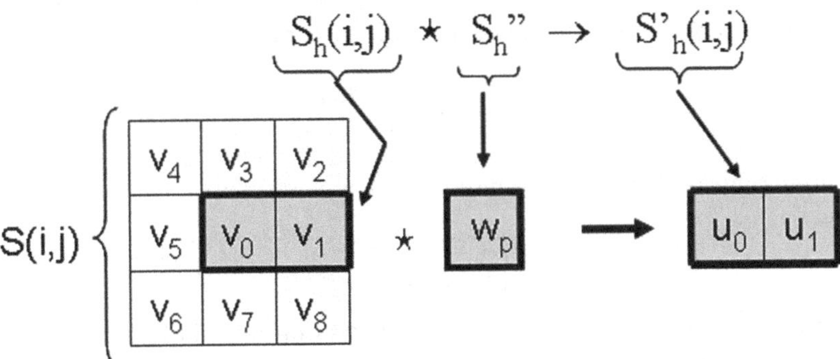

Fig. 1. Graphical representation of a local operator

where $S_h(m)$ is suggested to be included in Ω if any constant state cell $(a, m) \in S(m)$ belong also to Ω. If applicable, $\theta_h(m)$ updates next state configuration replacing the states of $V_h(m)$ by $U_h(m)$.

Attempting to apply all $\theta_h(m) \in \Theta$ to all $m \in M$ (may be unsuccessfully), makes $\Omega_B(t)$ to transit to the next global state $\Omega(t+1)$. Such a transition forms an *iteration*. A sequence

$$\Sigma(\Omega) = \Omega, \Omega(1), \ldots, \Omega(t), \Omega(t+1), \ldots, \Omega(T), \tag{8}$$

resulting of CA-model iterative functioning is called a *CA evolution*. Evolution of a CA is a discrete representation of a space-time process. Evolution has a termination, if there exists $t = T$, such that $\Omega(T) = \Omega(T+1) = \Omega(T+2) = \ldots$. Otherwise the CA simulates oscillatory or chaotic behavior [11]. In practice the number of iterations T is considered to be given a priori.

In order to make a CA-model consistent, it is useful to add a means performing a controlling function, representing them in the form of one cell local operators $\vartheta_1, \ldots, \vartheta_s$ as follows

$$\vartheta_l : (X, m_l) \to (w_l(X), m_l), \quad l \in \{1, \ldots, s\}, \tag{9}$$

where X - a set of external data. There are several types of controlling contexts, which differ mainly in transition function $w_l(X)$. The most used are three controlling context types.

Probabilistic context ϑ_p used when $\theta_h(m)$ corresponds to a random event,with a transition function

$$w_p(rand) = \begin{cases} 1 \texttt{ if} & rand < p, \\ 0 \texttt{ if} & rand \geq p, \end{cases} \tag{10}$$

where *rand* is a random number in the interval [0,1], p - the simulated event probability. Probabilistic context may be inserted into $\theta_h(m)$ by removing S''_h and writing the probability value above the substitutional arrow instead.

$$\theta_h(m): \; S_h(m) \xrightarrow{p_r} S'_h(m). \tag{11}$$

Timing context ϑ_t which indicates at what time steps $\theta_h(m)$ may be applicable, a typical example uses the following transition function.

$$w_t(t) = \begin{cases} 1 \text{ if } & t_{mod_n} = 0, \\ 0 \text{ if } & t_{mod_n} = 1. \end{cases} \tag{12}$$

Spatial context ϑ_s is used for indicating cell coordinates, where $\theta_h(m)$ is applicable.

$$w_s(m) = \begin{cases} 1 \text{ if } & m \in M', \\ 0 \text{ if } & m \notin M', \end{cases}, \quad M' \subset M. \tag{13}$$

Composed context is used when several conditions should be met for $\theta_h(m)$ applicability. It is the case of synchronous Margolus' diffusion [12], where both time and space should be controlled.

$$w_{ts}(i,j) = \begin{cases} 1 \text{ if } & t_{mod_2} = 0 \ \& \ (i+j)_{mod_2} = 0, \\ 0 \text{ otherwise.} \end{cases}, \tag{14}$$

The controlling contexts are also used in providing correct common behavior in CA-models composed of several CA.

3 Functioning Modes and Correctness Conditions

Transition from $\Omega(t)$ to $\Omega(t+1)$ may be performed in different modes. A mode prescribes a certain order of local operation application. All of them are combinations of two canonical modes: synchronous and asynchronous. The *synchronous mode* prescribes the following sequence of actions to perform a single iteration. For all $\theta_h(m) \in \Theta$:

1) make an attempt to apply $\theta_h(m)$ to all $m \in M$ and compute next state $u_{h_k}(m) \in U_h(m)$ if applicable,
2) update the cells of $\Omega(t)$ for which $u_{h_k}(m)$ is obtained, replacing $V_h(m)$ by $U_h(m)$ according to (6).

If one imagine that each cell is implemented in hardware, and hence, all may act simultaneously, getting benefit from the CA potential inherent parallelism, then the transition to the next global state might require one time step. But, in a conventional computer, all local operators are executed sequentially saving the computed next states $U_{h_k}(m)$ in another array, which serves as a global state for the next iteration. So, the number of needed time steps for implementing one iteration is

$$\Delta t = |M| \cdot \nu \cdot \tau, \tag{15}$$

τ being a processor time of an elementary operation, ν - a mean number of elementary operation needed to execute a local operator.

The *asynchronous mode* of operation prescribes complete randomness of local operators application. At each iteration the following should be done.

1) chose randomly an operator $\theta_h(m)$ out of Θ,
2) chose randomly a cell m and make an attempt to apply $\theta_h(m)$ to it,

3) if application is successful compute $u_{h_k}(m) \in U_h$ and make state replacement immediately,

4) the iteration is completed when all $\theta_h(m)$ are applied to all $m \in M$. Due to the succession of all events the iteration time for asynchronous mode is the same than in the synchronous one. It is important to notice, that two CA-models, differing only in modes of operation may have quite different evolutions, simulating quite different behavior of a physicochemical process under simulation.

Correctness of a CA-model is thought of as such an organization of the computational process, that has a termination when all needed operations are executed and it terminates without data loss. The correctness conditions for CA are well known [10], but may be easily violated when complex CA-models interact, and also when the computation is decomposed for multiprocessor implementation. The main condition for correct behavior is *determinacy*, which is formulated for CA-models as follows. Each local operator $\theta_h(m)$ is allowed to change simultaneously (synchronously) the cells of a set $\Pi \subseteq \Omega(t)$, containing not a single pair of cells $\{(v_k, m_k), (v_l, m_l)\}$ with $\{v_k, v_l\}$ belonging to one and the same $V_h(m)$, $m \in M$, $h = 1, \ldots, q$, which is formally expressed as follows.

$$T_h(m_g) \bigcap T_l(m_k) = \emptyset \qquad \forall (m_g, m_h) \in M, \ \text{and} \ \forall (h,l) \in \{1, \ldots, n\}. \tag{16}$$

Determinacy condition guarantees the absence of such a situations, when different transition functions, say, $F_h(m)$ and $F_l(\phi_g(m))$ attempt to change the state of one and the same cell. The condition (16) is not satisfied in synchronous CA, if more than one cell state is to be changed simultaneously It is always satisfied for classical CA where all local operators change only one cell state. It is also satisfied in asynchronous CA, where all local operators are applied sequentially.

4 Kinetic CA-Models

In simulation of physicochemical processes great experience is accumulated on investigating chemical reactions on a catalyst surfaces. So, it is the type of processes being taken as the typical example for illustrating practical use of CA-models and method of their parallelization. Although large variety of kinetic CA-models are known, a limited set of elementary local operators are used in them, which correspond to the following elementary actions.

Let $A = \{a, b, c, \emptyset\}$ be a set of cell states, where a, b, c indicate existence of a certain particle, the symbol $\emptyset$ corresponds to a free site. $M = \{(i, j) : i, j, = 0, \ldots, N\}$ is the set of sites on a 2D surface, each site corresponding to a cell name. The particles are allowed to interact in the site locality determined by a template (1).

A typical set of actions forming the kinetic surface reaction are represented by the following local operators.

Absorption is an act of appearing of a particle a on an empty site with probability p_a.

$$\theta_a(i,j) : \ (\emptyset, (i,j)) \xrightarrow{p_a} (a, (i,j)). \tag{17}$$

Dissociation is a break down of a particle into several parts. For example, a molecule of hydrogen H_2, when contacting a catalyst, dissociates into two atoms $b+b$, requiring two sites. The action happens with probability p_d.

$$\begin{aligned} \theta_{d_k}(i,j):\ & \{(\emptyset,(i,j)),(\emptyset,\phi_k(i,j))\} \star \{(1,m_d(p))(k,m_s)\} \\ & \to \{(b,(i,j)),(b,\phi_k(i,j))\}, \\ \vartheta_d &: (rand, m_d) \to (w_p(rand), m_d), \\ \vartheta_k &: (rand, m_k) \to (w_k(rand), m_k), \end{aligned} \tag{18}$$

where $w_p(rand)$ is computed according to (10), and

$$w_k(rand) = \begin{cases} k, & \texttt{if} \quad (k-1)/q < rand \leq k/q \\ 0 & \texttt{otherwise} \end{cases}, \tag{19}$$

$k \in \{1,\ldots,q\}$, q being the number of cells adjacent to a cell (i,j).

Sublimation or desorption is an act inverse to absorption, which means removing a particle a out of a cell with probability p_s

$$\theta_s(i,j):\ (a,(i,j)) \xrightarrow{p_s} (\emptyset,(i,j)). \tag{20}$$

Chemical reaction is an event like "$a+b \to c$" happening with a probability p_r, where at least two substances take part. Hence, the reactants should be in adjacent sites, while for the reaction product there may be several variants. The most simple is when the reaction is accompanied by sublimation.

$$\begin{aligned} \theta_{r_k}(i,j):\ & \{(a,(i,j)),(b,\phi_k(i,j))\} \star \{(1,m_d(p))(k,m_s)\} \\ & \to \{(\emptyset,(i,j))(\emptyset,\phi_k(i,j))\}, \\ \vartheta_r &: (rand, m_d) \to (w_r(rand), m_d), \\ \vartheta_k &: (rand, m_k) \to (w_k(rand), m_k), \end{aligned} \tag{21}$$

where $w_r(rand)$ and $w_k(rand)$ are as in (10) and (19), respectively.

Diffusion is a process of particles random walking aiming to equalize spatial concentration. The most natural way of simulating the process is to use a probabilistic asynchronous CA, introduced in [12] as a *naive diffusion model* with diffusion coefficient corresponding to probability p_e. a local operator

$$\begin{aligned} \theta_e(i,j) &: \{(u,(i,j)),(u_k,\phi_k(i,j))\} \star \{(i,m_i),(j,m_j),(k,m_k)(1,m_e\}, \\ & \to \{(u_k,(i,j)),(u,\phi_k(i,j))\}, \\ \vartheta_e &: (rand, m_e) \to (w_e(rand), m_e), \\ \vartheta_k &: (rand, m_k) \to (w_k(rand), m_k), \\ \vartheta_i &: (rand, m_i) \to (i(rand), m_i), \\ \vartheta j &: (rand, m_j) \to (j(rand), m_j), \end{aligned} \tag{22}$$

where $w_e(rand)$ and $w_k(rand)$ are computed according to (10) and (19), respectively, and

$$i(rand) = rand \times N, j(rand) = rand \times N. \tag{23}$$

5 Multiprocessor Implementation of Kinetic CA–Models

Fine-grained parallelism predetermines the method of domain decomposition for parallel implementation of CA algorithms. So, the cellular array is partitioned into n equal sized domains, $|Dom| = |M|/n$, each being allocated on a processor. It is well known that high parallelization efficiency may be achieved if

$$\tau \cdot |Dom| >> L \cdot t_{ex}, \tag{24}$$

where τ is time of a local operation, t_{ex} is the time needed for a single interprocessor exchange, L – the number of exchanges per iteration.

In synchronous CA $L = 1$, and, hence, with $t_{ex}/\tau \simeq 10^3$ the condition (24) is easily satisfied when $|Dom| \geq 10^4$.

When asynchronous CA-model is implemented on a multiprocessor each domain boundary cell updating should be accompanied by an act of exchange. It is necessary for correctness condition (16) is satisfied. So, $L = P$ exchanges per iteration is needed, P being domain perimeter, from what it follows, that with $t_{ex}/\tau \simeq 10^3$ the acceptable efficiency ($> 70\%$) is achieved with $|Dom| \geq 10^7$, which makes parallelization unreasonable. The problem is solved [10] by transforming the mode of CA functioning into a block-synchronous one by the following procedure.

1. A partition $\Pi = \{M_1, \ldots, M_q\}$ of M is induced by the main template $T(m)$ (1), such that $|M_k| = |m|/(q+1)$ for all $k \in \{0, \ldots, q\}$, $\bigcup_{k=0}^{q} M_k = M$, and $M_k \bigcap M_l = \emptyset$ forall $(k,l) \in \{0, \ldots, q\}$,

$$\bigcup_{m \in M_k} T(m) = M, \qquad T(m_g) \bigcap T(m_l) = \emptyset, \qquad \forall (m_g, m_l) \in M_k. \tag{25}$$

2. The transition $\Omega(t) \rightarrow \Omega(t+1)$ is performed during $q + 1 = |T(m)|$ stages. Each $k-$th stage comprises application of all local operators to all cells from a subset $M - k \in \Pi$ synchronously, and then exchange boundary cell states with the domains in adjacent processors. So,the number of exchanges per iteration is $L = q + 1$, which allows to obtain high efficiency for acceptable domain size.

Example. In a hydrogen fuel cell chemical reaction energy is transformed into electricity as follows. Hydrogen is infused into a carbon anode activated by a catalyst (platinum). Occurring in touch with catalyst hydrogen dissociates into protons and electrons according to formula

$$H_2 \rightarrow 2H^+ + 2e^-. \tag{26}$$

The protons walk to the cathode, where they join oxygen turning into water. The electrons go to the electrical circuit creating electrical current. Hydrogen dissociation process determines the fuel cell power, and, hence it is studied in details. Simulation of the process allows to obtain the dependence of generated current on hydrogen pressure and on the structure of catalyst cover in anode pores. The CA-model is a probabilistic CA $\aleph_a = \langle A, M, T, \Theta \rangle$ $A = \{\emptyset, H, e\}$, $M = (i,j) : i,j \in 0, 1 \ldots, N \cup \{E, m_s, m_p\}$, where E is electrons counter, $T(i,j) = \{(i,j),(i,j+1),(i-1,j),(i,j-1),(i+1,j)\}$, m_s, m_p - context cells names. Predicates in (w_s, m_s), and (w_p, m_p) correspond to the probabilities of hydrogen atom falling onto a platinum p_s, which is equal to the density of activated pore surface, and probability of hydrogen dissociation reaction p_H, which is equal to its partial pressure. The local operator set contains five CA local operators and two probability context operators of the type (10), $\Theta = \{\theta_e, \theta_{H_k} : k = 1,2,3,4\} \bigcup \{\vartheta_s, \vartheta_H\}$

$$\begin{array}{ll} \theta_e : & \{(H,(i,j)),(e,E)\} \rightarrow \{(\emptyset,(i,j)),(e+1,E)\} \\ \theta_H(k) : & \{(\emptyset,(i,j)),(\emptyset,\phi_k(i,j))\} \stackrel{p_H,p_s}{\longrightarrow} \{(H,(i,j)),(H,\phi_k(i,j))\}, \quad k = 1,2,3.4, \end{array} \tag{27}$$

The model was firstly tested on a single processor for $|M| = 10000 \times 10000$ that correspond to pore surface area of the anode equal to $S_a = 10^{-8}$ mm^2. The time needed for performing 100 iterations was $t = 64.4$ sec. The required data for $S_a \simeq 1$ mkm^2 were obtained using 36 processors with the same domain size. To achieve acceptable parallelization efficiency the CA was transformed into block-synchronous form with $|T(m)| = 3 \times 3$. Thus, 9 stages with synchronous exchange were used.

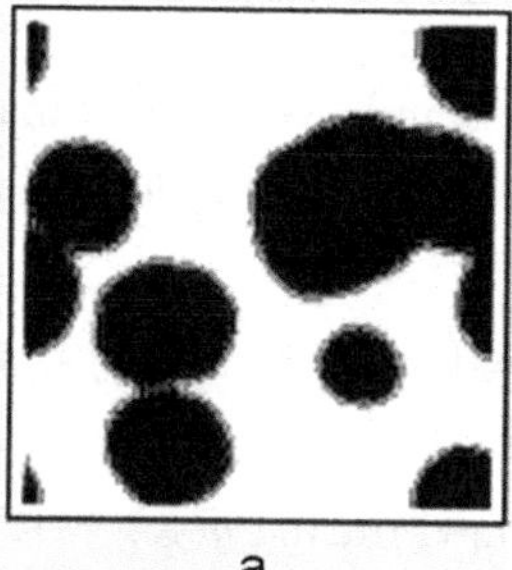

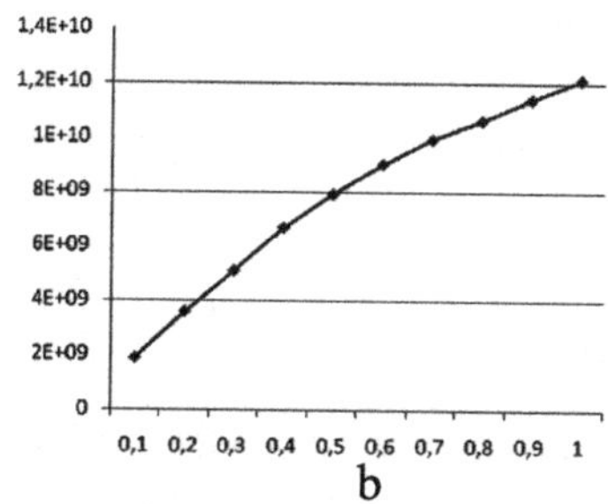

Fig. 2. Simulation of hydrogen dissociation in the anode of a hydrogen fuel element: a) platinum coverage with $p_s = 0.5$, b) dependence of electron number on partial pressure of incoming hydrogen

Simulation was performed on the cluster MVS-1000 (Alpha 21264A, 667MHz). Time for performing 1000 iterations was 1080 sec, which corresponds to parallelization efficiency $\simeq 0,6$.

6 Conclusion

It is shown that discrete CA models allow to study physicochemical processes "on the inside" by imitation the elementary events (chemical and phase transitions, particles displacements) on micro- and nano- levels. The proposed methods are alternative to the Partial Differential Equation solving, they allow to widen essentially the capabilities of mathematical and numerical modeling, admitting nonlinear and discontinuous phenomena to be simulated.

References

1. Toffoli, T.: Cellular Automata as an Alternative to (rather than Approximation of) Differential Equations in Modeling Physics. Physica D 10, 117–127 (1984)
2. Wolfram, S.: Statistical mechanics of Cellular automata. Review of Modern Physics 55, 607–664 (1993)
3. Bandman, O.: Composing Fine-Grained Parallel Algorithms for Spatial Dynamics Simulation. In: Malyshkin, V.E. (ed.) PaCT 2005. LNCS, vol. 3606, pp. 99–113. Springer, Heidelberg (2005)
4. Makeev, A.G.: Coarse bifurcation analysis of kinetic Monte Carlo simulations: a lattice-gas model with lateral interactions. Journ. of chemical physics 117(18), 8229–8240 (2002)
5. Elokhin, V., Latkin, E., Matveev, A., Gorodetskii, V.: Application of Statistical Lattice Models to the Analysis of Oscillatory and Autowave Processes on the Reaction of Carbon Monoxide Oxidation over Platinum and Palladium Surfaces. Kinetics and Catalysis 44(5), 672–700 (2003)
6. Neizvestny, I.G., Shwartz, N.L., Yanovitskaya, Z.S., Zverev, A.V.: 3D-model of epitaxial growth on porous 111 and 100 Si surfaces. Comp. Phys. Communications 147, 272–275 (2002)
7. Betz, G., Husinsky, W.: Surface erosion and film growth studied by a combined molecular dynamics and kinetic Monte Carlo code, Izvestia of Russian Academy of Sciences. Physical Series 66(4), 585–587 (2002)
8. Bandman, O.: Synchronous versus asynchronous cellular automata for simulating nano-systems kinetics. Bull Nov.Comp.Center, series Comp. Science (25), 1–12 (2006)
9. Achasova, S., Bandman, O., Markova, V., Piskunov, S.: Parallel Substitution Algorithm. Theory and Application. World Scientific, Singapore (1994)
10. Bandman, O.: Parallel Simulation of Asynchronous Cellular Automata Evolution. In: El Yacoubi, S., Chopard, B., Bandini, S. (eds.) ACRI 2006. LNCS, vol. 4173, pp. 41–48. Springer, Heidelberg (2006)
11. Wolfram, S.: A new kind of science - Champain, Ill. Wolfram Media Inc., USA (2002)
12. Toffoli, T., Margolus, N.: Cellular Automata Machine. MIT Press, USA (1987)

A Fast General Parser for Automatic Code Generation

Wuu Yang

National Chiao-Tung University, Taiwan, Republic of China

Abstract. The code generator in a compiler attempts to match a subject tree against a collection of tree-shaped patterns for generating instructions. Tree-pattern matching may be considered as a generalization of string parsing. We propose a new *generalized LR* (GLR) parser, which extends the LR parser stack with a parser *cactus*. GLR explores all plausible parsing steps to find the least-cost matching. GLR is fast due to two properties: (1) duplicate parsing steps are eliminated and (2) partial parse trees that will not lead to a least-cost matching are discarded as early as possible.[1]

Keywords: ambiguous grammar, code generator, compiler, context-free grammar, parsing, general parsing, pattern matching, tree pattern, tree-pattern matching.

1 Introduction

Many modern compilers consist of a front end that produces a machine-independent intermediate representation (IR) for the input program and a code generator that emits binary code for a specific computer architecture. IR usually is a tree structure and the code generator uses pattern matching to produce binary instructions. The patterns are designed according to the instruction set of the target architecture. For example, when the four tree patterns, shown in Figure 1 (a), are used to match the IR, shown in Figure 1 (b), two possible matches are shown in Figure 1 (c). We will say the IR tree is *covered* by the patterns in two different ways. Covering a tree with patterns can be regarded as parsing a string with a context-free grammar. A tree pattern can be regarded as the right-hand side of a production rule after the pattern is transformed to a linear form in the prefix order. The subject tree to be matched is similarly transformed into a string. For example, the four patterns in Figure 1 (a) can be regarded as the following four rules:

$\mathtt{r} \rightarrow \mathtt{c}$
$\mathtt{r} \rightarrow \mathtt{-\ c}$
$\mathtt{r} \rightarrow \mathtt{+\ -\ r\ r}$
$\mathtt{r} \rightarrow \mathtt{+\ r\ c}$

[1] The work reported in this paper is partially supported by National Science Council, Taiwan, Republic of China, under grants NSC-96-2628-E-009-014-MY3 and NSC 98-2220-E-009-051-.

C.H. Hsu and V. Malyshkin (Eds.): MTPP 2010, LNCS 6083, pp. 30–39, 2010.

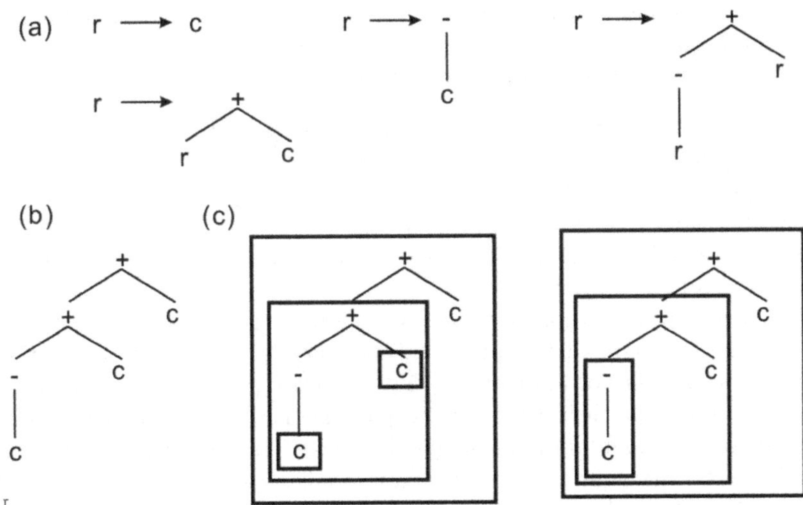

Fig. 1. (a) Four tree patterns. (b) The IR tree. (c) Two covers for the subject tree in (b).

Because tree patterns are derived from the instruction set, the resulting context-free grammar is called the *instruction grammar*. The subject tree in Figure 1 (b) is transformed into the string "`+ + - c c c`", which will be parsed with the above instruction grammar.

For most popular processor architectures, their instruction grammars are often ambiguous. That is, a task can usually be accomplished with two or more instruction sequences. The optimizer attempts to find the least-cost instruction sequence.

Since the instruction grammars are usually ambiguous, a subject tree could often be covered in two or more ways. For instance, the subject tree in Figure 1 (a) has two different covers according to the four tree patterns. We usually associate a *cost* for each production rule. The cost of a cover is the sum of the costs of the production rules used in the cover. Then our objective is to find the least-cost cover. For example, the Just-In-Time compiler in CVM [12] incorporates such a parser in its code generator.

In this paper we present a general parsing algorithm that would decide the least-cost cover. The algorithm, which is called the *GLR* parser, is a generalized LR parer. It enumerates all plausible candidate parse trees. It also attempts to skip repeated parsing steps and to discard partial parse trees as soon as it is clear that a partial tree will not result in a least-cost cover.

A sentence in an ambiguous grammar may have two or more parse trees. A straightforward representation of multiple parse trees is a collection of independent parse trees. However, this tree-collection representation suffers two weaknesses:

1. It wastes a lot of storage because the tree representations do not share common parts.
2. It does not highlight the differences between two parse trees.

We propose a new representation for multiple parse trees for the same input sentence. In the new representation, parse trees for the same input sentence will share many nodes and their differences will be highlighted with special *alternative* nodes. This new representation is built with a modified bottom-up parser. For example, Figure 2 (c) shows a composite tree that represents the two trees in Figure 2 (a) and (b).

The new representation is useful not only for ambiguous grammars. It is also useful for grammars that require a general parser. It is known that there are grammars, though not ambiguous, that require a parser to keep track of more than one potential derivation during parsing. For this kind of grammar, the parser needs to maintain multiple partial parse trees during parsing.

The GLR parser proposed in this paper makes use of this new representation to parse its input. Its output is a composite parse tree that represents one or more standard parse trees.

The rest of this paper is organized as follows: Related work is discussed in the next section. In Section 3, we will propose the general parser. We will show how to add the *alternative* nodes during parsing. In the last section, we will conclude this paper.

2 Related Work

Glanville and Graham [8] makes use of a standard LR parser for the code generator. In contrast, we use a generalized LR parser. Their method resolves conflicts by favoring shorter instruction sequences. A shift operation is always preferred in case of a shift-reduce conflict. Reduce-reduce conflicts are resolved by the semantic restrictions [9].

Christopher et al. [3] also suggests the use of general parsers, such as Earley's parser [4], to generate all possible parse trees and then to choose the one with the least cost.

Madhavan et al. [10,11] allows only tree patterns whose heights are 2 but it also allows wildcards. Our algorithm does not impose such a restriction.

Most code generation algorithms based on pattern matching relies on choosing the least-cost match among all possible matches. Some cut down the search space with various heuristics [3].

Because the instruction grammar is often ambiguous, some algorithms may not always terminate. For example, Madhavan [10] (p. 9) also mentioned that their algorithm may not always terminate and consequently has to revert to dynamic cost computations. In our case, we can use a table of pairs $(state, index)$. Whenever a repeating pair appears (which means cyclic derivation $state_1 \rightarrow state_2 \rightarrow \ldots \rightarrow state_1$ without actual progress in parsing), that branch can be abandoned without further exploration. This table can be implemented with a hash function.

The optimization phase of a traditional compiler attempts to find the least-cost among the equivalent instruction sequences. Many algorithms [8,1,6] makes use dynamic programming to generate such an optimal instruction sequence.

In our tree-pattern matching, we may adopt a parser that could enumerate all covers of a subject tree (efficiently) and the least-cost cover could be chosen. Earley's parser [4] could be a candidate parser.

If we use a general parser that could produce all possible covers for a subject tree, the least-cost cover could be easily chosen. The processing time is essentially that of the general parsers, which is no more than $O(n^3)$ for Earley's parser. On the other hand, algorithms based on dynamic programming usually assume a *contiguous execution* property [1] in the original machine instruction.

Aho et al. [1] constructs a finite state machine (similar to LR(0) machine) from the path strings. Our approach is similar to theirs.

3 GLR(1) Parsers

Similar to an LR(1) parser, a GLR(1) parser consists of a goto table and an action table [2,5]. Since we allow grammars with conflicts, an entry in the action table is a set of shift and/or reduce actions. (Note that in an LR(1) action table, an entry is either a shift action or a reduce action. No conflicts are allowed.) The goto table in the GLR(1) parser still encodes the LR(1) finite-state machine.

The parser stack in an LR(1) parser is implemented as an *ordered list of entries*. In contrast, the GLR parser is equipped with a *parser cactus*, which represents several parser stacks simultaneously. (A *cactus* is actually a tree structure. We choose the word "cactus" in order to avoid confusion with the term *parse trees* widely used in the parsing theory.) A path from the root to a leaf in the parser cactus represents a parser stack. Each parser stack in the cactus denotes a possible parsing process.

A node in the parser cactus corresponds to an entry in the traditional parser stack. The nodes are linked together with pointers. A leaf node denotes the top of a stack. A push operation is implemented as adding a node to a leaf; a pop operation is implemented as forking a new node at the appropriate node. All the leaf nodes are collected in the *Frontier* set. (In the algorithm below, we will leave popped nodes in the cactus in order to simplify the algorithm.)

The input string, which is the pre-order traversal of the subject tree, is stored in an 1-dimensional array *input*.

When a node A is added to the *children* set of another node B, we assume that there is an implicit pointer stored A that points to B. We do not show this pointer the following algorithm. See lines 17 and 28 in the *GLR* parsing algorithm below.

A *node* denotes an entry in the parse stack, which is declared as follows:

```
record
state: an LR(1) state
index: the position of the input symbol at this state
children: the set of child nodes
end record
```

A node's *state* is the parser state when the node is on the top of the parse stack. A node's *index* is the position of the input symbol when the node is on the top of the parse stack. A node's *children* is the set of all children of the node in the cactus. The *children* field is always initialized as the empty set when a node is allocated.

In line 22 in the *GLR* parsing algorithm below, we used the term *m-th ancestor*, which is defined as follows: a node's parent in the cactus is the *1st ancestor* of that node; the parent of a node's parent is the *2nd ancestor* of that node; etc. In LR parsing, to pop off m symbols from the stack is to locate the m-th ancestor of the stack-top node in the cactus. In order to accommodate λ-productions, we follow the convention that the 0-th ancestor of a node is the node itself.

```
    Algorithm GLR parser
1.    allocate a new node Root;
2.    Root.state := start state;
3.    Root.index := 1;
4.    Root.children := { }; /* an empty set initially */
5.    Frontier := {Root }; /* a set consisting of a single node */
6.    Accepted := { }; /* an empty set initially */
7.    while Frontier is not empty do {
8.        x := pick the first node from Frontier;
9.        Frontier := Frontier − {x};
10.       for each action actn in action(x.state, input[x.index]) do {
11.           switch(actn) {
12.           case shift:
13.               allocate a new node N;
14.               N.state := goto(x.state, input[x.index]);
15.               N.index := x.index + 1;
16.               N.children := { };
17.               x.children := x.children ∪ {N};
18.               Frontier := Frontier ∪ {N};
19.               break;
20.           case reduce i:
21.               m := length of the right-hand side of rule i;
22.               s := the m-th ancestor of node x;
23.               t := left-hand-side nonterminal of rule i;
24.               allocate a new node N;
25.               N.state := goto(s.state, t);
26.               N.index := x.index; /* do not consume input. */
27.               N.children := { };
28.               s.children := s.children ∪ {N};/* fork a new child */
29.               Frontier := Frontier ∪ {N};
30.               break;
31.           case accept:
32.               Accepted := Accepted ∪ {x};
33.               break;
```

```
            case error: /* simply throw away the node x. */
            } /* end of switch */
        } /* end of the for loop */
    } /* end of the while loop */
```

After the *GLR* parser terminates, each node in the *Accepted* set corresponds to a derivation, which is a cover of the subject tree with the given tree patterns. If the *Accepted* set is empty, the input string cannot be generated from the grammar. This means that the subject tree cannot be covered by the tree-patterns.

Example. Consider the following (ambiguous) grammar:

$S \rightarrow A\ B\ C$
$A \rightarrow u$
$C \rightarrow v$
$B \rightarrow P\ y$
$B \rightarrow Q\ y$
$P \rightarrow x$
$Q \rightarrow x$

Figure 2 (a) and (b) shows two parse trees for the input `uxyv`. Figure 2 (c) shows the composite parse tree. The square node "`Alt(B)`" means that there are two candidates for the subtree rooted at `B`. Figure 3 shows the parsing steps for the input `uxyv`. Node N0 contains the initial state `Init`. The first three steps are shift, reduce, and shift, respectively. At steps d and e, the terminal `x` is reduced to `P` and `Q`, respectively. There are two branches (i.e., two stacks) on the parse cactus at this step. At steps f and g, a shift operation is performed on each of the two stacks. At steps h and i, a reduce operation is performed on each of the two stacks. Note that at step i, the new node—N9—and the existing node

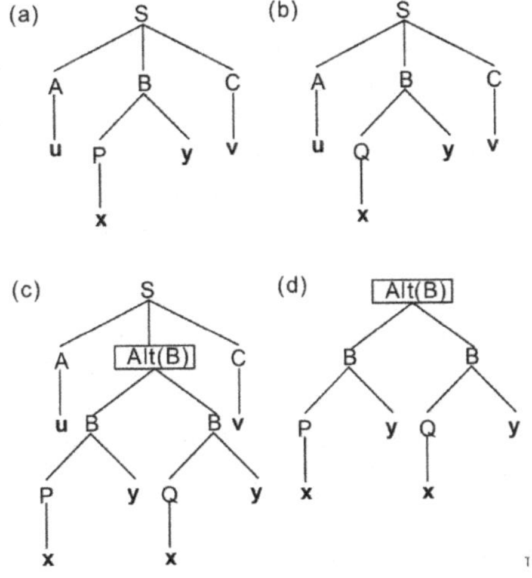

Fig. 2. Two parse trees for the sample input `uxyv`

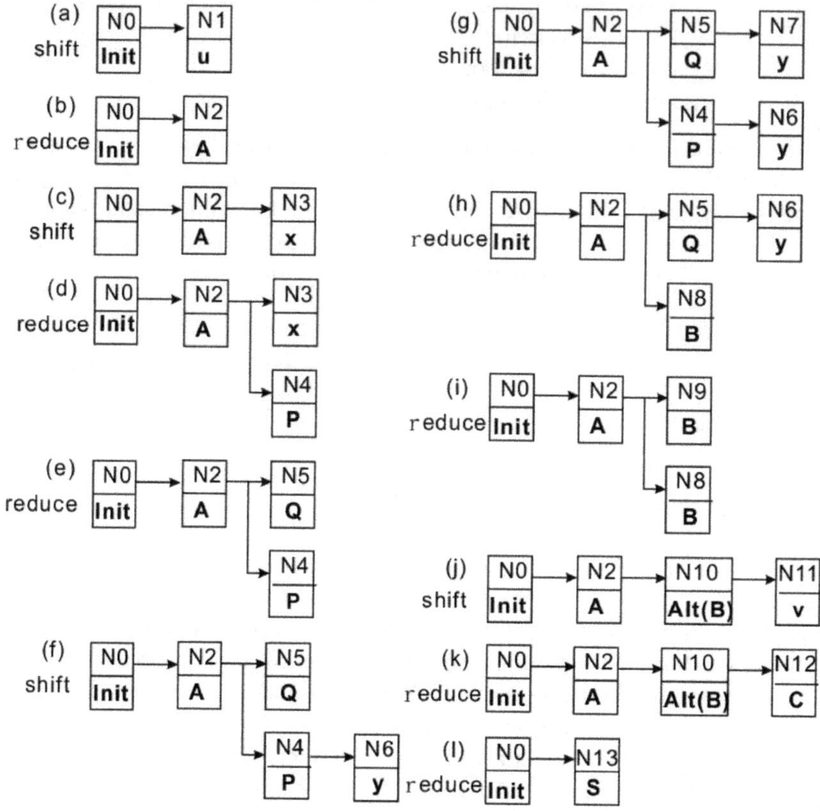

Fig. 3. The parser cactus during the parsing steps. N0 is the initial state `Init`.

N8 are combined together (thus creating the `alternative` node N10) because they both contain the same parser state and they represent the same segment of the input (that is, the segment `xy`). The subtree stored in node N10 is shown in Figure 2 (d). Afterwards, `v` is shifted into the stack and then is reduced to the nonterminal `C`. Finally, the sentential form `ABC` is reduced to the start symbol `S`. The tree stored in node N13 is the complete composite parse tree, which is shown in Figure 2 (c). □

There are several assumptions in the above algorithm: First, we assume nodes in the *Frontier* set are ordered by their *index* fields. When a node N is added to *Frontier* (in line 18 and line 29), it is implicitly assumed to be inserted to correct place in *Frontier*. Thus, in line 8, "pick the first node from *Frontier*", a node with the smallest *index* field is picked up.

3.1 Introducing the `Alternative` Nodes

The above algorithm demonstrates only the parsing part. A collection of parse trees can be built during parsing, as follows. Every node in the cactus contains the subtree rooted at that node. When a shift operation is performed (in line 12), the new node contains a tree that has the newly shifted terminal symbol only. When a reduce operation is performed (in line 20), a tree is created whose root is the nonterminal t (in line 23) and whose children are the m popped-off nodes. This tree is stored in the new node N (line 24).

When a reduce operation is performed, a new node is created. Before forking a new branch, we check if the new node has a sibling that contains the same parser state and represents the same segment of the input. If so, the two nodes will be combined together and a new `alternative` node will be created to represent the two nodes. We will introduce the following definition:

Definition. Two nodes in the parser cactus are called *twins* if they satisfy the following three conditions:

1. They are siblings, i.e., they have the same parent in the cactus.
2. They represent the same segment of input.
3. They are in the same parser state.

The second condition is easy to evaluate since nodes in the *Frontier* set is ordered by their input positions (i.e., indices). The other two conditions are also easy to evaluate.

Twins can be combined into an `alternative` node. For example, in Figure 3 (i), nodes N8 and N9 are twins because they contains the same state and represents the same segment of the input (that is, `xy`). Thus, N8 and N9 are combined, creating the node N10, which is an `alternative` node, shown in Figure 3 (j). The node N10 contains the subtree shown in Figure 2 (d). Combining two nodes means that further derivations from the two nodes are identical. Parsing time is also reduced by avoiding repeated derivations.

The `alternative` nodes are effective in that the conflicts in the instruction grammars are usually restricted to certain subtrees. Even though a subject tree may have more than one cover, these covers share many common nodes. The `alternative` nodes can reduce memory usage and the parsing time.

3.2 Eliminating Ineligible Parser Stacks

We will also add a *cost* field to every node. The *cost* of a leaf node is 0. When a `reduce` i action is performed (line 20), the cost of a newly allocated node N is the sum of the cost of the m nodes corresponding to the right-hand side of production i plus the cost of production i. The cost of a production rule is defined by the tree-pattern designer. The cost could indicate the number of registers used in the subtree, the number of instructions generated for the subtree, the number of memory accesses needed during the execution of the instructions translated from from the subtree, or the execution time for subtree, etc. The computation of the cost can be formally specified with an attribute grammar, with a single synthesized attribute *cost*.

In a code generator, we are interested in the *least-cost* parse tree(s). All other parse trees are discarded. In the GLR parser, we can remove a parse tree before it is completely built. After we have decided two nodes are twins, instead of combining them, the one with the larger cost can be deleted immediately, saving the time and space required for further exploration.

Note that, in an ambiguous grammar, an input sentence may have an infinite number of parse trees. This is due to recursive null productions, as follows:

$$\mathtt{A} \rightarrow \mathtt{B}$$
$$\mathtt{B} \rightarrow \mathtt{A}$$
$$\mathtt{B} \rightarrow \lambda$$

By adding the additional requirement that the cost of a production rule is always positive, the *GLR* parser will always terminate when the cost is more than that of the currently optimal parse trees.

4 Conclusion

The code generator is an important component in a compiler. Pattern matching is frequently used to cover a subject tree with patterns in a code generator. Tree pattern matching can be regarded as a general parsing problem, with an often ambiguous instruction grammar. We designed a general parsing algorithm that can find a least-cost cover for a subject tree. We also developed a data structure that can represent multiple parse trees.

References

1. Aho, A.V., Ganapathi, M., Tjian, S.W.K.: Code Generation Using Tree Matching and Dynamic Programming. ACM Transactions on Programming Languages and Systems 11(4), 491–516 (1989)
2. Aho, A.V., Lam, M.S., Sethi, R., Ullman, J.D.: Compilers, Principles, Techniques, and Tools, 2nd edn. Addison-Wesley, New York (2007)
3. Christopher, T.W., Hatcher, P.J., Kukuk, R.C.: Using Dynamic Programming to Generate Optimized Code in a Graham-Glanville-Style Code Generator. In: Proceedings of the 1984 SIGPLAN symposium on Compiler construction, ACM SIGPLAN Notices, vol. 84, pp. 25–36. ACM, New York (1984)
4. Earley, J.: An efficient context-free parsing algorithm. ACM Comm. 13(2), 94–102 (1970)
5. Fischer, C.N., LeBlanc Jr., R.J.: Crafting a Compiler with C. Benjamin/Cummings, Reading (1991)
6. Fraser, C.W., Hanson, D.R., Proebsting, T.A.: Engineering a Simple, Efficient Code Generator Generator. ACM Letters on Programming Languages and Systems 1(3), 213–226 (1992)
7. Glanville, R.S.: A machine independent algorithm for code generation and its use in retargetable compilers. Ph.D. dissertation, University of California, Berkeley (December 1977)
8. Steven Glanville, R., Graham, S.L.: A new method for compiler code generation. In: Proceedings of 5th ACM SIGACT-SIGPLAN Symposium on Principles of Programming Languages, Tucson, Arizona, pp. 231–254 (1978)

9. Nigel Horspool, R.: An Alternative to the Graham-Glanville Code-Generation Method. IEEE Computers 20, 33–39 (1987)
10. Madhavan, M., Shankar, P.: Optimal regular tree pattern matching using pushdown automata. In: Arvind, V., Sarukkai, S. (eds.) FST TCS 1998. LNCS, vol. 1530, pp. 122–133. Springer, Heidelberg (1998)
11. Madhavan, M., Shankar, P., Rai, S., Ramakrishna, U.: Extending Graham-Glanville Techniques for Optimal Code Generation. ACM Transactions on Programming Languages and Systems 22(6), 973–1001 (2000)
12. Sun Microsystems, Connected, Limited Device Configuration, Specification Version 1.0a, Java 2 Platform Micro Edition 1.0a, CA, USA (May 19, 2000)

A Multi-core Software API for Embedded MPSoC Environments

Jia-Jhe Li, Shao-Chung Wang, Po-Chun Hsu, Po-Yu Chen, and Jenq Kuen Lee

Department of Computer Science
National Tsing Hua University, Hsinchu, Taiwan
{jjli,scwang,pchsu,pychen}@pllab.cs.nthu.edu.tw, jklee@cs.nthu.edu.tw

Abstract. Heterogeneous multi-core architectures are the mainstream of processor designs for high-end embedded systems. Although such architectures promise high performance and low power consumption, challenges are raised for how to program such devices. This paper presents "Multi-core Software APIs" (MSA) to address these issues. MSA is a library-based framework based asynchronous remote procedure call (RPC) mechanism. Aiming at distributed memory architectures, which is common in embedded systems, MSA supplies a function-offloading programming model. MSA consists of three modules, RPC module, message module, and streaming module, to provide task offloading, data transmission, and streaming data transmission, respectively. Furthermore, this paper provides two case studies, π calculation and stereo vision, to show how MSA works on building multi-core applications.

Keywords: multi-core, programming, remote procedure call, streaming.

1 Introduction

In recent years, heterogeneous multi-core architectures have become the mainstream of processor designs for high-end embedded systems [18,14]. Such architectures consist of a group of different processors connected with interprocessor communication mechanisms such as directed memory access, share memory, and interrupts. Exploiting the unique characteristics of different cores and associated hardware is then critical to benefit from such architectures. In heterogeneous multi-core systems, applications are usually broken into smaller tasks, which are distributed to different cores for utilizing certain edges of a core. As a result, providing a paradigm to manage tasks for migration and communication raises challenges of programming such architectures.

One of the most promising programming models to overcome these issues is remote procedure call (RPC). RPC treats each component on different cores as a remote procedure, and allows remote procedures to be invoked by other procedures on other cores. Based on this model, we present a Multi-core Software API, or MSA, to ease programming challenges of heterogeneous multi-core systems. MSA is a library-based framework for heterogeneous multi-core architecture. Due to the nature of constrained resource on embedded systems, MSA is

C.H. Hsu and V. Malyshkin (Eds.): MTPP 2010, LNCS 6083, pp. 40–50, 2010.

designed as compact as possible by providing only the necessary functionalities. For offloading and managing tasks on heterogeneous multi-core environments, MSA provides a set of RPC APIs for programmers. For communications between different tasks across cores, MSA supplies message passing and data streaming module to handle control data and streaming data. Beyond the functionality on programming embedded multi-core system, the high-level library design allows software developers to deliver their applications efficiently to different platforms and shorten the time to market.

The rest of this paper is organized as follows. Section 2 describes the architecture that we target. Section 3 provides an overview and design of MSA. Section 4 exhibits two case studies and experimental results. Section 5 examines related work and section 6 concludes our work.

2 Heterogeneous Multi-core System

The heterogeneous multi-core system this paper experimented is based on a configurable SID-based Multi-core Simulator [7], which is shown in Figure 1. It simulates a virtual platform consisting of one master core, four slave cores, and peripherals like LCD controller and camera. The master core is a RISC processor aiming for high-end applications, and slaves cores are served with PAC DSPs [2], which are heterogeneous five-way issue VLIW digital signal processors.

The virtual platform employs a distributed memory sub-system, where the main memory connects to cores in the system and private memories dedicate to their associated DSP. While the virtual platform allows MPU to access main memory and private memories of DSPs seamlessly, it restricts DSPs to access their own private memory and inflicts serious access latency when they step into main memory. On the other hand, MPU equips a L1 data and instruction cache which causes issues on memory coherence between MPU and DSPs. All these cores are set to 300 MHz for performance measurement. All components are connected with an interconnection model to provide accurate timing information.

The virtual platform provides a linux-based environment for software development, where the version of linux kernel is 2.6.18. We developed a device driver for the DSP to provide underlying mechanisms like activates DSP and deactivates DSP. Furthermore, GCC 3.4.4 is used to compile all software executing on the master core, and pacc32 [10] is used to compile those programs running on slave cores. All compilation is done with optimization level at O2.

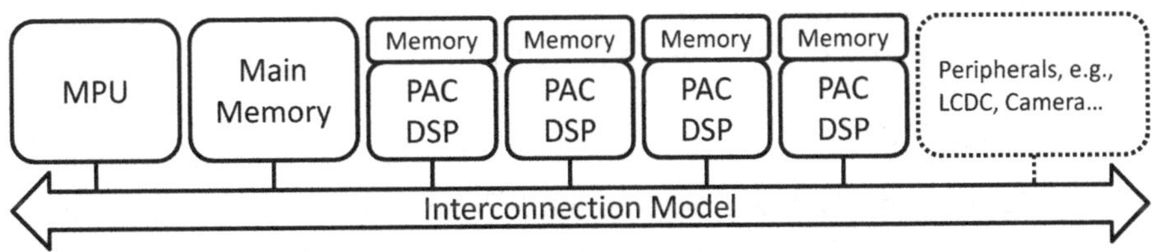

Fig. 1. Experimental Environment

Table 1. Highlight of Multi-core Software API

RPC Module	
msa_load_procset(char *i, CoreID c)	load executable image "i" to "c" core
msa_RPC_start(char *p, CoreID c)	start procedure "p" on "c" core
Message Module	
msa_msg_send(MsgID m, void *p, MsgAttr *a)	send data from pointer "p" to message box "m" with given attribute "a"
msa_msg_recv(MsgID m, void *p)	receive data from message box "m" to pointer "p"
Stream Module	
msa_stream_send(StrID s, void *p, StrAttr *a)	send data from pointer "p" to streaming channel "s" with given attribute "a"
msa_stream_recv(StrID s, void *p)	receive data from streaming channel "s" to pointer "p"

3 The Design of Multi-core Software API

This paper presents "Multi-core Software API," MSA, to enable programming in embedded systems with heterogeneous multi-core. With MSA, a software developer breaks an application into a main application and several procedure sets. While the main application contains the part of the program going to be executed on the master core, a procedure set contains one or more procedures going to be executed on a slave core. Note that each core may equip a different instruction set, and, as a result, each procedure set has to be compiled and linked separately with corresponding tool-chains. Procedures in a procedure set can be identified as two categories, remote procedures and local procedures. The different is that a remote procedure can be invoked by the main application, but a local procedure can only be executed by procedures in the same procedure set. In such paradigm, the main application manages the control flow of the whole application, and procedure sets wait for orders from main application to do the computation. The main application and procedure sets exchange data by message and streaming mechanisms provided by MSA, where message is for occasional data and streaming is for continuous data. Furthermore, both mechanisms support collective operations for communication-intensive programs. The rest of this section discusses the detail of them.

3.1 Using MSA

Task migration in MSA is done through RPC module. RPC module maintains a "core ID table" and a "remote procedure table" to manage slave cores and in-between procedure sets. The core ID table statically contains unique IDs and physical information such as memory maps of cores in the system; the remote procedure table holds names and corresponding addresses of remote procedures.

To activate a slave cores, `msa_load_proceset` in Table 1 loads an executable image onto a specific core. This function not only copies instructions and data section from image `i` to the slave processor `c`, but also parses the symbol table for function information to construct the remote procedure table. Once an image has been loaded to the associated core, the master core can then use `msa_RPC_start` to invoke remote function `p` on core `c`. Note that `msa_RPC_start` invokes remote procedures in the asynchronous manner, and returns to its caller immediately. MSA also provides APIs to suspend, resume, and stop remote procedures as well, although they are omitted in table 1 due to limit of paper length.

On the other hand, exchanging data between procedures distributed in different cores could be daunting. This is because that each core in system could provide different architecture designs, for example, cache and non-cache, physical and virtual memory, and big-endian and little-endian. Furthermore, these issues make designing scalable and portable applications more challenging on heterogeneous multi-core environments. To hide these hardware details from programmers, MSA provides communication modules for two types of data with a centralized mechanism.

One of the communication modules is streaming module. It is designed for transmitting continuous data between procedures. Aiming at hiding buffer management and synchronization between cores, streaming module refers to [6] to encapsulate these tedious details to improve programmability. This module defines a set of "streaming channels," and a sender uses `msa_stream_send` to send data to a streaming channel. Similar to send operation, a receiver calls `msa_stream_recv` to receive data from a channel. Once a channel is full (empty) because of asymmetrical speeds of senders and receivers, streaming module puts the sender (receiver) to sleep to allow execution of other tasks.

Another communication module is message module, which is designed for transmitting occasional data. This module transmits such data efficiently by excluding the overhead of buffer managements. Similar to streaming module, message module maintains a set of "message boxes" to provide a central view of communications between procedures. Each message box consists of a message ID (MID), data attributes, and associated data. A sender invokes `msa_msg_send` to explicitly assign a MID to store the data, and a receiver requests data by

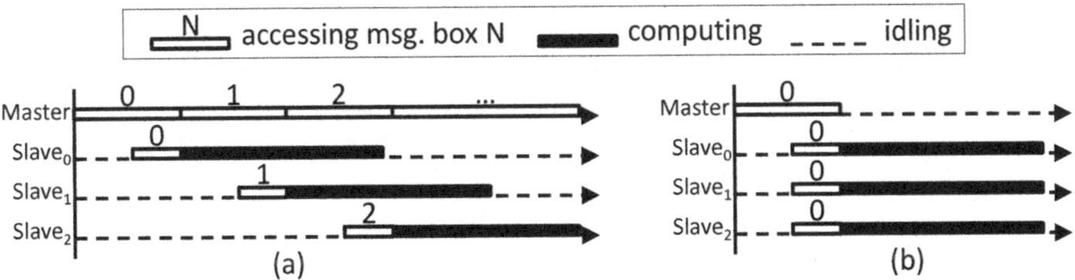

Fig. 2. Broadcasting without/with Availability

msa_msg_recv with the same MID. Note that senders and receivers does not directly communicate with each other, and therefore hide hardware-variant characteristics from programmers.

3.2 Collective Communication

Collective communications, such as broadcasting (one to all), multicasting (one to many) and gathering (many to one), are in urgent need of many parallel programming patterns. For example, in the π-calculation program which will be discussed in section 4.1, the master has to inform all slaves the number of trails, and slaves return their result to the master. Although the concept of collective communication is transparent, it is delicate to implement these operations on MSA framework.

Figure 2(a) instances a broadcast operation where a master sends a piece of data to slaves. Since the master can only do one thing at a time, it first sends the data to message box 0, and waits for slave_0 to retrieve it. After slave_0 gets the data, the master sends the data the message box 1 and waits for slave_1. This process continues until all slaves got the data. One can find that each slave spent most of its time for waiting the master's data, which makes the master the bottleneck once the number of slaves increases. This is because that the master has to send the same data multiple times. In the other word, easing the overhead of the master is critical to improve the efficiency of collective communication.

To tackle this issue, MSA defines an attribute called "availability" for each message box and streaming channel. Availability stands for times that a piece of data can be retrieved. By defining this attribute, the sender only transmits the same data to a message box for once, but allows the same data to be retrieved for multiple times. Figure 2(b) shows the same broadcast scenario but set the availability to three. In this case, the master still sends the data to message box 0. However, each slave can retrieve the data from message box 0 directly without further delay. The benefit of this design is significant especially when the core number is large, because that the average idle time in (a) increases as well as the core number increases, while the idle time in (b) is almost a constant.

```
int main() {
  int hits = 0, trials = 10000;
  float pi;

  hits = hit_calculation(trials);

  pi = (float)4*hits/trials;

  printf("PI is %f\n", pi);
  return 0;
}
```

```
int hit_calc(int trials) {
  int i, hits = 0;
  float x, y;

  for (i=0; i<trials; i++) {
    x = rand();
    y = rand();
    if ((x*x + y*y) <= 1.0) hits++;
  }

  return hits;
}
```

Fig. 3. Sequential Code Example of the Monte Carlo Calculation of Π

4 Case Study

4.1 *Π* Calculation

To demonstrate how MSA can be applied to RPC and message modules, this section presents a π calculation program with the Monte Carlo method as an example. The concept of this method is to randomly generate coordinates that uniformly distributes on an 1×1 square, and compute the ratio of coordinates within the quarter circle in the same square. This ratio should be π / 4. While this is a trivial example, it covers how task offloading and communication work with MSA.

Figure 3 shows the single core program for calculating π [4]. The program consists of two functions, `main` and `hit_calc`. `main` first initializes necessary variables, then it invokes `hit_calc` and supplies a parameter indicating how many coordinates should be generated and tested. After `hit_calc` completes, it returns to `main` with the number of coordinates dropping within the circle. `main` then accumulates the hit ratio and calculates π. One can observe that each iteration of the for loop in `hit_calc` is independent from each others. Which means that `hit_calc` can be duplicated and distributed to different cores for improving the performance.

With this observation in mind, the program can be rewritten as shown in Figure 4. In the revised version, the `main` function executes at the master core, and `_RPC_hit_calc` is going to be compiled for different architectures and distributed to slave cores for execution. The first `for` loop in `main` loads image files to different cores with `msa_load_procset` and invokes `_RPC_hit_calc` on different slave cores with `msa_RPC_start`.

After invocations, `main` sends the number of trails to message box 0 to inform slave cores, while each slave core is blocked for that. Note that `msa_msg_send` set the availability of message 0 to exact core number for broadcasting. `main` then snoops message boxes to wait for result of slave cores, and calculates the result of π before the program stops.

Table 2. Execution Time of *Π* Calculation

	MPU	MPU + 1 DSP	MPU + 2 DSPs	MPU + 4 DSPs
Execution Time	10.48	5.68	3.14	2.12

Table 2 shows the experimental result with 10000 trails. The execution time on single core (MPU) scenario was 10.48 seconds, which was measured with the program in figure 3. Once we offloaded the function `hit_calc` as showed in figure 4, the execution time dropped dramatically. On one MPU and one DSP scenario, the execution time was 5.68 seconds, which was faster than the single core scenario by 4.8 seconds. Furthermore, the experiment also showed the well scalability of MSA, where the execution time of three cores and five cores environments were 3.14 and 2.12 seconds respectively.

```
int main() {
  int hits = 0, trials = 10000;
  int partial_hits = 0;
  float pi;
  CoreID cid;
  int local_trials = trials / NUMCORE;
  MsgAttr m = {sizeof(int), NUMCORE};

  msa_msg_init();

  for (cid=0; cid<NUMDSP; cid++) {
    msa_load_procset(image[cid], cid);
    msa_RPC_start("_RPC_hit_calc", cid);
  }

  msa_msg_send(0, &local_trials, &m);

  for (cid=0; cid<NUMCORE; cid++) {
    msa_msg_recv(cid, &partial_hits);
    hits += partial_hits;
  }
  pi = (float)4*hits/trials;

  printf("PI is %f\n", pi);
  return 0;
}
```

```
void _RPC_hit_calc() {
  int trials, i, hits = 0;
  float x, y;
  MsgAttr m = {sizeof(int), 1};

  msa_msg_recv(0, &trails);

  for(i=0; i<trials; i++) {
    x = rand();
    y = rand();
    if ((x*x + y*y) <= 1.0) hits++;
  }

  msa_msg_send(COREID, &hits, &m);
  return;
}
```

Fig. 4. MSA Example of the Monte Carlo Calculation of Π

4.2 Stereo Vision with Belief Propagation Method

This paper enabled a stereo vision application on the SID ESL platform to demonstrate how MSA works in real world applications. Stereo vision has been extensively investigated in the recent years to provide high-quality applications. One of the most advanced applications is to apply the technique in inferring the 3-D position of an object. By computing the depth and disparity of referenced images through matching the images in the same plane of different view position, the range information of the environment can help robots to adapt to the real world. One of the most significant approximate algorithms is belief propagation (BP) that gathers information from the neighborhoods of each pixel in an image to find the minimum matching cost of the local point and its neighborhoods [17]. Although having highly accurate results with outstanding quality, BP requires significant computation power that makes it less practicable in real-time application domains.

This paper enabled a BP application based on [9]. Figure 5 shows the execution flow of BP. BP can be recognized as three stages, where the first stage initializes the execution with two images from different viewpoints, the second stage gathers and updates the messages iteratively from the neighboring nodes for disparity cost, and the third stage summarizes the distance information. Among all of these stages, the second stage is an obvious target of optimization for the iterative execution. In the second stage, each node gathers the message

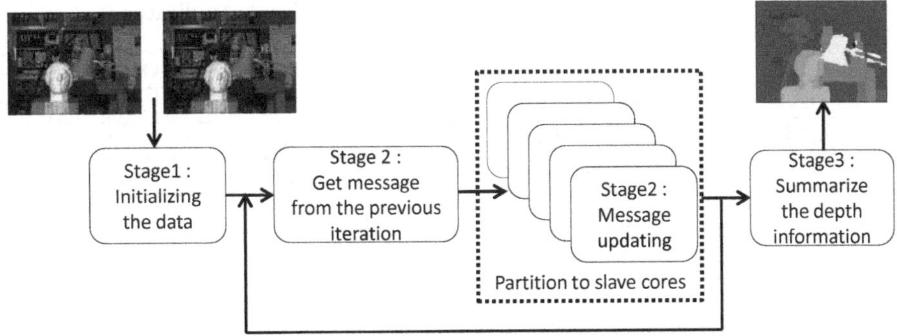

Fig. 5. Execution Flow of Belief Propagation

computed in the previous iteration, and then updates the message stored in up, down, left, right direction nodes in each iteration.

To utilize such observation, we extracted stage two procedures from the main application, and packaged these procedures as a procedure set. Executing the modified BP first offloads the second stage to slave cores, and then partitions and sends the input images according to number of slave cores. The main application then checks assigned message boxes to gather results of every slave cores and update the distance information with the results. This process repeats until reached the requested iteration number of stage two, and then starts stage three to produce the final distance information.

Table 3 shows the breakdown analysis of BP on different configurations from gprof on MPU. The input data we used are two images and each one contains 96×72 pixels. Furthermore, we limited the stage two to execute just once to shorten the time of experiments. We further classified procedures in BP into three groups: MPU computation, major computation, and MSA overhead. Major computation stands for the part of procedures to be offloaded to DSPs; MPU computation is the group of computation remaining on MPU; MSA overhead consists of MSA functions. Table 3 exhibits those information based on report from gprof. Each group in the table includes two numbers, where the "Time" is the number reported by gprof in seconds, and "Ratio" shows the time relating to the total execution time of MPU-only configuration.

This analysis shows that MPU computation takes 18.56 seconds in MPU-only scenario. The times measured in other scenarios are similar to MPU-only with negligible differences that were caused by error of time measuring. On the other hand, major computation spent 35.34 seconds on MPU-only scenario. Offloading the part reduced more than 60% of execution time in MPU + 1 DSP by utilizing the edge of the DSP. Cooperating multiple DSPs further reduced the execution time of major computation. Experimental result indicated the execution time of major computation on one DSP, two DSPs, and four DSPs scenarios are 3.04, 1.16, and 0.94 seconds, respectively. Observation indicated that the major reason causing the non-linear improvement was again error of time measuring.

Table 3. Breakdown Analysis of Stereo Vision Application

	MPU Computation		Major Computation		MSA Overhead		Total
	Time	Ratio	Time	Ratio	Time	Ratio	Time
MPU only	18.56	34.4%	35.34	65.6%	0	0%	53.9
MPU + 1 DSP	18.48	34.3%	3.04	5.6%	1.5	2.8%	23.02
MPU + 2 DSPs	18.69	34.7%	1.16	2.2%	1.48	2.7%	21.33
MPU + 4 DSPs	18.51	34.3%	0.94	1.7%	1.4	2.6%	20.85

Experience of measuring MPU computation showed that the error could up to 0.21 seconds, which made the execution time of major computation on multi-core scenarios reasonable.

Finally, the single core version of BP method on MPU spent 53.9 seconds to complete an iteration. Partitioning `msg` function to a DSP immediately cut off 30.88 seconds, which made the execution time of one MPU and one DSP version to be 23.02 seconds. Three cores and five cores environments further reduced execution times to 21.33 and 20.85 seconds respectively. On the other hand, experimental result showed that MSA only took about 1.5 seconds on each configuration. The overhead are relatively small, less than 3%, when comparing to the total execution time. All these experiments show that MSA is a lightweight yet efficient library for embedded multi-core systems.

5 Related Work

Many programming models are proposed to relieve the complexity of developing multi-core applications. Models like [16,1,3] supply the multi-thread programming model to express parallelism on homogeneous multi-core systems. Since such architectures consist of cores with identical instruction set architectures, these models lack of considering characteristics of different cores. MSA addresses this issue by providing a RPC-like mechanism to explicitly offload computations to specific cores and manage tasks among different cores.

While there are several languages proposed to either hide the underlying details of cores from programmers [19,12] or expose those cores for allowing explicit manipulation [15], they all require compiler and/or runtime supports to achieve that. Which is not desirable for industries of embedded systems because developing and verifying compilers and optimizations for new languages are expensive. Furthermore, porting legacy programs to utilize features of a new language is tedious and error-prone. On the other hand, MSA provides a library-based mechanism and therefore cooperates with any existing compilers.

Streaming RPC [6] is the most related to our work. It enables library-based design flow to support asymmetric streaming-function off-loading. While it provides a formal mathematical model in building such framework, the original form is more on dual-core design. Our work in this paper provides the linguistics for multi-core and multi-cast version of streaming RPC. There are other parallel libraries aiming for programming multi-core applications. MapReduce [5] allows users to assign a "map" function to process given key/value pairs on cluster environments, and a "reduce" function to merge results returning from distributed map functions. Similar to MapReduce, [8,13,11] all provide structured interfaces for software developers to design their parallel programs. However, all these works are designed for cluster environments and not suitable on embedded system with heterogeneous multi-core and distributed memory sub-system.

6 Conclusion

This paper presented and implemented a "Multi-core Software API" to enable easy, scalable, and portable programming for heterogeneous multi-core systems with distributed memory architectures. In addition to introduction of MSA, we also provided two case studies to show how to apply MSA on the "π calculation" problem with the Monte Carlo method, and a "stereo vision" application with belief propagation method. While the result suggested that MSA is a lightweight and efficient solution for programming heterogeneous multi-core systems, there are still challenges ahead. For example, how to support more programming paradigms, such as map-reduce, on heterogeneous multi-core embedded systems with scalability and portability; how to lower the communication overhead in such centralized environment.

References

1. Butenhof, D.R.: Programming with POSIX threads. Addison-Wesley, Reading (1997)
2. Chang, D.C.-W.: PAC digital signal processor. In: Proceedings of Fall Microprocessor Forum (2006)
3. Chrysanthakopoulos, G., Singh, S.: An asynchronous messaging library for c#. In: Synchronization and Concurrency in Object-Oriented Languages, SCOOL (2005)
4. Culler, D.E., Dusseau, A., Goldstein, S.C., Krishnamurthy, A., Lumetta, S., von Eicken, T., Yelick, K.: Introduction to split-c. Tech. rep., University of California–Berkeley (April 1995), `http://www.cs.cmu.edu/~seth/papers/dusseau-tr92.pdf`
5. Dean, J., Ghemawat, S.: Mapreduce: simplified data processing on large clusters. In: OSDI'04: Proceedings of the 6th conference on Symposium on Opearting Systems Design & Implementation, p. 10. USENIX Association, Berkeley (2004)
6. Hsieh, K.Y., Liu, Y.C., Wu, P.W., Chang, S.W., Lee, J.K.: Enabling streaming remoting on embedded dual-core processors. In: 37th International Conference on Parallel Processing, ICPP '08, September 2008, pp. 35–42 (2008)
7. Huang, C.W., Shih, W.K., Hsu, Y., Lee, J.K.: Configurable sid-based multi-core simulators for embedded system education. In: Workshop on Embedded Systems Education'09, Grenoble, France (2009)
8. Kuchen, H.: A skeleton library. In: Monien, B., Feldmann, R.L. (eds.) Euro-Par 2002. LNCS, vol. 2400, pp. 620–629. Springer, Heidelberg (2002)

9. Lai, C., Hsieh, K., Lai, S., Lee, J.: Parallelization of belief propagation method on embedded multicore processors for stereo vision. In: IEEE/ACM/IFIP Workshop on Embedded Systems for Real-Time Multimedia, ESTImedia 2008, pp. 39–44 (2008)
10. Lin, Y.C., Tang, C.L., Wu, C.J., Hung, M.Y., You, Y.P., Moo, Y.C., Chen, S.Y., Lee, J.K.: Compiler supports and optimizations for PAC VLIW DSP processors. In: Ayguadé, E., Baumgartner, G., Ramanujam, J., Sadayappan, P. (eds.) LCPC 2005. LNCS, vol. 4339, pp. 466–474. Springer, Heidelberg (2006)
11. Matsuzaki, K., Iwasaki, H.: A library of constructive skeletons for sequential style of parallel programming. In: InfoScale' 06: Proceedings of the 1st international conference on Scalable information systems. ACM International Conference Proceeding Series, vol. 152, p. 13. ACM Press, New York (2006)
12. McCool, M.D., Wadleigh, K., Henderson, B., Lin, H.Y.: Performance evaluation of gpus using the rapidmind development platform. In: SC '06: Proceedings of the 2006 ACM/IEEE conference on Supercomputing, p. 181. ACM, New York (2006)
13. Murray, A.B., Cole, M., Gilmore, S., Hillston, J.: Flexible skeletal programming with eskel. In: Cunha, J.C., Medeiros, P.D. (eds.) Euro-Par 2005. LNCS, vol. 3648, pp. 761–770. Springer, Heidelberg (2005)
14. Qualcomm: The snapdragon platform (2010), `http://www.qctconnect.com/products/snapdragon.html`
15. Reid, A.D., Flautner, K., Evans, E.G., Lin, Y.: Soc-c: efficient programming abstractions for heterogeneous multicore systems on chip. In: CASES '08: Proceedings of the 2008 international conference on Compilers, architectures and synthesis for embedded systems, pp. 95–104. ACM, New York (2008)
16. Reinders, J.: Intel Threading Building Blocks. O'Reilly, Sebastopol (2007)
17. Sun, J., Zheng, N., Shum, H.: Stereo matching using belief propagation. IEEE Transactions on Pattern Analysis and Machine Intelligence 25(7), 787–800 (2003)
18. Texas Instruments: Omap^{TM}4 mobile applications platform (2009)
19. Thies, W., Karczmarek, M., Amarasinghe, S.: Streamit: A language for streaming applications. In: Horspool, R.N. (ed.) CC 2002. LNCS, vol. 2304, pp. 49–84. Springer, Heidelberg (2002)

A VLIW-Based Post Compilation Framework for Multimedia Embedded DSPs with Hardware Specific Optimizations

Meng-Hsuan Cheng, Kenn Slagter, Tai-Wen Lung, and Yeh-Ching Chung

System Software Laboratory
Department of Computer Science
National Tsing Hua University
Hsinchu, Taiwan 30013, R.O.C
{luse,kennslagter,lungkaiser}@sslab.cs.nthu.edu.tw,
ychung@cs.nthu.edu.tw

Abstract. In high performance and low power multimedia embedded system design, VLIW-based embedded DSPs compilers that exploit ILP have become popular and play an important role today. For this reason, we need optimizing embedded DSP compilers that can both generate capable and efficient code in terms of performance, power, size, and productivity. In this paper, we show a post-compilation framework that can further optimize programs that have already been compiled and optimized by another compiler, by using runtime information and exploiting hardware specific features of DSPs. Finally, we show in our simulation results, that even programs compiled at the best optimization level, can obtain significant improvement through the use of this framework.

Keywords: VLIW Compiler optimization, DSP Compiler optimization, Post optimization.

1 Introduction

In multimedia embedded system design, it is desirable for the system to be high in performance and low in cost. To achieve both these goals, VLIW-based DSPs use some hardware specific features, such as zero-overhead loops, vector and pixel sub-words operations, heterogeneous processing units, and compiler-supported branch prediction. These hardware specific features are used to increase computing efficiency instead of using dynamic scheduling logic that would increase hardware complexity and cost. In traditional VLIW-based compiler design, the most important optimization technique is done by using ILP (Instruction-Level Parallelism). For the inquisitive an example of this technique in use can be seen in the IMPACT VLIW compiler framework [4]. However, since there is a tendency for the number of instructions in a basic block of a multimedia program to be small, ILP in a multimedia program that has specialized algorithms and program structure tends to be rather limited. Due to this problem, it is important that a VLIW-based DSP compiler can capitalize on hardware specific features, when it is optimizing an application program. Since hardware specific features

C.H. Hsu and V. Malyshkin (Eds.): MTPP 2010, LNCS 6083, pp. 51–58, 2010.

tend to be application-friendly not compiler-friendly, most VLIW-based DSP compilers cannot take the advantage of those specialized features effectively. Moreover, in VLIW-based DSPs, the connectivity between computation units and storage units is restricted in order to minimize hardware and interconnection cost. The partial connection of registers and functional units is also an obstacle for VLIW-based DSP compilers to select and schedule instructions effectively. Without an efficient VLIW-based DSP compiler, designers of high performance and low cost multimedia embedded systems are forced to use fully handwritten assembly codes in order to get better performance and code density. However, handwriting assembly code is not an acceptable solution as it tends to result in long development time and it lacks portability.

In general, off the shelf VLIW-based DSP compilers cannot use hardware specific features effectively due to design trade-offs. In this paper, we propose a VLIW-based post compilation framework for multimedia embedded DSPs with hardware specific optimizations. The main purpose of our framework is to enhance the performance of the executable code generated by other VLIW-based DSP compilers. Traditionally, post compilation framework instrumentations are used to provide methods that allow low-level language code such as machine code or assembly code to be optimized. During runtime, a multimedia application program can have its code separated into regions and have each region classified as having either cold region code or hot region code according to the execution time of that region. With this runtime information, compilers can focus their more aggressive optimizations on the hot region codes [17]. By focusing on hot region codes and less on cold region codes compilers can achieve better overall performance. Therefore, most post compilation frameworks tend to focus on specific optimizations such as instruction rescheduling, register reallocation, speculative execution [21], post-pass power optimization and post-pass loop optimization, to enhance the machine code generated by other compilers based on this type of runtime information.

Our proposed framework focuses on instruction rescheduling and post-pass loop optimizations. It consists of six parts, a frontend parser, a run-time information collector, a profiling database synthesizer, a hardware database synthesizer, a hardware specific optimizer, and a code generator. The frontend parser is used to parse the codes generated by other VLIW-based DSP compilers and transform them into the intermediate representation (IR) of our framework. The run-time information collector is used to collect the necessary run-time information. The profiling database synthesizer and the hardware database synthesizer generate useful runtime information about a program as well as hardware related information to help the hardware specific optimizer to optimize a program. The hardware specific optimizer contains two optimization techniques, hardware specific instruction optimizations and hardware specific loop optimizations. The hardware specific instruction optimizations include specific instruction rematch optimizations, instruction rescheduling and recourse reallocation optimization. These optimizations can increase the computation performance, exploit more ILP, and use a compiler-supported profiling-based branch predictor to improve branch performance. For hardware specific loop optimizations, we combine the zero-overhead loop, vector operations, pixel operations, and simple software pipelining techniques to improve the loop performance that dominates multimedia embedded programs. Finally, the code generator is used to generate the enhanced machine code.

To evaluate the performance of the proposed post compilation framework, we implemented a post compilation framework for the ADI Blackfin DSP [1] and used the Blackfin GCC 3.4 [8] and VDSP++ 4.5 [23] as the frontend compilers. Two benchmarks, the DSP Stone [18] benchmark and the JM 9.8 H.264 reference code [10], were used as test samples and were then simulated by the Blackfin cycle-accurate simulator in order to collect runtime information. Experimental results showed that, for the DSP Stone benchmark, our framework on average was able to get 17.5% and 9% performance gain with the codes generated by the Blackfin GCC 3.4 and VDSP++ 4.5 respectively, when using optimization level 3. For the JM9.8 H.264 reference code, which is an optimized DSP library that has been hand-tuned, our framework was able to get a 5.8% performance gain.

The organization of the rest of the paper is as follows. In Section 2, we describe our VLIW-based post compilation framework in more detail. In Section 3, we give more details about the target experimental platform and the results. Finally, in Section 4 we present our conclusions.

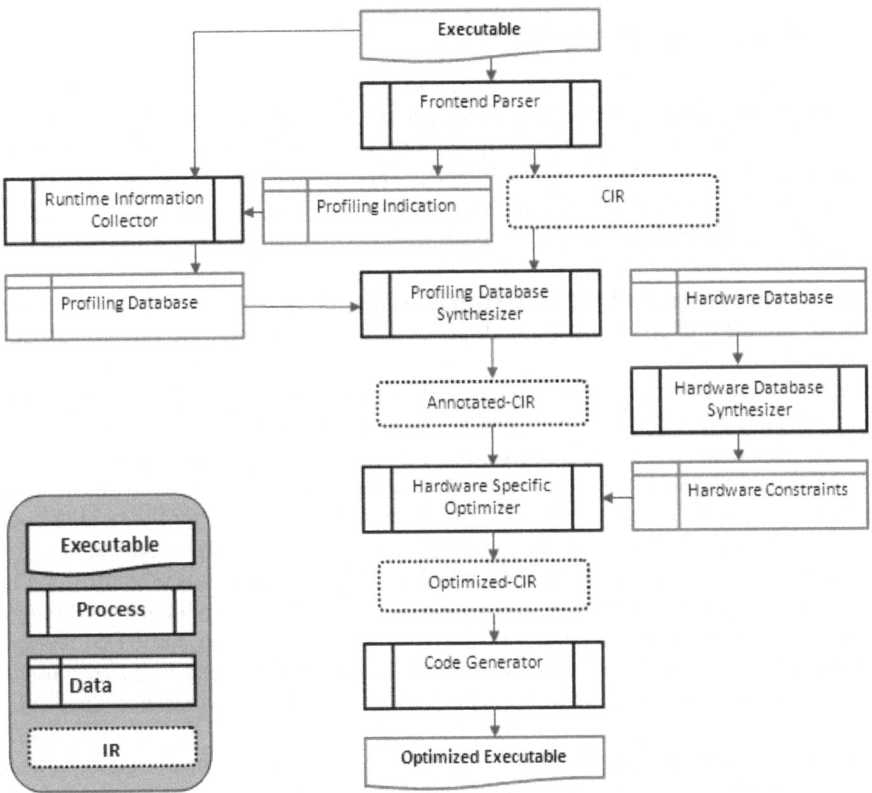

Fig. 1. Compilation flow of the post-compilation framework

2 The Post-Compilation Framework

The architecture of the proposed post compilation framework is shown in Figure 1. As is shown in the figure, the framework consists of six components, a frontend parser, a runtime information collector, a profiling database synthesizer, a hardware database synthesizer, a hardware specific optimizer, and a code generator. The frontend parser is used to parse the machine code generated by a frontend compiler into the common intermediate representation (CIR) and generates profiling indicators that provides hints for the runtime information collector. The runtime information collector is used to record the runtime information, needed for the hardware specific optimizer, into a database and is based on the profiling indicators and the execution of the machine code. The profiling database synthesizer is used to annotate the CIR according to the profiling database. The hardware database synthesizer is used to generate useful hardware information according to the hardware database. The hardware specific optimizer is used to optimize the annotated-CIR with specific hardware features based on any useful hardware information generated by the hardware database synthesizer. The final optimized executable is then generated by the code generator. In the following subsection, we will describe the optimization stage in more detail.

2.1 Hardware Specific Optimizations

The hardware specific optimizer contains two components, a hardware specific instruction optimizer and the hardware specific loop optimizer. The hardware specific instruction optimizer handles optimizations with multiply-accumulate calculations (MAC), ILP scheduling, and with branch prediction. The hardware specific loop optimizer also handles optimization of the zero-overhead loop buffer, the vector-unit and pixel-unit, and in the software pipeline

2.1.1 The Hardware Specific Instruction Optimizer

Since multimedia programs use multiply-accumulate operations frequently, especially in some matrix computing, the multiply-accumulate unit is an important feature of DSPs that can improve the performance of programs that perform multiply-accumulate calculations (MACs). The MAC instruction of most DSPs use specific registers (such as accumulate register). Since defects in a compiler intermediate representation may result in a compiler generating MAC instructions ineffectively, we used IBURG [6] [7] style algorithms, in our framework, to generate MAC tree pattern-matching code in case the core compiler (frontend compiler) lost some MAC instruction matching.

Due to the multiple-issues involved in VLIW-based DSPs, it is often difficult to do scheduling optimizations that can exploit ILP. Since ILP exploited scheduling is a complicated problem that cannot be solved in polynomial time, most DSP compilers cannot generate multiple issue instructions effectively. In our work, we used the artificial resource assignment (ARA) algorithm in [14] [15] combined that with our generated runtime information to solve this problem. This augmented algorithm allowed us to reschedule the result of the frontend compiler. Since register utilization can be increased by using our runtime information to reallocate registers, the artificial resource assignment algorithm is used to help schedule instructions efficiently.

Branch prediction is an important feature in modern deep pipeline design because a stall in the pipeline can affect a systems performance a lot. In order to minimize hardware and power consumption, most DSPs use compiler supported branch prediction (also known as static branch prediction), instead of dynamic branch prediction a mechanism that is implemented in the hardware. For example, the ADI Blackfin DSP supports a special instruction called the branch prediction appendix (bp). The branch prediction appendix helps a compiler improve its branch performance by predicting if a branch is to be taken or non-taken. In general, compilers will predict a backward-going branch as taken and a forward-going branch as non-taken. In the proposed framework, we use a profile based branch predictor [11] to support the branch prediction appendix. In early studies [11], the profile based branch predictor was able to improve prediction rates from 3% to 24% in SPEC92.

2.1.2 The Hardware Specific Loop Optimizer

Zero-Overhead Loop Buffer Optimization

Most DSPs have a zero-overhead loop buffer, which is a hardware technique that can minimize loop overhead without the penalty of increasing code size. The zero-overhead loop buffer is a specific instruction buffer used for low cost counter-based loops. Without decrementing a counter, evaluating a loop condition, then calculating and branching to a new target address, the zero-overhead loop instruction can save 2 to 6 instructions in one loop depending on the structure of the loop. In addition, the zero-overhead loop instruction can also reduce power consumption and increase performance in a low instruction cache utilization environment since it is not accessing the instruction memory. Moreover, when external memory is used in an embedded system, the synchronization of an external memory bus tends to be costly and the data transfer of the external memory bus can consume a lot of energy. With a zero-overhead loop optimization, an embedded system can execute a loop efficiently and use less power.

However, the zero-overhead loop instruction has some restrictions, as it cannot be used with branch instructions or call/return instructions. In the proposed framework, the hardware specific loop optimizer finds loops and tries to optimize them with the zero-overhead loop instruction. Any loop that does not contain a call/return or branch instruction in it can be optimized with a zero-overhead loop instruction easily. However, to optimize loops that contain call/return or branch instructions, the framework has to use other methodologies such as function inlining, loop unswitching[3], and speculative execution in order to take advantage of the zero-overhead loop instruction.

Vector-Unit and Pixel-Unit Optimization

In order to increase the parallelism of a program, some small precision operations can be computed in parallel. This is called data-level parallelism (DLP) [12] and it is independent of the ILP. In modern DSP design, a SIMD operation (also known as a vector operation) is the methodology used to exploit data-level parallelism. SIMD operations are very suitable for several media-processing applications, such as audio, video and image processing. The pixel unit is designed for multimedia applications to take advantage of these instructions to align bytes, perform dual 16 bit and quad 8 bit addition (or subtraction) operations, and for pixel averaging operations. Moreover, a program can have good code-density if it takes advanatge of vector and pixel operations.

In our proposed framework, with run-time value range information, the hardware specific optimizer can identify a variable that is declared as a word (32-bit) but is instead used as a sub-word (8-bit or 16-bit) during runtime. Vector and pixel optimizations can then be performed on these variables by using the algorithm in [24].

Software Pipelining Optimization

In the proposed framework, the hardware specific loop optimizer integrates all the optimizations described above into one simple software pipeline that can construct several loop iterations and combine them into one new loop iteration. Since the new loop can compute several loop iterations and pack them into a single loop iteration, we can issue multiple instructions in parallel and allows us to optimize the software pipeline.

3 Performance Evaluation

To evaluate the performance of the proposed post compilation framework, we implemented a post compilation framework for the ADI Blackfin DSP [1] and used the Blackfin GCC 3.4 [8] and VDSP++ 4.5 [23] as the frontend compilers. Two benchmarks, the DSP Stone [18] benchmark and the JM 9.8 H.264 reference code [10], were used as test samples and were then simulated by the Blackfin cycle-accurate simulator in order to collect runtime information. Tables 1 and 2 show the execution results of the proposed framework for the DSP Stone benchmark, our framework on average was able to get 17.5% and 9% performance gain with the codes generated by the Blackfin GCC 3.4 and VDSP++ 4.5 respectively, when using optimization level 3.

In Table 3, a comparison of the execution cycles on the JM9.8 H.264 reference code is obtained by decoding three frames using a baseline profile with and without the hand-tuned DSP library. From Table 3, we observe that, with the post-optimizing framework, we were able to get an additional 15.16 % performance improvement.

Table. 1. Execution cycles in DSP Stone on Blackfin GCC 3.4

DSP Stone File Name	Execution	Cycles		Performance	Improve
Optimization Level	**None**	**O3**	**Post**	**41.35%**	**17.56%**
complex_multiply	1060	930	719	32.17%	22.69%
convolution	1905	1391	1131	40.63%	18.69%
dot_product	989	942	771	22.04%	18.15%
fir	2391	1556	1305	45.42%	16.13%
fir2dim	9582	5439	4755	50.38%	12.58%
iir_biquad_N_sections	2507	1675	1326	47.11%	20.84%
iir_biquad_one_section	1022	977	834	18.40%	14.64%
lms	3045	1899	1503	50.64%	20.85%
matrix (matrix1)	49926	25377	20283	59.37%	20.07%
matrix(matrix2)	47328	25518	20029	57.68%	21.51%
matrix1x3	1266	1097	974	23.06%	11.21%
n_complex_updates	5689	3856	3445	39.44%	10.66%
n_real_updates	3079	1884	1460	52.58%	22.51%
real_update	935	890	736	21.28%	17.30%
startup	5383	2545	2149	60.08%	15.56%

Table. 2. Execution cycles in DSP Stone on Blackfin GCC 3.4

DSP Stone File Name	Execution	Cycles		Performance	Improve
Optimization Level	None	Opt	Post	69.39%	8.80%
complex_multiply	456	366	326	28.51%	10.93%
convolution	2257	416	383	83.03%	7.93%
dot_product	474	323	311	34.39%	3.72%
fir	25160	2712	2523	89.97%	6.97%
fir2dim	3460	622	585	83.09%	5.95%
iir_biquad_N_sections	3636	564	514	85.86%	8.87%
iir_biquad_one_section	424	342	312	26.42%	8.77%
lms	4575	1035	705	84.59%	31.88%
matrix (matrix1)	155253	14345	12194	92.15%	14.99%
matrix (matrix2)	147221	15053	12344	91.62%	18.00%
matrix1x3	1147	324	302	73.67%	6.79%
n_complex_updates	11573	1521	1521	86.86%	0.00%
n_real_updates	4633	741	741	84.01%	0.00%
real_update	371	320	320	13.75%	0.00%
startup	14567	2677	2485	82.94%	7.17%

Table 3. Execution cycles on JM 9.8 H.264 decoder using a baseline profile

DSP Stone File Name	Execution	Cycles			Performance	Improve
Optimization Level	O3 Only	O3+DSPLib	Post - O3	Post - DSPLib		
JM 9.8 H264 Decoder	450011213	184517090	381791064	173817381	15.16%	5.80%

When the code did not use the hand-tuned DSP library an improvement of 5.8% was obtained when we did use the hand-tuned DSP library. These results indicate that the framework can work well with real multimedia programs even when they use hand-tuned optimizations.

4 Conclusion

In this paper, we proposed a VLIW-based post compilation framework for multimedia embedded DSPs with hardware specific optimizations. The proposed framework was able to better optimize a program by using runtime information and exploiting specific hardware features of an embedded DSP. The hardware specific optimizer in our VLIW-Based post compilation framework was divided into two parts. One was the hardware specific instruction optimizer and the other was the hardware specific loop optimizer. These post-optimizations were shown to have a significant impact on a programs performance. Finally, the simulation results showed that it is possible for our proposed framework to work well with real multimedia programs even if they have had hand-tuned optimizations.

References

1. The Analog Devices, Inc. Website (1995), `http://www.analog.com/en/`
2. Aho, A.V., Lam, M.S., Sethi, R., Ullman, J.D.: Compilers Principles, Techniques, and Tools, 2nd edn. Addison-Wesley, Reading (2006)

3. Bacon, D.F., Graham, S.L., Sharp, O.J.: Compiler Transformations for High-Performance Computing. ACM Computing Surveys (December 1994)
4. Chang, P.P., Mahlke, S.A., Chen, W.Y., Warter, N.J., Hwu, W.W.: IMPACT: An architectural framework for multiple-instruction-issue processors. In: Proc. 18th. Int. Symp. Computer Architecutre (1996)
5. Falk, H.: Control Flow Optimization by Loop Nest Splitting at the Source Code Level, Research Report No 773 (October 2002)
6. Fisher, J.A., Faraboschi, P., Young, C.: Embedded Computing: a VLIW approach to architecture, compilers and tools. Morgan Kaufmann, San Francisco (2005)
7. Fraser, C.W., Hanson, D.R., Proebsting, T.A.: Engineering a simple, efficient code-generator generator. ACM Letters on Programming Languages and Systems, 213–226
8. The GCC - the gnu compiler collection (1987), `http://gcc.gnu.org/`
9. Gyllenhaal, J.C., Hwu, W.M., Rau, B.R.: Hmdes version 2.0 specification, Univ., Illinois, Urbana, IL, Tech. Rep. IMPACT (1996)
10. The H.264/AVC JM Reference Software, The Image Processing HHI (2006), `http://iphome.hhi.de/suehring/tml/`
11. Hennessy, J.L., Patterson, D.A.: Computer Architecture: A quantitative approach, 4th edn. Morgan Kaufmann, San Francisco (2006)
12. Kozyrakis, C.E., Patterson, D.A.: Scalable Vector Processors for Embedded Systems. IEEE Computer Society Press, Los Alamitos (2003)
13. Marwedel, Goosens, G. (eds.): Code Generation for Embedded Processors. Kluwer, Norwell (1995)
14. Rajagopalan, S., Rajan, S.P., Malik, S., Rigo, S., Araujo, G., Takayama, K.: A Retargetable VLIW Compiler Framework for DSPs With Instruction-Level Parallelism. IEEE Transactions on CAD of IC and System 20(11) (November 2001)
15. Rajagopalan, S., Vachharajani, M., Malik, S.: Handling irregular ILP within conventional VLIW schedulers using artificial resource constraints. In: Proc. Int. Conf. Compilers, Architecture, and Sysnthesis for Embedded Systems, November 2000, pp. 157–164 (2000)
16. Padua, D.A., Wolfe, M.J.: Advanced Compiler Optimizations for Supercomputers. Communication of the ACM (December 1986)
17. Zhang, K., Zhang, T., Pande, S.: Binary Translation to Improve Energy Efficiency through Post-pass Register Re-allocation. In: Proceedings of the 4th ACM international conference on Embedded software (2004)
18. Zivojnovic, V., Velarde, J.M., Schläger, C., Meyer, H.: DSP-stone: A DSP-oriented benchmarking methodology. In: Proc. Int. Conf. Signal Processing Applications and Technology, October 1994, pp. 715–720 (1994)
19. Saghir, M.A.R., Chow, P., Lee, C.G.: Application-driven design of DSP architectures and compilers, Acoustics, Speech, and Signal Processing. In: ICASSP-94 (1994)
20. Kumar, R., Gupta, A., Pankaj, B.S., Ghosh, M., Chakrabarti, P.P.: Post-Compilation Optimization for Multiple Gains with Pattern Matching. ACM SIGPLAN Notices (2005)
21. Liao, S.S., Wang, P.H., Wang, H., Hoflehner, G., Lavery, D., Shen, J.P.: Post-Pass Binary Adaptation for Software-Based Speculative Precomputation. In: ACM PLDI'02 (June 2002)
22. Angiolini, F., Menichelli, F., Ferrero, A., Benini, L., Oliveri, M.: A Post-Compiler Approach to Scratchpad Mapping of Code. In: International Conference on Compilers, Architectures and Synthesis of Embedded Systems CASES 2004 (September 2004)
23. The Analog Devices, Visual DSP++, Website (1995), `http://www.analog.com/en/`
24. Suzuki, M., Fujinami, N., Fukuoka, T., Watanabe, T., Nakata, I.: SIMD Optimization in COINS Compiler Infrastructure. In: Proceedings of the Innovative Architecture for Future Generation High-Performance Processors and Systems (2005)

Parallelization of Motion JPEG Decoder on TILE64 Many-Core Platform

Xuan-Yi Lin[1], Chung-Yu Huang[1], Pei-Man Yang[2], Tai-Wen Lung[1], Shau-Yin Tseng[2], and Yeh-Ching Chung[1]

[1] Department of Computer Science, National Tsing Hua University
Hsinchu, Taiwan 30013, R.O.C.
{xylin,ychung}@cs.nthu.edu.tw
[2] Information & Communications Research Laboratories
Industrial Technology Research Institute
Hsinchu, Taiwan 310, R.O.C.
{peimanyang,tseng}@itri.org.tw

Abstract. The ubiquity of many-core architectures poses challenges to software developers to make scalable software. To parallelize data-intensive applications on a many-core platform, one has to consider both hardware architecture and software characteristics when writing parallel codes. In this paper, we take Motion JPEG decoder as an example data-intensive application and take TILE64 as an example many-core platform. We parallelize the decoder with two different strategies and observe their impact on program performance and scalability. We design two algorithms, *READ* and *WRITE*, which differ in the direction of data movement between processor cores. Experimental results show that *READ* algorithm outperforms *WRITE* algorithm by 217% when decoding 1080P video on the TILE64 platform. It indicates that the arrangement of data flows in a data-intensive parallel program can have huge impact on program performance and scalability on a many-core platform.

Keywords: many-core architecture, parallel processing, Motion JPEG.

1 Introduction

With rapid industry development of many-core architectures, mass-produced processors now contain tens to hundreds of cores in a single chip. While the trend of processor making is to increase core count rather than processor frequency, software developers can no longer rely on the so called "free lunch" [1] that automatically makes their program run faster on processors clocked at higher frequencies.

For application developers, in order to make the performance of their programs scale well with the number of available cores on many-core architectures, existing software needs to be modified or re-written from ground up. The effort required to adapt existing software to a new many-core processor is directly correlated with the programming language and programming model used. Well understanding of both hardware architecture and software characteristics is also crucial to build scalable software on a many-core platform.

C.H. Hsu and V. Malyshkin (Eds.): MTPP 2010, LNCS 6083, pp. 59–68, 2010.

When in the course of parallelize a data-intensive application for a many-core platform, data flow should be considered with hardware architecture in mind. Arrangement of the flow of workloads among processors will have direct impact on the performance and scalability of the adapted program.

In this paper, we explore the method of parallelizing a data-intensive application on a many-core system and observe its impact on program performance and scalability. We take Motion JPEG decoder as an example data-intensive application and TILE64 as an example many-core system. We designed two shared-memory based algorithms, *WRITE* and *READ*, to parallelize a Motion JPEG decoder on the TILE64 platform. *WRITE* is a straightforward algorithm and is easier to implement compared to *READ*. We apply both *WRITE* and *READ* algorithms to an open-source Motion JPEG decoder to evaluate their performance. Benchmark result shows that although the *READ* algorithm requires extra effort and time to implement, it scales far better than the *WRITE* algorithm. The decoder runs as much as 3.17 times faster when adopting the *READ* algorithm instead of the *WRITE* algorithm.

This paper is organized as follows. Section 2 provides background knowledge for TILE64 processor and Motion JPEG files. The *WRITE* and *READ* algorithms are introduced in Section 3 and benchmarked in Section 4. Conclusions of this work are given in Section 5.

2 Preliminaries

2.1 The TILE64 Processor

TILE64 is a general purpose many-core processor made by Tilera [2]. It has an array of 64 identical processor cores (each referred to as a *tile*) interconnected via on-chip two-dimensional mesh networks [tile ref]. TILE64 is fully programmable using standard ANSI C under Linux environment. In addition to standard Linux C, TILE64 can also be programmed using proprietary API called iLib. The iLib library supports two communication mechanisms, shared memory and streaming, for processes running on different cores to communicate. Software developer can use both communication primitives in a program.

Fig. 1 illustrates the architecture overview of a TILE64 processor. There are four memory controllers located at the four corners of processor array. These on-chip memory controllers provide access to an external memory system that is accessible by all tiles. The interface to the memory networks provides access to other tiles and to the DDR2 memory.

To use shared memory mechanisms in a program, the process which is sharing information can call *malloc_shard()* function of the iLib to get an address pointing to a block of shared memory. Then the sharing process notifies other processes the location of shared memory by sending them the pointer to shared memory.

2.2 Motion JPEG

A Motion JPEG (M-JPEG) file is basically a large file containing a sequence of independent JPEG frames. Fig. 2 shows structure of a typical M-JPEG file. There is

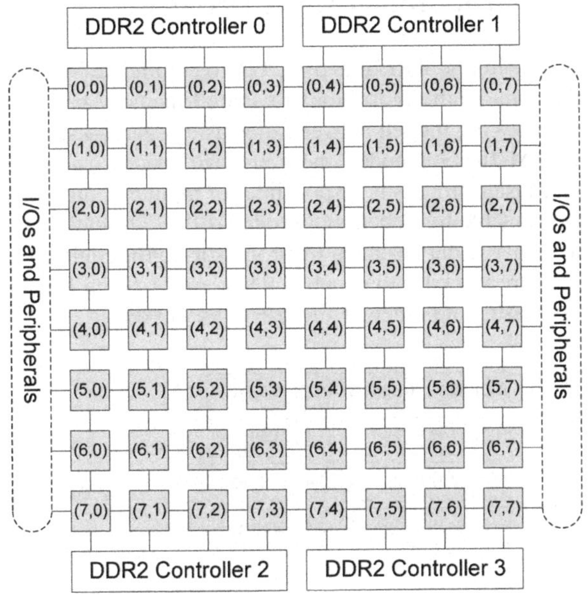

Fig. 1. TILE64 processor architecture overview

no data dependence between frames within an M-JPEG file, thus it is inherently parallel at inter-frame level. The inherent parallelism of an M-JPEG file makes it easy to parallelize an M-JPEG decoder by instructing processors to decode different frames concurrently.

Data size of JPEG frames in an M-JPEG file will vary based on the complexity of individual frames. Decoded YUV frames, however, are equally sized. Fig. 3 illustrates decoding of an M-JPEG file. Because JPEG has high compression rate, size of decoded YUV data is significantly larger than original JPEG data.

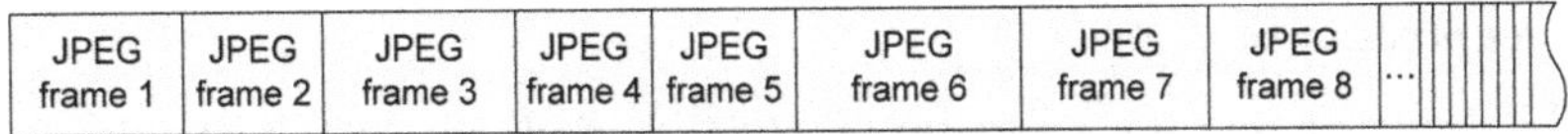

Fig. 2. A Motion JPEG file

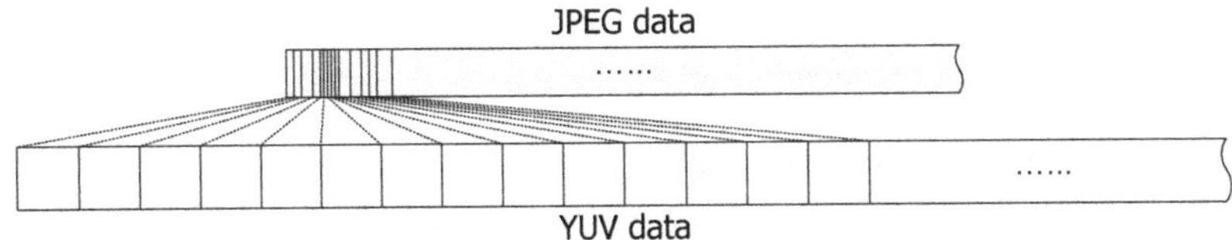

Fig. 3. Decoding of an M-JPEG file into YUV video sequence

3 Parallelization of Motion JPEG Decoder

We design two algorithms, *WRITE* and *READ*, to parallelize Motion JPEG decoder on the TILE64 platform. Both algorithms are shared memory based. JPEG data frames and YUV data frames are moved between tiles using shared-memory mechanism.

In the parallel M-JPEG decoder, there are two process roles, master process and worker process. Master process is responsible for input and output operations. Worker processes are responsible for decoding individual JPEG frames.

Fig. 4 shows a particular instance of processor configuration for both algorithms. In Fig. 4, 32 tiles are working together to decode a M-JPEG file, among the 32 tiles, *tile (0, 0)* acts as master and other 31 tiles serve as workers.

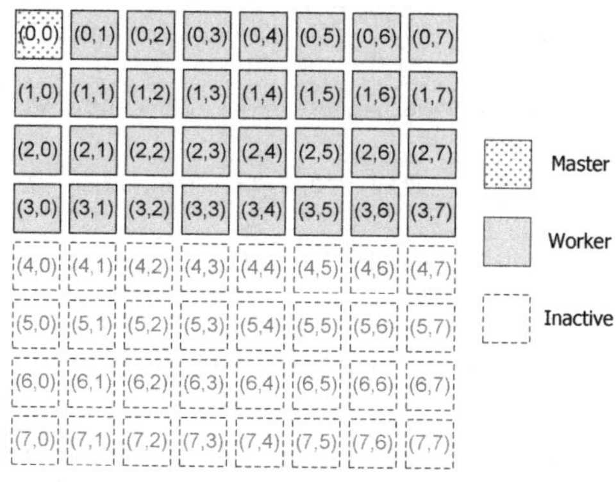

Fig. 4. Decoder configuration

3.1 The *WRITE* Algorithm

Following is the program pseudo code of *WRITE* algorithm for master and worker process. The *WRITE* is a straightforward algorithm. Illustration of the algorithm is given in Fig. 5.

Master process:

```
Initialize shared memory space, JPEG_buffer[] and YUV_buffer[].
Broadcast address pointers of JPEG_buffer[] and YUV_buffer[] to all worker
processes.
Open and parse input M-JPEG file, mjpegFile.
output_frame_num = 0;
while( frames_to_decode != 0 )
{
  if ( JPEG_buffer[] is not full )
  Fetch next JPEG frame in mjpegFile and enqueue it to JPEG_buffer[].
  if ( YUV_buffer[] is not empty )
  {
  if ( YUV_frame(output_frame_num + 1) is available and valid )
```

12. {
13. Output(*YUV_frame(output_frame_num* + 1));
14. *output_frame_num++;*
15. *frames_to_decode – –;*
16. }
17. }
18. }

Worker process:

1. Receive address pointers of *JPEG_buffer[]* and *YUV_buffer[]* from master process.
2. while(*frames_to_decode* != 0)
3. {
4. if (*JPEG_buffer[]* is not empty)
5. {
6. Move first JPEG frame in *JPEG_buffer[]* to private JPEG buffer.
7. *private_YUV_buffer* = DecodeJPEGframe (*private_jpeg_buffer*);
8. Copy *private_YUV_buffer* to corresponding position in *YUV_buffer[]*.
9. Set the validity of the YUV frame to valid.
10. }
11. }

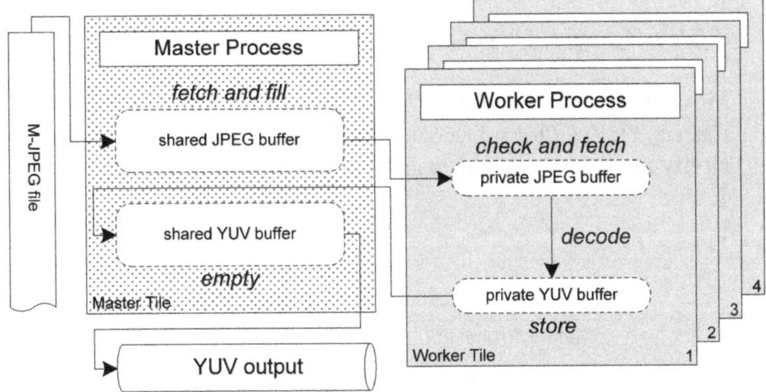

Fig. 5. Illustration of the *WRITE* algorithm

3.2 The *READ* Algorithm

In the *READ* algorithm, every worker process allocates YUV buffer as shared, so the YUV buffer is accessible by master process. Illustration of the algorithm is given in Fig. 6.

Master process:

1. Initialize shared memory space, *JPEG_buffer[]*.
2. Broadcast address pointers of *JPEG_buffer[]* to all worker processes.

```
Receive address pointers of shared_YUV_buffer from all worker processes.
Open and parse input M-JPEG file, mjpegFile.
output_frame_num = 0;
while( frames_to_decode != 0 )
{
if ( JPEG_buffer[] is not full )
Fetch next JPEG frame in mjpegFile and enqueue it to JPEG_buffer[].
if ( received notification from worker process )
{
Fetch YUV_frame(output_frame_num + 1) from the worker process.
Output( YUV_frame(output_frame_num + 1) );
output_frame_num++;
frames_to_decode – –;
}
}
```

Worker process:

```
Receive address pointers of JPEG_buffer[] and YUV_buffer[] from master
process.
Initialize shared memory space, shared_YUV_buffer.
Send address pointer of shared_YUV_buffer to master process.
while( frames_to_decode != 0 )
{
if ( JPEG_buffer[] is not empty )
{
Move first JPEG frame in JPEG_buffer[] to private JPEG buffer.
shared_YUV_buffer = DecodeJPEGframe (private_jpeg_buffer);
Notify master process the availability of private_YUV_buffer.
}
}
```

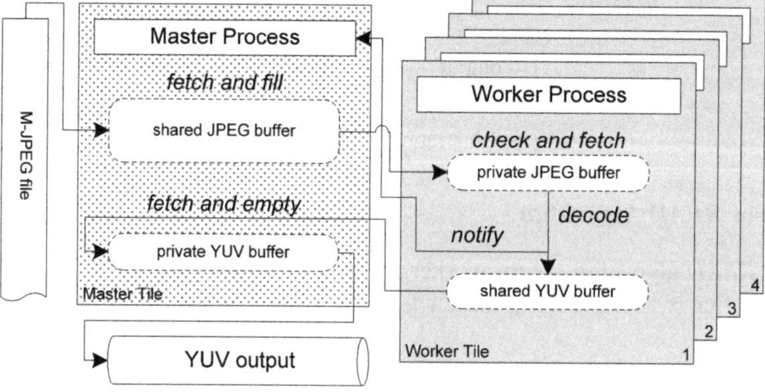

Fig. 6. Illustration of the READ algorithm

4 Experimental Results

We apply *WRITE* and *READ* algorithms to an open source Motion JPEG decoder, *MJPEG Tools* [3] and run the parallelized M-JPEG decoder on TILE64 platform to observe performance and scalability of the decoder. We use the parallel decoder to decode four videos of different resolution. Table 1 lists the test files used.

Table 1. Motion JPEG test files used

	deadline	*city*	*stockholm*	*factory*
Format	CIF	4CIF	720P	1080P
Resolution	352x288	704x576	1280x720	1920x1088
Frames	1374	600	604	1339

4.1 Performance of *WRITE*

Fig. 7 shows speedup of parallel M-JPEG decoder with *WRITE* algorithm using different number of tiles. Number of tiles used shown in the figure, for example 1+15, represents one master process and 15 worker processes.

From the results we can see that the performance of *WRITE* algorithm does not scale beyond 1+15 tiles. To better understand the scalability problem, we also record throughput information of individual tiles and present it visually in Fig. 8 and Fig. 9. Fig. 8 and Fig. 9 show per-tile decoding throughput with master process running on *tile (0, 0)* and *tile (3, 3)* respectively. From Fig 8 and Fig. 9 we can see that worker processes with physical location closer to master process have higher performance. That is because it takes a lot more time for further tiles to write data to the master tile.

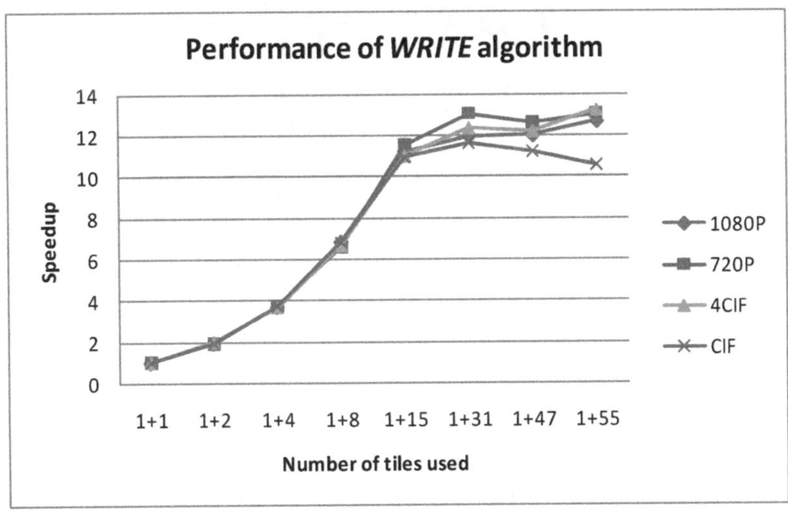

Fig. 7. Decoding performance of parallel M-JPEG decoder using WRITE algorithm

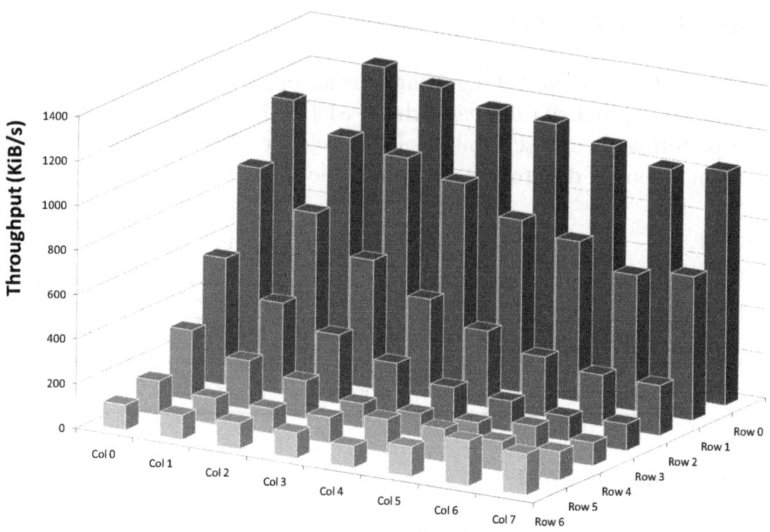

Fig. 8. *WRITE* algorithm per-tile decoding throughput under 1080P workload

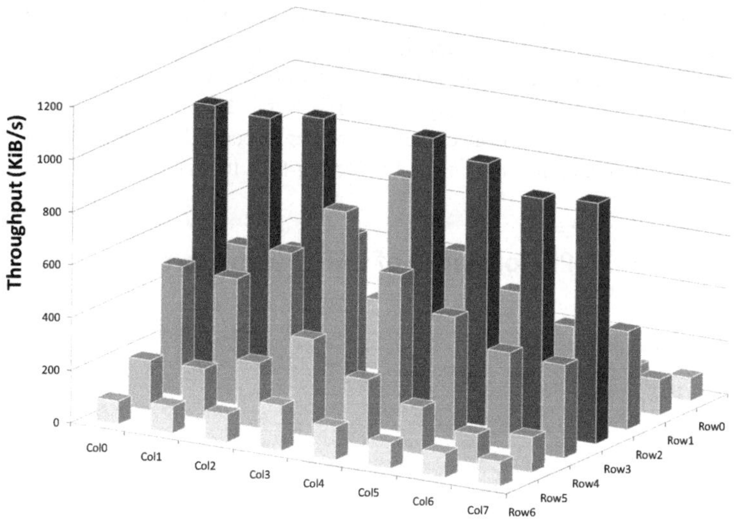

Fig. 9. *WRITE* algorithm per-tile decoding throughput under 1080P workload with master process running on *tile* (3,3)

4.2 Performance of *READ*

Performance of *READ* algorithm is shown in Fig. 10 and Fig. 11. It shows that *READ* algorithm scales beyond 1+31 tiles when decoding a 1080P video file. Throughput data shows that latency of read operation is barely affected by distance between tiles.

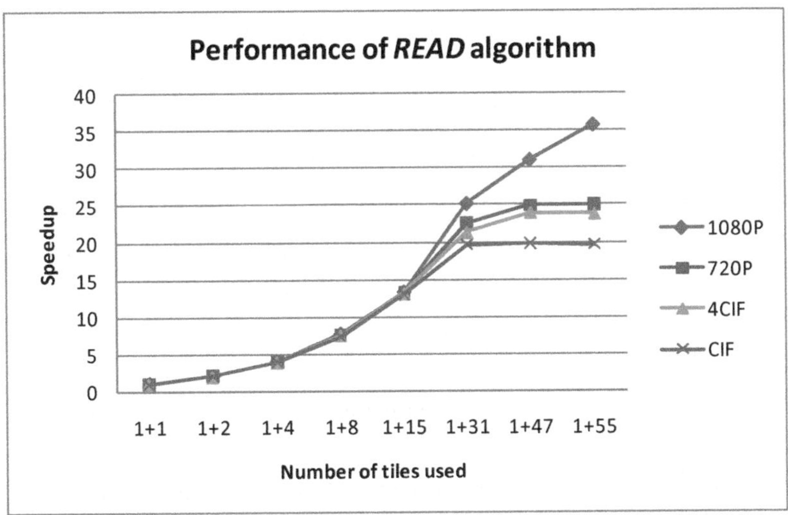

Fig. 10. Decoding performance of parallel M-JPEG decoder using *READ* algorithm

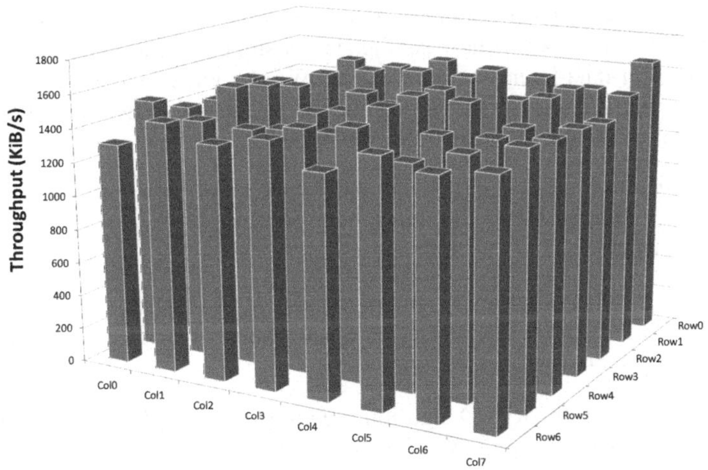

Fig. 11. *READ* algorithm per-tile decoding throughput under 1080P workload

4.3 Performance Advantage of *READ* over *WRITE*

Fig. 12 shows the performance advantage of *READ* algorithm over *WRITE* algorithm. The greatest performance gain can be observed at the configuration of using 1+55 tiles to decode a 1080P video file. It has a performance improvement of 217%. It means that on the TILE64 platform, M-JPEG decoder using *READ* algorithm runs 3.17 times faster than using *WRITE* algorithm when decoding a 1080P video file.

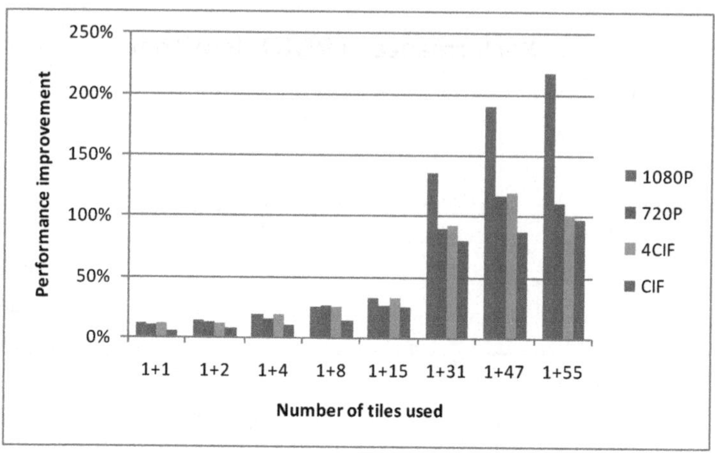

Fig. 12. Performance improvement of *READ* algorithm over *WRITE* algorithm

5 Conclusion

In this paper, we conduct parallelization of Motion JPEG decoder on the TILE64 platform. We want to know how parallelization strategies can impact scalability and performance on data-intensive applications. We designed two share-memory based algorithms, *WRITE* and *READ* to parallelize a Motion JPEG decoder. From the experimental results we have the following remarks:

Remark 1. Parallelization strategy with consideration of both hardware and software characteristics is necessary in building high performance and scalable software on many-core platforms.

Remark 2. On TILE64, latency of write operations to shared memory addresses increases with the distance between sharing tile and writing tile. Read operations are not affected by such overhead.

Remark 3. Although the READ algorithm requires extra implementation overhead, it scales far better than that of the WRITE algorithm.

References

1. Sutter, H.: The Free Lunch Is Over: A Fundamental Turn Toward Concurrency in Software. Dr. Dobb's Journal 30(3) (2005)
2. Tilera Corporation, `http://www.tilera.com`
3. MJPEG Tools, `http://mjpeg.sourceforge.net`

A Compound Scheduling Strategy for Irregular Array Redistribution in Cluster Based Parallel System

Shih-Chang Chen[1], Ching-Hsien Hsu[2], Tai-Lung Chen[1], Kun-Ming Yu[2], Hsi-Ya Chang[3], and Chih-Hsun Chou[2,*]

[1] College of Engineering
[2] Department of Computer Science and Information Engineering
Chung Hua University, Hsinchu, Taiwan 300, R.O.C., Fax: +886-5186416
[3] National Center for High-Performance Computing, Hsinchu 30076, Taiwan
{scc,robert,tai}@grid.chu.edu.tw, yu@chu.edu.tw, jerry@nchc.org.tw, chc@chu.edu.tw

Abstract. With the advancement of network and techniques of clusters, joining clusters to construct a wide parallel system becomes a trend. Irregular array redistribution employs generalized blocks to help utilize the resource while executing scientific application on such platforms. Research for irregular array redistribution is focused on scheduling heuristics because communication cost could be saved if this operation follows an efficient schedule. In this paper, a two-step communication cost modification (T2CM) and a synchronization delay-aware scheduling heuristic (SDSH) are proposed to normalize the communication cost and reduce transmission delay in algorithm level. The performance evaluations show the contributions of proposed method for irregular array redistribution.

1 Introduction

Scientific application executing on parallel systems with multiple phases requires appropriate data distribution schemes. Each scheme describes the data quantity for every node in each phase. Therefore, performing data redistribution operations among nodes help enhance the data locality.

Generally, data redistribution is classified into regular and irregular redistributions. `BLOCK`, `CYCLIC` and `BLOCK-CYCLIC(`*c*`)` are used to specify array decomposition for the former while user-defined function, such as `GEN_BLOCK`, is used to specify array decomposition for the latter. High Performance Fortran version 2 provides `GEN_BLOCK` directive to facilitate the data redistribution for user-defined function. To perform array redistribution efficiently, it is important to follow a schedule with low communication cost.

With the advancement of network and the popularizing of cluster computing research in campus, it is a trend to join clusters in different regions to construct a complex parallel system. To performing array redistribution on this platform, new techniques are required instead of existing methods.

[*] Corresponding author.

C.H. Hsu and V. Malyshkin (Eds.): MTPP 2010, LNCS 6083, pp. 69–77, 2010.

Schedules illustrate time steps for data segments (messages) to be transmitted in appropriate time. The cost of schedules given by scheduling heuristics is the summation of cost of every time steps while cost of each time step is dominated by the message with largest cost. A phenomenon is observed that most local transmissions, which are happened in a node, do not dominate the cost of each step although they are in algorithm level for existing methods. In other words, they are overestimated. Since a node can send and receive only one message in the same time step [5], the arranged position of each message becomes important. Therefore, a *two-step communication cost modification* (*T2CM*) and a *synchronization delay-aware scheduling heuristic* (*SDSH*) are proposed to deal with the overestimate problems, reduce overall communication cost and avoid synchronization of schedules in algorithm level.

The rest of this paper is organized as follows: Section 2 gives a survey of existing works related to array redistribution. Section 3 gives notations, terminology and examples to explain each parts of scheduling heuristics. The proposed techniques are described in section 4. Section 5 presents the results of the comparative evaluation, while section 6 concludes the paper.

2 Related Work

Array redistribution techniques have been developed for regular array redistribution and GEN_BLOCK redistribution in many papers. Both kinds of redistribution issues require at least two sorts of techniques. One is communication sets identification which decomposes array for nodes; the other one is communication scheduling method which derives schedules to shorten the overall transmission cost for redistributions.

ScaLAPACK [9] was proposed to identify communication sets for regular array redistribution. Guo *et al.* [2] proposed a symbolic analysis method to help generate messages for GEN_BLOCK redistribution. Hsu *et al.* [3] proposed the *Generalized Basic-Cycle Calculation* method to shorten the communication for generalized cases. The research on prototype framework for distributed memory platforms is proposed by Sundarsan *et al.* [11] who developed a method to distribute multidimensional block-cyclic arrays on processor grids. Karwande *et al.* [8] presented *CC-MPI* with the compiled communication technique to optimize collective communication routines. Huang *et al.* [6] proposed a flexible processor mapping technique to reduce the number of data element exchanging among processors and enhance the data locality. To reduce indexing cost, a processor replacement scheme was proposed [4]. With local matrix and compressed *CRS* vectors transposition schemes the communication cost can be reduced significantly. Combining the advantages of relocation scheduling algorithm and divide-and-conquer scheduling algorithm, Wang *et al.* [12] proposed a method with two phases for GEN_BLOCK redistribution. The first phase acts like relocation algorithm, but the contentions avoidance mechanism of second phase will not be proceeded immediately while contentions happened. To minimize the total communication time, Cohen *et al.* [1] supposed that at most k communication can be performed at the same time and proposed two algorithms with low complexity and fast heuristics. A study [7] focusing on the cases of local redistributions and inter-cluster redistribution was given by Jeannot and Wagner. It compared existing scheduling methods and described the difference among them. Rauber and Runger [10] presented a data-re-distribution library to deal with composed data structures which are distributed to one or more processor groups for executing multiprocessor task on distributed memory machines or cluster platforms. Hsu *et al.* [5] proposed a two-phase degree-reduction scheduling heuristic to minimize the overall communication cost. The proposed method derives each time step of a complete schedule by performing degree reduction technique while the number of messages of each node representing the degree of each vertex in algorithm level.

3 Preliminary

Following are notations, terminology and examples to explain each parts of scheduling heuristics for GEN_BLOCK redistribution. To improve data locality, multi-phase scientific problems require appropriate data distribution schemes for specific phases. For example, to distribute array for two different phases on six nodes, which are indexed from 0 to 5, two strings, {13, 20, 17, 17, 12, 21} and {16, 18, 13, 16, 29, 8}, are given, where the array size is 100 units. These two strings provide necessary information for nodes to generate messages to be transmitted among them. Fig. 1 shows these messages marked from m_1 to m_{11} and are with information such as data size, source node and destination node in the relative rows.

Scheduling heuristics are developed for providing solutions of time steps to reduce total communication cost for a GEN_BLOCK redistribution operation. In each step, there are several messages which are suggested to be transmitted in the same time step. To help perform an efficient redistribution, scheduling methods should avoid node contention, synchronization delay and redundant transmission cost. It is also important to follow policies of messages arrangement, i.e. with the same source nodes, messages should not be in the same step; with the same destination nodes, messages should be in different step; a node can only deal with one message while playing whether source node or destination node. These messages that cannot be scheduled together called conflict tuples, for example, a conflict tuple is formed with messages m_1 and m_2. Note that if a node can only deal with a message while it is a source/destination node, the number of steps for a schedule must be the equal to or more than the number of messages from/to these nodes. In other words, the minimal number of time steps is equal to the maximal number of messages in a conflict tuple, CT_{max}.

Fig. 2 gives a schedule with low communication cost and arranges messages in the number of minimal steps. In this result, there are three time steps with messages sent/received to/from different nodes. The values beside $m_{1\sim11}$ are data size, the cost of each step is dominated by the largest one. Thus, m_3, m_1 and m_8 dominate step 1, 2 and 3, and the estimated cost are 17, 13 and 4, respectively. To avoid node contentions, messages m_1 and m_2 are in separate steps due to destination nodes of both messages are the same. Based on same argument, m_2 and m_3 are in separate steps due to both messages are members of a conflict tuple. The total cost which represents the performance of a schedule is the summation of all cost of steps. In other words, a schedule with lower cost is better than another one with higher cost in terms of performance.

Information of messages			
No. of message	Data size	Source node	Destination node
m_1	13	0	0
m_2	3	1	0
m_3	17	1	1
m_4	1	2	1
m_5	13	2	2
m_6	3	2	3
m_7	13	3	3
m_8	4	3	4
m_9	12	4	4
m_{10}	13	5	4
m_{11}	8	5	5

Fig. 1. Information of messages generated from given schemes to be transmitted on six nodes which are indexed from 0 to 5

A result of scheduling heuristics		
No. of step	No. of message	Cost of step
Step 1	$m_3(17)$, $m_5(13)$, $m_7(13)$, $m_{10}(13)$	17
Step 2	$m_1(13)$, $m_6(3)$, $m_9(12)$, $m_{11}(8)$	13
Step 3	$m_2(3)$, $m_4(1)$, $m_8(4)$	4
Total cost		34

Fig. 2. A result of scheduling messages with low communication cost and minimal steps

The result in Fig. 2 schedules messages in three steps, which is the number of minimal steps or CT_{max}. The total cost is small which representing low communication cost due to messages with larger cost and messages with smaller cost are in separate steps. However, the schedule can still be better by providing a cost normalization method and a new scheduling technique to avoid synchronization delay among nodes during message transmissions in next section.

4 The Proposed Method

In this paper, a *two-step communication cost modification* (*T2CM*) and a *synchronization delay-aware scheduling heuristic* (*SDSH*) are proposed to normalize the communication cost of messages and reduce transmission delay in algorithm level. The first step of *T2CM* is a *local reduction* operation, which deal with the message happened in local memory. In other words, candidates are transmissions whose source node and destination node are the same node. For example, m_1, m_3, m_5, m_7, m_9 and m_{11}

are such kind of transmissions which happened inside nodes. The second step is a *inter amplification* method, which is responsible for transmissions happened across clusters. Assumed there are two clusters, and node 0~2 are in cluster 1, other nodes are in cluster 2. Then m_6 is such message which is transmitted from cluster 1 to cluster 2. Both operations are responsible for different kind of transmissions due to the heterogeneity of network bandwidth. The *local reduction* operation reduces simulated cost of messages to 1/8 which is evaluated from PC clusters that connected with 100Mbps layer-2 switch. On same argument, *inter amplification* operation increases cost of messages five times. The cost then becomes more practical for real machines when scheduling heuristics try to give a perfect schedule with low communication cost. For previous research, the difference does not exist in algorithm level of scheduling heuristics in and could result in erroneous judgments and high communication cost.

Fig. 3 gives the results of *local reduction* and *inter amplification* operations modifying data size for messages $m_{1\sim11}$. The given schedule in Fig. 2 becomes the results in Fig. 4. Difference of Fig. 2 and Fig. 4 shows the schedule could be improved and explains the explain the erroneous judgments. First, the dominators in step 1 and 2 are changed to others whose estimated cost is larger in Fig. 4. For example, the m_3 and m_1 are replaced by m_{10} and m_6 for both steps, respectively. Second, the cost of step 1 and step 2 are changed due to new dominators are chosen in both steps. Furthermore, the synchronization delay is small in algorithm level but results in more node idle time in practical. For instance, the cost of m_3, m_5, m_7 and m_{10} are 17, 13, 13 and 13 are close to each other in step 1 in Fig. 2. But it is quite different in practical in Fig. 4, they should be 2.125, 1.625, 1.625 and 13, respectively. Node 1, 2 and 3 must wait for node 4 and 5 to proceed next step because when the transmissions of m_3, m_5 and m_7 are finished, the transmission of m_{10} is still on the way.

Information of messages			
No. of message	Data size	Source node	Destination node
m_1	**1.625**	0	0
m_2	**3**	1	0
m_3	**2.125**	1	1
m_4	**1**	2	1
m_5	**1.625**	2	2
m_6	**15**	2	3
m_7	**1.625**	3	3
m_8	**4**	3	4
m_9	**1.5**	4	4
m_{10}	**13**	5	4
m_{11}	**1**	5	5

Fig. 3. The *local reduction* and *inter amplification* operations derive new data size for messages $m_{1\sim11}$

A result of scheduling heuristics		
No. of step	No. of message	Cost of step
Step 1	$m_3(2.125)$, $m_5(1.625)$, $m_7(1.625)$, $m_{10}(13)$	13
Step 2	$m_1(1.625)$, $m_6(15)$, $m_9(1.5)$, $m_{11}(1)$	15
Step 3	$m_2(3)$, $m_4(1)$, $m_8(4)$	4
Total cost		32

Fig. 4. The results with new dominators and cost

The proposed *synchronization delay-aware scheduling heuristic* is a novel and efficient method to avoid delay among clusters and shorten communication cost while performing GEN_BLOCK redistribution. To avoid synchronization delay, the transmissions happened in local memory are scheduled together in one single step instead of separating them among time steps like the results in Fig. 4. Other messages are pre-proceeded by *inter amplification* and then scheduled by a low cost scheduling method which selects messages with smaller cost to shorten the cost of a step and avoid the node contentions. Fig. 5 shows the results of *SDSH* which is with low synchronization delay and is contention free. There are two reasons making the results in Fig. 5 better than the results in Fig. 4. First, *SDSH* successfully avoids synchronization delay by congregating m_1, m_3, m_5, m_7, m_9 and m_{11} in step 3. It also helps reduce the cost of a step. Second, messages m_6 and m_{10} are the most important transmissions in the schedule due to their communication cost can dominate any steps. It is a pity that they are separated in two steps in Fig. 4 due to the node contentions. For example, it is impossible to move m_6 to step 1 to shorten the cost of step 2 due to m_5 and m_7. The message m_5 owns node 2 as source node and so does m_6. Both messages cannot be scheduled in the same step. Similarly, m_6 and m_7 cannot be scheduled together due to destination node. On same argument, it is impossible to move m_{10} to step 2 due to m_9 and m_{11}. If m_5, m_7, m_9 and m_{11} can be placed in other step, it would be possible to place m_6 and m_{10} together to minimize the communication cost of the results. *SDSH* successfully places them in step 3 and then schedules m_6 and m_{10} in step 1 to shorten the cost of other steps. This operation also successfully avoids node contentions that happened in Fig. 4.

A result of the proposed method		
No. of step	No. of message	Cost of step
Step 1	$m_2(3)$, $m_6(15)$, $m_{10}(13)$	15
Step 2	$m_4(1)$, $m_8(4)$	4
Step 3	$m_1(1.625)$, $m_3(2.125)$, $m_5(1.625)$, $m_7(1.625)$, $m_9(1.5)$, $m_{11}(1)$	2.125
Total cost		21.125

Fig. 5. A result of proposed method with low synchronization delay and contention free

5 Performance Evaluation

To evaluate the proposed method, it is compared with a scheduling method, *TPDR* [5]. The simulator generates schemes (strings) for 8, 16, 32, 64 and 128 nodes, and there are three nodes in a cluster. To constrain the data size of each node, the lower bound and upper bound of each value in the strings are 1 and the value that array size divided by the number of nodes, where the array size is 10,000. If the array is distributed on eight nodes, the lower bound and the upper bound of data size are 1 and 1250 for each node, respectively.

Fig. 6 shows the results of comparisons between *SDSH* and *TPDR*. For each set of node, the number on the right side represents the cases that *SDSH* performs better, *TPDR* performs better or tie cases. In the simulation results for 8 nodes, the proposed method wins 813 cases which is less than 90% because it is easy for both methods to find the same results when performing GEN_BLOCK redistribution on few number of nodes. Therefore, the number of tie cases is over than 10%, and is much more than the results of other sets. When performing GEN_BLOCK redistribution with more nodes, *SDSH* outperforms *TPDR*, and *TPDR* loses over 92% cases in the rest of the comparisons. Note that the proposed method always find the best results in over 93% cases including the tie cases in all comparisons. It also shows the contribution of *SDSH* for shortening transmission cost and avoiding synchronization delay.

Results of evaluations			
Num. of nodes	*SDSH*	*TPDR*	Same
8	813	76	111
16	946	43	11
32	950	48	2
64	914	79	7
128	903	96	1
Percentage	90.52%	6.84%	2.64%
Total	4526	342	132

Fig. 6. The results of both methods on five sets of nodes with 5,000 cases in total

Attributes of given cases				
Num. of nodes	CT_{max}	Average CT_{max}	Cost of 1,000 cases	
			SDSH	*TPDR*
8	6	3.271	6733580	7953932
16	8	3.762	5733523	6983753
32	10	4.246	3564076	4354899
64	10	4.661	2412444	2781670
128	11	5.009	1282008	1520884

Fig. 7. Attributes of given cases for five set of nodes

The attributes of generated cases dependents on the number of nodes, for example, higher CT_{max} and lower communication cost are with higher number of nodes. It is hard to find the same schedules for two scheduling heuristics with larger number of nodes. Fig. 7 shows the information of cases which are used to evaluate the *SDSH* and *TPDR*.

CT_{max} of results with 128 nodes is 11 which is almost two times larger than the CT_{max} of results with 8 nodes. The average CT_{max} also grows with higher number of nodes. The total cost of schedules given by both methods for 1000 cases with different number of nodes explains the contribution of *SDSH* in Fig. 6. The proposed method provides better schedules and the improves the communication cost about 15% while comparing to *TPDR*. It also explains how *SDSH* outperforms its competitor. Overall speaking, *SDSH* is a novel, efficient and simple method to provide solutions for scheduling communications of GEN_BLOCK redistribution.

6 Conclusions

To perform GEN_BLOCK redistribution efficiently, research focused on developing scheduling heuristic to shorten communication cost in algorithm level. In this paper, a *two-step communication cost modification* (*T2CM*) and a *synchronization delay-aware scheduling heuristic* (*SDSH*) are proposed to normalize the transmission cost and reduce synchronization delay. The *two-step communication cost modification* provides *local reduction* and *inter amplification* operations to enhance the importance of messages. The *SDHC* deal with messages separately to avoid synchronization delay and reduce the cost. The performance evaluation shows that the proposed methods outperforms its competitor in 92% cases and improves about 15% on overall communication cost.

References

1. Cohen, J., Jeannot, E., Padoy, N., Wagner, F.: Messages Scheduling for Parallel Data Redistribution between Clusters. IEEE Transactions on Parallel and Distributed Systems 17(10), 1163–1175 (2006)
2. Guo, M., Pan, Y., Liu, Z.: Symbolic Communication Set Generation for Irregular Parallel Applications. The Journal of Supercomputing 25(3), 199–214 (2003)
3. Hsu, C.-H., Bai, S.-W., Chung, Y.-C., Yang, C.-S.: A Generalized Basic-Cycle Calculation Method for Efficient Array Redistribution. IEEE Transactions on Parallel and Distributed Systems 11(12), 1201–1216 (2000)
4. Hsu, C.-H., Chen, M.-H., Yang, C.-T., Li, K.-C.: Optimizing Communications of Dynamic Data Redistribution on Symmetrical Matrices in Parallelizing Compilers. IEEE Transactions on Parallel and Distributed Systems 17(11) (2006)
5. Hsu, C.-H., Chen, S.-C., Lan, C.-Y.: Scheduling Contention-Free Irregular Redistribution in Parallelizing Compilers. The Journal of Supercomputing 40(3), 229–247 (2007)
6. Huang, J.-W., Chu, C.-P.: A flexible processor mapping technique toward data localization for block-cyclic data redistribution. The Journal of Supercomputing 45(2), 151–172 (2008)
7. Jeannot, E., Wagner, F.: Scheduling Messages For Data Redistribution: An Experimental Study. The International Journal of High Performance Computing Applications 20(4), 443–454 (2006)

8. Karwande, A., Yuan, X., Lowenthal, D.K.: An MPI prototype for compiled communication on ethernet switched clusters. Journal of Parallel and Distributed Computing 65(10), 1123–1133 (2005)
9. Prylli, L., Touranchean, B.: Fast runtime block cyclic data redistribution on multiprocessors. Journal of Parallel and Distributed Computing 45(1), 63–72 (1997)
10. Rauber, T., Rünger, G.: A Data Re-Distribution Library for Multi-Processor Task Programming. International Journal of Foundations of Computer Science 17(2), 251–270 (2006)
11. Sudarsan, R., Ribbens, C.J.: Efficient Multidimensional Data Redistribution for Resizable Parallel Computations. In: Stojmenovic, I., Thulasiram, R.K., Yang, L.T., Jia, W., Guo, M., de Mello, R.F. (eds.) ISPA 2007. LNCS, vol. 4742, pp. 182–194. Springer, Heidelberg (2007)
12. Wang, H., Guo, M., Wei, D.: Message Scheduling for Irregular Data Redistribution in Parallelizing Compilers. IEICE Transactions on Information and Sysmtes E89-D(2), 418–424 (2006)

Optimization of Intercluster Communications in the NumGRID

Maxim Gorodnichev, Sergey Kireev, and Victor Malyshkin

Institute of Computational Mathematics and Mathematical Geophysics,
Pr. Lavrentjeva, 6, 630090, Novosibirsk, Russia
{maxim,kireev,malysh}@ssd.sscc.ru
http://ssd.sscc.ru

Abstract. The NumGRID is a middleware for joining geographically distributed computational clusters in order to run large-scale scientific applications that use MPI standards for communication between processes. Intercluster communication system of the NumGRID provides a single communication environment for the processes located on internal nodes of joined clusters. The paper discusses shortcoming of the NumGRID v.1 implementations, suggests new principles for the NumGRID v.2 implementation and presents an experimental evaluation of this principles.

Keywords: NumGRID, computational grid, MPI.

1 Introduction

The NumGRID is a joint Russian-French project of the M2P2 laboratory (UMR 6181 CNRS-Universits d'Aix-Marseille et Ecole Centrale, Marseille) and ICM&MG SB RAS, Novosibirsk. The NumGRID provides middleware for joining geographically distributed computational clusters in order to run large-scale scientific applications (such a joint computational resource is also referred to as a NumGRID further on). Other important goals are:

- to prolong lifetime of the old clusters by consolidating them with newer systems or with other obsolescent clusters;
- to allow for multi-part applications where each part requires specialized hardware or system software. The NumGRID will join specialized clusters and place parts of the application according to their specific requirements.

An overview of the NumGRID architecture is given in the Section 2, the Section 3 discusses limitations of the NumGRID v.1 implementation and develops principles for the NumGRID v.2 intercluster architecture, the Section 4 presents an experimental evaluation of the new principles, and the conclusion is given in the Section 5.

C.H. Hsu and V. Malyshkin (Eds.): MTPP 2010, LNCS 6083, pp. 78–85, 2010.

2 Overview of the NumGRID Architecture and Mode of Use

The development of the NumGRID started [1] with the following basic requirements:

- clusters should be joined on a basis of common communication layer for the processes located on worker nodes,
- communication layer should be based on MPI [2] standards specifications,
- NumGRID should enable running distributed jobs without major changes to the local cluster administrative policies,
- clusters can be heterogenious (different CPU performance, different memory capacity, etc),
- each cluster is composed (see Fig. 1) of the head node and a number of worker nodes. Worker nodes are used for running jobs and are connected to each other via high performance network. Also, they are connected to each other and to a head node via less capable TCP/IP private network which is used for shared file system and job control. A head node has another TCP/IP network interface to accept user connections from outer world (internet). A head node is used to compile, queue and monitor jobs.

The NumGRID is composed of the NumGRID-MPI and the NumGRID-UI modules. The NumGRID-MPI partialy implements MPI standards for the heterogeneous environment of joined clusters and thus provides communication layer for the joint computational resource. NumGRID-UI provides a tool for configuration of the NumGRID environment and running jobs. The NumGRID-MPI includes (see Fig. 1 the NumGRID gateway and the NumGRID-MPI library. In order to run an application, the NumGRID gateways should be started on the head nodes of the clusters and gateways should be connected into a connected graph via TCP/IP networks. User application and the library are supposed to be compiled for every cluster involved in the NumGRID. Application is linked against the NumGRID-MPI library and is run as local MPI job on each cluster. We will refer to such a local job as a sub-job further. Each process of the local MPI job then connects via TCP/IP network automatically to the NumGRID-MPI gateway located on the head node of its cluster and thus, through the gateways, each process can communicate to any of the other processes located on other clusters. All the processes together constitute MPI_COMM_WORLD communicator. Local MPI installation library is used for intra-cluster communications thus enabling use of the internal high performance networks to the full extent. User should define connections between gateways, port numbers on each gateway listen, number of the processes to be run on each of the clusters and ordering of the processes. Basically, this requires user to open ssh sessions for each of the clusters, build a gateway executable and run it on a clusters head node. Here is an example of the command a user should issue to run a gateway:

```
./numgrid_net --internalport 10001 --externalport 10002 \
--localquorum 4 --globalquorum 8 --name gw2 \
--connect cluster2.example.dom:20002}
```

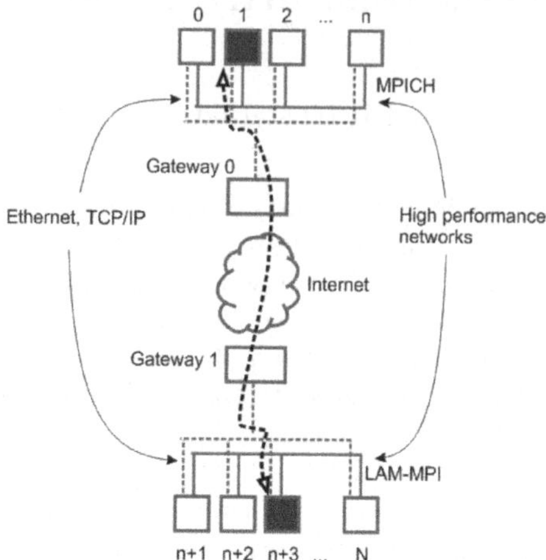

Fig. 1. The NumGRID basic layout

The `--internalport`is a number of a port to which sub-job processes should connect, the `--externalport`is open for other gateway to connect to this one, the `--localquorum`is a number of processes in the sub-job destined to this cluster, the `--globalquorum`is a the total number of processes in a job, the `--name`gives a name to a gateway instance, and the `--connect`is used to indicate gateways to which a given one is to be connected (hostname or IP and port number). Then user queues sub-jobs on each of the cluster. Each job is given a number of command line options defining the place of the sub-job in the job and gateway parameters. For example, a user might run:

```
mpiexec -n 4 /home/maxim/numgrid/test.exe \
--gateway-address 10.0.1.1 --gateway-port 10002 \
--subjob-id 0 --rank-base 0
```

Command line options provide a sub-job with the details of the gateway, tell the sub-job its number. The `--rank-base`gives a lowest global rank of the the processes in the sub-job. Graphical user interface (the NumGRID-GUI) is provided to make a single button that will start a job without a user having to type in all these commands. The NumGRID-GUI makes it easy to change parameters of the job and try again as much as needed. The NumGRID jobs do not make a difference with other jobs running on the clusters and thus do not disturb other users and cluster administration. This may result in unsynchronized sub-job readiness (due to different wait time in cluster queues). Such a mode of use is supposed to be followed for the purpose of application development and for running normal calculations. However, if there is a need to exploit all resources of the clusters involved simultaneously, it is supposed that user makes an agreement with administrations of the clusters and gets an exclusive access right.

3 Optimization of Inter-cluster Communications

3.1 Limitations of the NumGRID-MPI v.1 Architecture and Implementation

Present work has revised the NumGRID-MPI v.1 implementation and has revealed several shortcomings. Following are given the assumptions from which the NumGRID-MPI v.1 implementation proceeded and their limitations are explained.

- Gateways can connect each other directly. The fact that direct physical links can be absent was ignored. Hence, there was no space for routing optimizations. Proper mapping of virtual topologies to hardware cannot be done without taking properties of physical links into account.
- Gateways is not bound by memory limitations and can store messages of any length. Message transfer between processes located in different clusters is carried on in three steps. First, the source process sends the whole message to the gateway of this cluster, and then message is transferred as a whole to the gateway of the destination cluster. Finally, message is delivered to the destination process. This can incur unpredictable load on the cluster head nodes. Another disadvantage of such an approach is the high message delivery latency.
- Cluster cannot be added dynamically to the running NumGRID job and processes cannot be added to the job within one cluster. Thus, a user cannot supply more resources to a running job even if there are a reason and opportunity to do so. The MPI-2.0 standard specification declares functionality that makes dynamic process management possible. However this functionality cannot be implemented in the NumGRID-MPI v.1.

There are also some implementation drawbacks like an only message queue for all types of messages and all destinations at a gateway instance. Thus, delivery of a message is highly dependent on the other messages in the system. Together with the previous assumptions, all the architecture of the NumGRID v.1 disables implementation of effective nonblocking communications that are critical for application performance in heterogeneous environments.

3.2 New Principles

Analysis of the NumGRID-MPI v.1 bottlenecks leads to the following basic principles laid for the new implementation of inter-cluster communication system:

- Packet switching [3] transfer method should be used. It will a) resolve the problem with limited memory on the head nodes, b) allow short messages to be delivered while the longer messages are on transfer, c) decrease latency, d) allow for multi-path routing , e) allow for memory allocation optimization.
- User should be able to describe how the clusters are really connected and the properties of each link. It will enable intelligent routing policies.

- Messages can be assigned a priority [4]. This can be used in high level programming systems where run-time system messages might need a prioritized delivery. Users should be warned that message prioritization is out of the scope of the MPI standards.
- NumGRID construction schemes should allow for dynamic resource and process management.
- Nonblocking communication functions should be efficiently implemented. This requires thread-safe NumGRID-MPI implementation.

Gateways were redesigned to have designated buffers for each connection (gateway <=> process and gateway <=> gateway). Buffers can be extended to a limited size measured in packet sizes. Limits and packet size are configurable.

4 Experimental Evaluation

Experimental evaluation has been fulfilled in a test environment in which two virtual clusters were placed onto a real one. Figure 2 explains the configuration. Two gateways have been placed onto a head node of a real cluster and processes of an application have been located on worker nodes of the cluster. This setting excludes the factor of the network between two clusters and helps us understand the overhead incurred by the NumGRID implementation itself. In other respects, the situation does not differ from situations where two real clusters are involved.

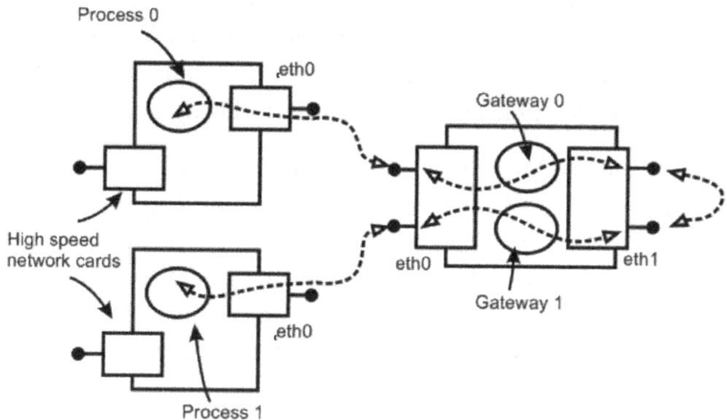

Fig. 2. Mapping of two virtual clusters onto a real one. Worker nodes are on the left, head node is on the right.

4.1 Ping-Pong Test

Overhead. The test application is composed of two processes. The process ranked 0 sends a message of a certain size to the processes ranked 1 and receives the message of the same size back. Test demonstrates how the time of the message roundtrip depends on the message size. The Fig. 3 presents the timings of the message exchange done with the NumGRID as described in Fig. 2 and local MPI installation. The later uses high performance network for message transfer. The NumGRID results presented for two packed sizes. The tests demonstrate that the NumGRID overhead grows with the size of a packet decreasing. However, the overhead is very high (10 times) even for relatively big packet size.

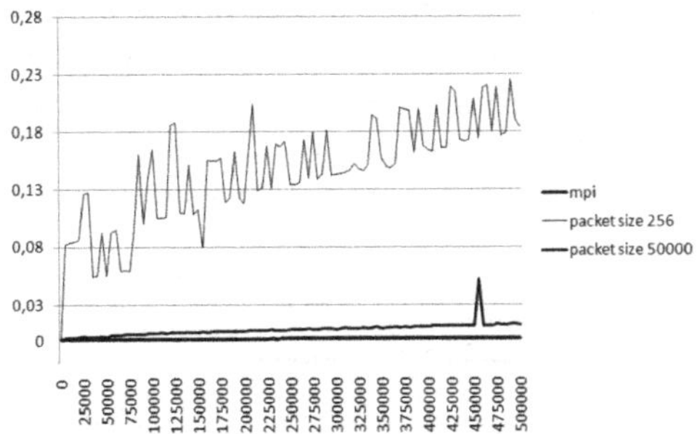

Fig. 3. The time, sec., the message needs to travel between two processes forward and back vs. message size, bytes

Small message with the long message in the background. The test (Fig. 4) demonstrates how the time of small message delivery (forward and back) depends on the packet size. Test is done in such a way that a small message delivery begins after a long message delivery between two clusters starts and small message delivery ends before the long message delivery ends.

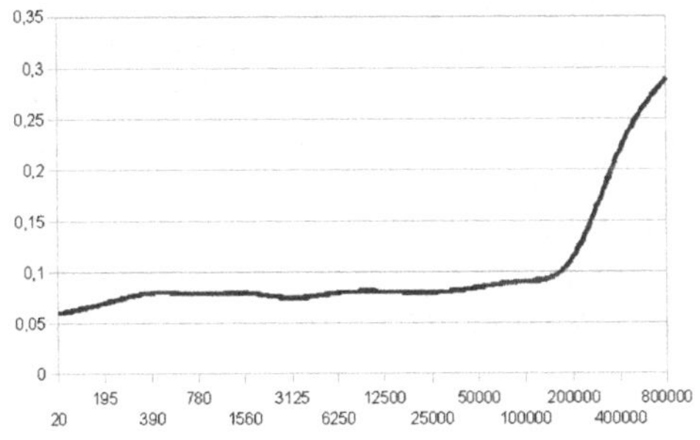

Fig. 4. Small message delivery with the long message in the background. Time, sec., of forward and back travel vs. the size of a packet, bytes.

4.2 Real Application: Solution of a Wave Equation

Following are the results of experiments with explicit wave equation solver by S. Kireev. An equation is solved in a rectangular domain. Paralleled implementation is done on the basis of domain decomposition onto a line of processes. Communications are required at the cross-sections at the end of each time step.

Packet size variation. This test explores the influence of the packet size on the performance of the application. We have measured execution time of 100 time step simulations for the 10000x10000 (2D) problem with two MPI processes (Fig. 2). Communications are done at the cut line of the domain and thus make a messages of the size $10000 * element\_size = 10000 * sizeof(double) = 80000B$. Figure 5 presents the execution time and the time that was taken by communications. Time of application run with local MPI library is given also for reference. One can notice that the lines of the local MPI implementation and NumGRID-MPI implementation do not differ substantially after a certain packet size value. It means that it is possible to choose a packet size that is less than the size of the messages but still does not much overhead as compared to the non-packeted transfers.

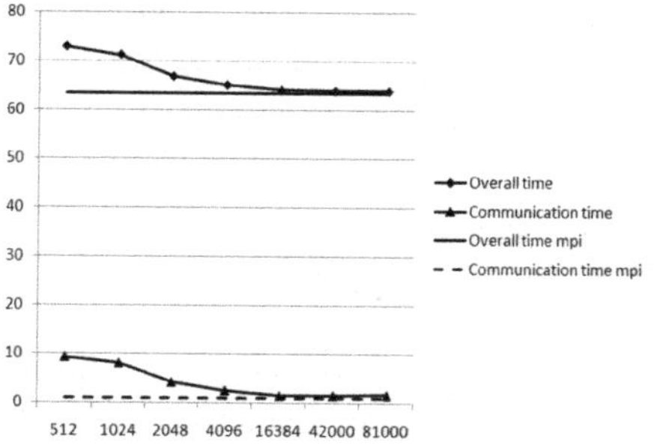

Fig. 5. Time, sec., to solve 100 time-steps of a 10000x10000 problem vz. fragment size, in sizeof(double)

Problem size variation. In this test (Fig. 6) the size of the problem changes. The NumGRID-MPI implementation with packet size equal to 8192 bytes is compared to the local MPI implementation. The results are presented in Fig. 6. The part *a* demonstrates variation of the size of the cut dimension. Variation of this size does not change the volume of communications. It explains why the NumGRID-MPI and local MPI implementations give almost equal results. Variation of the size of the uncut dimension changes size of the messages and thus the overhead of the NumGRID-MPI becomes visible.

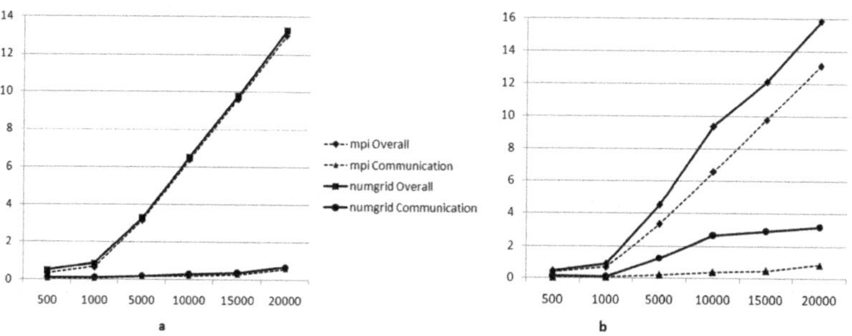

Fig. 6. Problem size variation. Part *a* varies the cut dimension, while the part *b* varies the size of the dimension that remains uncut. Execution time, sec., vs. problem size.

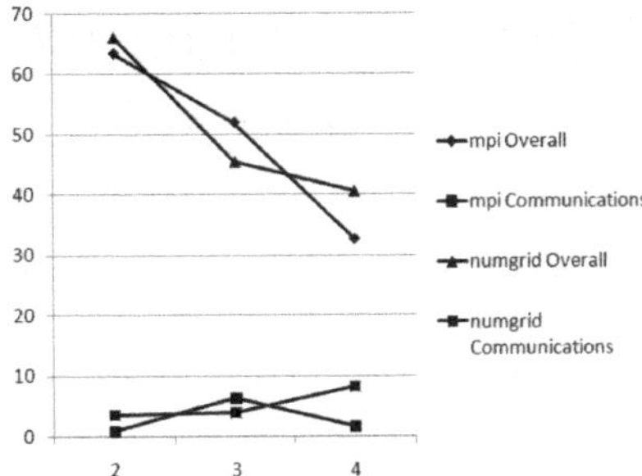

Fig. 7. Scaling. Time, sec., vs. the number of CPUs.

Scaling. This test (Fig. 7) compares local MPI implementation to the NumGRID implementation with 2, 3 and 4 cpus. The packet size is fixed to 8192 bytes. In the NumGRID-MPI environment, the 3 processes were split with 1 going to one cluster and other 2 to another one. A group of 4 processes was split so that each cluster gets 2 processes.

5 Conclusion

New design of the inter-cluster communications of the NumGRID satisfies to the requirements of the future development of the NumGRID. It enables intelligent optimizations of inter-cluster communications, implementation of the nonblocking communication routines, virtual topologies, dynamic process management and provides developers of applications and higher level parallel programming systems with a finer control over the inter-cluster communications in the NumGRID environments. However, the current implementation demonstrates a substantial overhead incurred by the NumGRID message-processing routines. On the other hand, the test reveals that there is still possible to achieve, in a heterogeneous environment, a performance comparable to the performance of local MPI implementations run within a high performance intra-cluster network. This is fulfilled by varying a packet size and mapping of the application to the hardware. Further work should be focused on optimization of message processing system, extension of the MPI support and running large scale computations in the NumGRID environment.

References

1. Fougere, D., Gorodnichev, M., Malyshkin, N., Malyshkin, V., Merkulov, A., Roux, B.: NumGrid middleware: MPI support for computational grids. In: Malyshkin, V.E. (ed.) PaCT 2005. LNCS, vol. 3606, pp. 313–320. Springer, Heidelberg (2005)
2. Message Passing Interface specifications, `http://www.mpi-forum.org`
3. Baran, P.: On Distributed Communications Networks. IEEE Transactions on Communications Systems (March 1964)
4. Real-Time Message Passing Specification based on MPI, `http://www.mpirt.org/`

Metacluster System for Managing the HPC Integrated Environment

Victor Gergel and Andrew Senin

N.I. Lobachevsky State University of Nizhni Novgorod, Russia
SeninAndrew@gmail.com

Abstract. Clusters became the de-facto standard in modern high-performance computing. At present it is rather often when a single organization has a few clusters and wants to connect them into a multicluster to benefit from reduced task waiting time and increased total available processing power. This paper studies one of possible approaches to the problem of uniting the computing resources which addresses exactly the case of owning clusters by a single proprietor. Such approach was implemented in Metacluster system of Nizhni Novgorod State University, Russia. Current state of Metacluster was reviewed. Key features, component-based architecture and main functions implementation details were described.

Keywords: Metacluster, cluster management system, tasks scheduling, computing resources monitoring, Microsoft High Performance Computing Server 2008.

1 Introduction

Cluster technologies play a leading role in high performance computing world ([1]). According to the current (as of November, 2009) TOP 500 list of supercomputing sites clusters occupy 417 positions which are 83% of installations ([8]). In most cases clusters offer an optimal ration of price/performance for solving a wide range of computing insensitive tasks such as finance math, physical modeling, new medicine discovering etc. Cluster systems are easy to upgrade, there are a lot of off-the-shelf hardware components available and administration can be done by technicians who do not have deep knowledge in the field of high performance computing. These factors have led to a massive spread of clusters in the scientific community, in industry, and even individual researchers often can afford to purchase or build their own a small cluster.

Increase in the number of cluster systems makes attractive the idea of clusters sharing to increase the total available processing power, as well as to reduce the average waiting time of tasks to start. The latter is achieved through the use of advanced scheduling strategies that take into account temporary fluctuations in workload of various clusters (if one of the clusters is idle then tasks from the most loaded cluster can be transferred to the idle one) and at the same time flexibility in allocation of resource (for example, tasks of cluster owners might have higher priority etc.).

C.H. Hsu and V. Malyshkin (Eds.): MTPP 2010, LNCS 6083, pp. 86–94, 2010.

The idea of clusters sharing is reflected in the concept of grid ([2]). There are currently a few popular technologies available which allow combining computing resources into the grid: Globus Toolkit, gLite, Unicore etc. Some of grid systems combine clusters located on different continents (eg. World Community Grid, [9]), consist of thousands of organizations and millions of computing nodes. But often there is a need to unite computing resources which belong to a single organization: a few clusters or labs of workstations. If these resources are physically close to each other, one can setup a dedicated connection between them with characteristics similar to network within a cluster. This allows the computing resources owner to expect from the derived *multicluster* to be more effective than when combining similar resources on the Internet. For solving the problem above one or another grid implementation can be used. But such solution will be paid by unnecessary complexity of administration and inconveniences for end-users. The concept of grid does not imply the availability of centralized management, which is natural in case the computing resources are owned by different organizations. But in the case of a single organization, this approach introduces additional cost and complicates tasks scheduling.

An alternative approach in creating a multicluster is to centralize the management of all connected computing resources. Such an approach may be inapplicable when managing thousands of resources all around the world. But in the case all clusters that make up a multicluster are owned by a single organization centralized management can be more effective and natural because it reflects the fact of owning the resources by a single proprietor. The high performance computing environment management system Metacluster developed at the University of Nizhni Novgorod is based on such principle. The main purpose of the system is more effective use of computing resources through load balancing between clusters, and effective scheduling strategies within each cluster.

2 Metacluster Overview

There were three consistently developed version of Metacluster during its project history. The first version of was Metacluster developed in 2002 and was focused on the organization of effective management of individual clusters while ensuring high reliability and fault tolerance when providing remote access to the cluster via the Internet. An important feature of the developed system was its installation on clusters operated under the family of operating systems Microsoft Windows. Selection of the operating system was due to the desire to simplify the problem of practical usage of high performance cluster systems by end-users – many application developers have more experience in Windows environment and their development of programs for Linux-based clusters could be quite complex. Besides, a large amount of computing resources which can be reused to build clusters (labs of workstations, student's terminals), as a rule, is running Windows. And finally, at the beginning of Metacluster development there were a rather limited number of affordable, reliable and easy to use cluster management systems working with Windows as practice of high performance computing on Windows had not have wide distribution. As shown by subsequent developments, the choice of Windows was justified, because it attracted to the subject of high performance computing a wide range of users. At present, most of the leading software manufacturers for cluster management also support Windows (Platform LSF, Condor, Microsoft HPC 2008 and others).

The next version of Metacluster was presented in 2005 and was aimed at supporting of multiclustering – the ability to manage simultaneously multiple clusters providing a single access point to all connected computing resources. This has opened great opportunities for users, as they can run the job on a wide range of computing resources united in a multicluster (or so-called integrated high performance computing environment) as well as to improve the overall efficiency of clusters due to dynamic load balancing between them.

The newest version of Metacluster was developed in 2008 with support of Linux clusters as well as integration with third-party cluster management systems including Microsoft High Performance Computing Server 2008.

The key features of Metacluster are the following: multicluster management, ability to work on different operating systems, integration with third-party cluster management systems. Consider each of the characteristics in detail.

2.1 Multicluster Management

Metacluster allows connecting to the integrated high performance computing environment an arbitrary number of clusters, possibly remote and not in the same network. The system takes over the problem of delivering necessary input files to the required cluster and to copy results of calculations on a virtual user desktop. This allows users not to think about where the tasks physically are being performed. In case of temporary unavailability of a particular cluster, as well as in the case of new computing resources being connected to Metacluster there is nothing changes for users: the system finds out itself which resource it is more effective to resend the tasks to. All the user needs to do is to specify requirements for a task: architecture and a number of processors, RAM, free space on hard disks and other characteristics. If necessary it is possible to explicitly specify a cluster and later its nodes. But in this case a user will have to restart the tasks by him-/herself in case the selected resources are disconnected.

2.2 Multiplatform Cluster Integration

Metacluster enables integration of clusters running different operating systems. This is achieved through specific implementation of platform-specific code and the use of open protocols for component communications. Currently major versions of operating systems Windows and Linux are supported. When formulating a problem user selects a target operating system and implementation of communication mechanism (for example a specific implementation of MPI) required by the application. The task of Metacluster is to choose appropriate resources and to launch the application in a specific for selected operating system and MPI implementation manner.

Flexibility in selecting of operating system allows extending the range of Metacluster users and spectrum of applications. However Metacluster partially retains its focus on work in Windows. Thus, the integrator (distribution of tasks across clusters) and the remote access component work only in Windows.

2.3 Cooperation with Third-Party Cluster Management Systems

Metacluster can interact with cluster management systems of third party developers. When connecting of such a cluster Metacluster takes over work on the interaction with the cluster management system: adding of tasks, tracking task statuses, monitoring of computing resources, etc. This allows connecting of clusters to the integrated high performance computing environment management system without disturbing existing users, who can continue operating in the familiar way. Currently integration with Microsoft High Performance Computing Server 2008 is implemented.

3 Metacluster Architecture

There are four main components of Metacluster: remote access manager, integrator of clusters, manager of a cluster and inspector of a node.

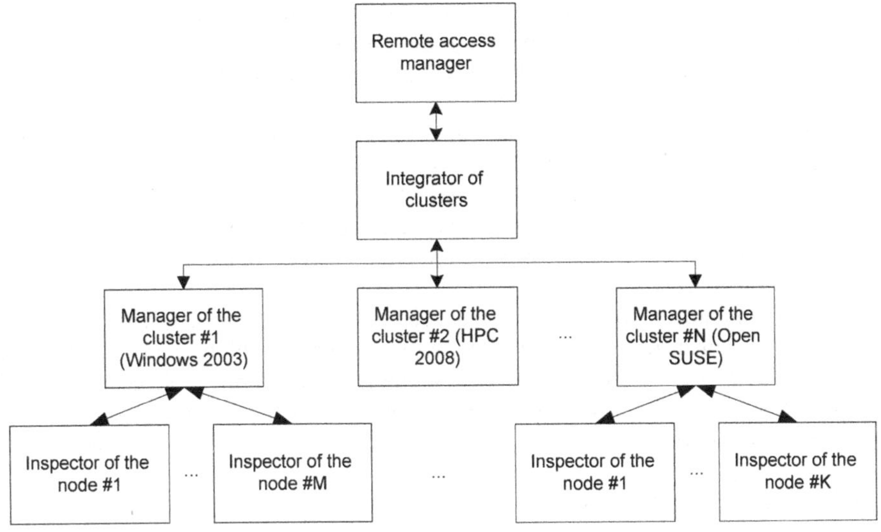

Fig. 1. Architecture of Metacluster

3.1 Remote Access Manager

The remote access manager is a single access point for all the resources under control of Metacluster. The access is possible either through a Web service ([10]) or through web interface. The web service allows integrating Metacluster with custom user applications that need to use high performance computing resources. The open standards the web service is based to (XML, WSDL, SOAP) help to interact with Metacluster using virtually any programming language and any operating system.

Metacluster web interface provides convenient and easy access to computing resources via the Internet and does not require installing any special software. After the authentication procedure a Metacluster user gets access to his/her a virtual desktop with uploaded applications and input files as well as results of previous calculations. The virtual desktop supports all basic file management operations: creating of directories, moving/ coping, renaming, zipping/ unzipping, loading/ downloading of files, etc. To add a new task a Metacluster user selects an executable file and sets properties of the computing problem: required computing resources, input parameters, standard output file, task priority, etc. After adding the task its state is shown in the task queue. The web interface can also be used to draw different statistical graphics and reports on the cluster usage during selected periods of time. Almost all of these operations can be performed from a user's program through the utilization of web services - in fact, the web interface is just one of the web service clients.

3.2 Integrator of Clusters

The integrator of clusters is the central point of multicluster control. The integrator receives tasks from the remote access manage, distributes them among the clusters and monitors task states. The integrator liaises with each of the clusters included in Metacluster. It is not required for cluster nodes and the integrator to be in the same subnet, only communication between the integrator and the cluster manager is necessary. The integrator periodically checks the availability of each cluster. In the case of connection loss the integrator may decide to move jobs to another cluster. The integrator stores a queue of all tasks running on the multicluster, monitors their states and transmits the information to the remote access manager.

The integrator selects only a cluster which runs the tasks; nodes are selected by the appropriate cluster manager. However, in case user explicitly sets a cluster and nodes where he/she wants the task to run the integrator simply translates this information to the cluster manager.

For communication of the cluster manager and the remote access manager MS .NET remoting mechanism is used which assures high reliability and speed of data transmission. The presence of an open source cross-platform implementation of MS .NET remoting ([11]) allows connecting the integrator with the cluster manager running on Linux and other operating systems. The integrator itself only runs on Windows.

3.3 Manager of a Cluster

The cluster manager component is a cluster management system for a single cluster. Job of the component is to distribute tasks on the cluster nodes in accordance with local planning strategy, to monitor of cluster computing resources and to provide fault tolerance of individual sites. The cluster manager has a local task queue of tasks from the integrator and tasks from local users. The cluster manager tracks states of running tasks and transfers this information to the integrator.

A specific cluster manager implementation can be used to connect clusters under control of different operating systems to Metacluster as well as clusters with third-party cluster management systems. In the latest case the cluster manager acts as a communicator of the integrator and the cluster management system by converting integrator command into the local cluster management system commands. Currently Windows and Linux native cluster managers are implemented and a communicator for Microsoft High Performance Computing Server 2008.

3.4 Inspector of a Node

The inspector of nodes component must be installed on all cluster nodes. Its job is to execute cluster manager command on remote nodes. The commands may include the following: start/stop of processes, restart of hung services, providing information of node workload and state as well as states of separate processes of interest. For communication with the cluster manager MS .NET remoting mechanism is used.

4 Metacluster Implementation Features

One of the most important challenges of cluster management is to ensure the efficient use of available computing resources. Only in this case the maximum possible workload of a computer system can be achieved, which, in turn, will minimize time of tasks execution. Thus, the scheduling and monitoring subsystems are two of the most important functions of cluster management. In case of a multicluster they are even more important as they are required ability to work in heterogeneous geographically distributed environment. Let's consider the implementation of these functions of Metacluster in more details.

4.1 Task Scheduling

Scheduling in Metacluster is carried out at two levels: at first the integrator selects the most appropriate cluster for the scheduling task, and then the corresponding cluster manager allocates computing nodes for the task. When choosing a cluster the integrator takes into account requirements specified by a user: necessary resources (number of processors, memory, disk space, etc.), operating system, installed libraries and packages. If several clusters simultaneously satisfy these requirements, then the least busy is chosen, that is the one on which the expected time of task completion is less.

The primary scheduling algorithm on the cluster manager level is backfilling ([3]) – one of the most efficient and widely used cluster scheduling algorithms. The algorithm is a modification of the FCFS (First Come First Served) method with priorities. The basic FCFS algorithm schedules tasks in order established by a weighted sum of several parameters: task waiting time, task priority, user priority etc. For the backfilling algorithm to work it is necessary to set up its expected execution time. With using this information the scheduler can plan less priority but small tasks to be executed before high priority major tasks by utilizing free time slots in the schedule. Thereby the average tasks waiting time is reduced.

The cluster manager can be integrated with Maui – a popular task scheduler on cluster systems. Maui allows an administrator to configure advanced settings of scheduling policies. Integration with Maui was accomplished by exchange of network sockets messages in WIKI protocol ([12]). Details of integration are presented for example in [4].

4.2 Monitoring of Computing Resources

The main jobs of the Metacluster monitoring subsystem are the following:

- Providing up-to-date information of dynamic computing nodes metrics: CPU workload, amount of free RAM, amount of HDD free space etc.,
- Graphics with statistics of cluster workload in the Metacluster web interface.

The monitoring subsystem must be cross-platform and work stable on heterogeneous geographically distributed multicluster system. We found that the Ganglia distributed system for monitoring of high performance computing resources ([5]) best suits to the listed requirements. It consists of the following main components:

- Gmond is a service of node monitoring which must be installed on each cluster node,
- Gmetad is a service of statistics gathering which is installed on cluster head node to collect and store statistics into a cyclic database. In case of having multiple clusters the service can accumulate statistics from different gmetad services,
- Web interface is a visual part for graphic display of statistics.

Ganglia is open source software written in C programming language. The node monitoring service has been ported to a wide range of different operating systems, but the service of statistics gathering is intended for use only on Unix-like systems. In particular, according to the developers gmetad does not compile on Windows. Nevertheless, as a part of integration of Metacluster and Ganglia the Metacluster developing team has ported gmetad to Windows – see [6] for more details.

The Metacluster scheduler receives monitoring information from a specially developed library which uses network sockets to communicate with gmetad (format of messages is XML). The Ganglia web interface provides convenient and intuitive means of displaying information about the workload of the cluster. Ganglia Web interface was built into the statistics page of Metacluster web interface to display cluster workload for selected intervals.

4.3 Managing of Different Task Types

Metacluster does not restrict types of user applications. Those may be sequential programs, parallel programs written with using different implementations of MPI, tasks which utilize packages installed on a cluster. Such flexibility is achieved by moving parts of the launching logic into Perl scripts, editable by a cluster administrator. An administrator determines what types of tasks are allowed to run on a cluster, for example: sequential executables, scripts in Python, MPICH2 programs, OpenMPI programs, Fluent tasks ([13]) etc. For each type of task an administrator binds its handling script.

The handling scripts accept the following parameters:

- Module name. File selected in the Metacluster web interface to run. In case of running self-written application it may be an executable or a script. In case of utilizing a package installed on a cluster this may be a name of input file with model description and settings,
- Command line arguments,
- Standard output and standard error redirection,
- Working directory. Full path to a directory which corresponds to the virtual folder from which the user started his/her application,

- Environment variables set by a user in the web interface,
- Nodes allocated by the scheduler to run the task.

With using given values the script must create a full command string to run the task. For example in case of MPICH2 application it may be the following command:

```
<MPI install path>\mpiexec -hosts <list of nodes> -env
<environment variable> <module name> <command line
arguments> > <output filename> (some parameters were
omitted for clarty).
```

In addition of forming a command line a script may execute additional service operations in order to prepare the task to run: checking correctness of passed parameters, preparing necessary environment, registering of environment variables, writing messages to a log file etc. An executable file which is actually started as result of executing the script command must be running until the task finishes. Otherwise Metacluster considers this as a completion of the task and starts the release of resources procedure which closes a whole tree of running processes started by the task.

5 Conclusions

The high performance computing environment management system Metacluster manages 3 clusters of the Supercomputing technologies center of Nizhni Novgorod state university: a cluster of 64 dual-core dual-processor servers based on Intel Xeon 3.2 GHz, 4 GB RAM and 2 miniclusters on 4 and 5 nodes, respectively. Gigabit Ethernet is the main data network inside the clusters.

Resources under control of Metacluster are actively used by a wide range of research workers and students of the University and some other users. There are some applications developed at the university which utilize high performance computing resources through integration with Metacluster: ParaLab, Global Expert. Metacluster was adapted to act as a part of an experimental grid segment of the SKIF-grid program ([14]). Metacluster is also used in the project “Developing of high performance software complex for the quantum mechanical calculations and modeling of nanoscale atomic-molecular systems and complexes” (see [7]).

There are more than 100 user registered in the system. During peak periods the workload of the system was more than 500 tasks per day. Metacluster proved to be easy to use, stable and extendable system.

Currently the Metacluster team works on using of multiple clusters in one MPI task as well as on improving and extending the remoting manager component options.

References

1. Gergel, V.P.: Theory and practice of parallel computing. Internet university of information technologies intuit.ru, Moscow (2007) (in Russian)
2. Foster, I., Kesselman, C., Nick, J., Tuecke, S.: The physiology of the grid: An open grid services architecture for distributed systems integration (2002)

3. Lee, C.: Parallel Job Scheduling Algorithms and Interfaces. Department of Computer Science and Engineering. University of California, San Diego (2004)
4. Kustikova, V.D., Senin, A.V.: Integration of clusters management system Metacluster with Maui scheduler. In: Proceeding of Technology Microsoft in theory and practice of programming, NNSU, N. Novgorod (2008) (in Russian)
5. Massie, M., Chun, B., Culler, D.: The Ganglia Distributed Monitoring System: Design, Implementation, and Experience. Parallel Computing 30 (2003)
6. Lozgachev, I.N., Senin, A.V.: Monitoring of high performance computing in Metacluster clusters management system. In: Proceeding of Technology Microsoft in theory and practice of programming, NNSU, N. Novgorod (2008) (in Russian)
7. Vasil'ev, V.N., Buhanovski, A.V., Kozlov, S.A., Maslov, V.G., Roganov, N.N.: High performance software complex for modeling of nanoscale atomic-molecular systems. In: Technologies of high performance computing and computer modeling, SPbSU ITMO. SPbSU ITMO, University telecommunications, Saint Petersburg, vol. 54 (2008) (in Russian)
8. TOP500 Highlights (November 2009), `http://www.top500.org/lists/2009/11/highlights`
9. World Community Grid official website, `http://www.worldcommunitygrid.org/`
10. Web Services Glossary, `http://www.w3.org/TR/ws-gloss/`
11. Mono project official site, `http://www.mono-project.com/Main_Page`
12. Wiki Interface Specification, version 1.2, `http://www.clusterresources.com/products/mwm/docs/wiki/wikiinterface.shtml`
13. ANSYS FLUENT Flow Modeling Software, `http://www.ansys.com/products/fluid-dynamics/fluent/`
14. SKIF-grid initiative official web site, `http://skif-grid.botik.ru/`

The ParaLab System for Investigating the Parallel Algorithms

Victor Gergel and Anna Labutina

Nizhni Novgorod State University
gergel@unn.ru, anna.labutina@cs.vmk.unn.ru

Abstract. In this paper we introduce a software system which allows to carry out and visualize computational experiments for studying and researching the parallel algorithms of solving complicated computational problems in imitation mode on one single sequential computer. User can "assemble" a parallel computational system of cluster type that consists of multiprocessor and multicore nodes connected with the network, set up the problem to be solved, carry out the parallel solving algorithm, collect and analyze the results of computational experiments. To estimate the execution time of parallel method on current hardware system we use the sophisticated models. For every implemented parallel method we proved the theoretical estimations of the execution time by comparing the real time of the execution on the NNSU high performance cluster with the time, that can be calculated using the model.

Keywords: high performance computing, parallel computing, parallel computations modeling, cluster, multiprocessor architecture, multicore architecture.

1 Introduction

The development of the computer architecture and network technologies, together with investigations of new time-consuming scientific and applied problems that demand massive computations showed high necessity of parallel computations, made high performance computing the cornerstone of programming and computational technology.

But despite of science needs and the actuality of parallel computations, so far they are not as widely used as it was predicted. One of the possible reasons is the necessity of developing new parallel algorithms to solve the new computationally intensive problems. It is well-known that the speedup of solving the task on parallel computational system can only be achieved when the algorithm is divided into set of independent processes that can be run simultaneously. The other reason is that the debugging of parallel code is a high complexity problem, which makes it necessary to fully understand the behavior of the system of computational processes run in parallel. That is why competence in modern high performance computational system design trends, in new tools developed to achieve parallelism, the ability to create models and methods for solving the problems in parallel are the major qualities for specialists in applied mathematics, computer science and information technologies.

C.H. Hsu and V. Malyshkin (Eds.): MTPP 2010, LNCS 6083, pp. 95–104, 2010.

2 Working with ParaLab

While working with ParaLab the user has an access to a wide range of tools to set the computational experiment parameters. She can model the computational system, chose the problem, carry out the parallel algorithm, collect and analyze the results of computational experiments.

2.1 Modeling the Parallel Computational System

ParaLab allows to simulate the parallel computational experiments execution on multiprocessor (*SMP*) and *multicore* architectures. The computational system appears to consist of the *computational nodes* (*computers*). Each node has one or more *processors*, and each processor has one or more *cores*. The ParaLab system architecture doesn't limit the maximum amount of cores in processor and processors in one node, but for the sake of visualization we limit the number of cores to be equal to 1, 2 or 4 and number of processors to be 1 or 2.

In order to simulate the computer system, it is necessary to determine the network topology, the number of computational nodes, the number of processors and cores in one node, the performance of each core, and the characteristics of the communication network (the latency, the bandwidth and the data communication method). It should be noted that the computer system is assumed to be homogeneous in the ParaLab system, i.e. all the computational nodes have the equal amount of processors, every processor consists of the same number of cores, cores possess equal performance, and all the communication lines have the same characteristics.

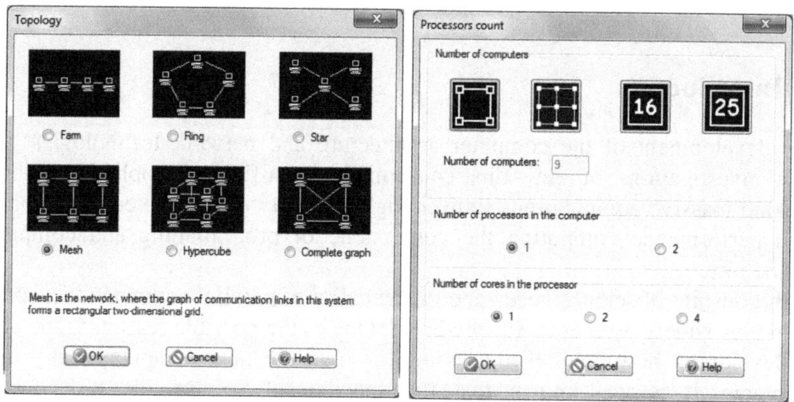

Fig. 1. Dialog windows to set up the computational system parameters

The data communication network *topology* is defined by the structure of communication lines among the computer system processors. The system ParaLab supports the following network topologies: farm, ring, star, grid, hypercube, full graph (clique).

The system ParaLab allows user to set the desirable number of nodes for the selected topology. The choice of the system configuration is performed in accordance with the type of the topology used.

The *performance* of the core in the ParaLab system is measured by the number of floating point operations per second (flops). It should be noted that to estimate the execution time of the experiment, it is assumed that all the computer instructions correspond to the same floating point operation.

The time of data transmission among the processors determines the *communication overhead* of the parallel algorithm execution in a multiprocessor system. The main set of parameters, which makes possible to estimate the data communication time, contains the following values:

–*latency* (α). It is the time, which characterizes the duration of preparing a message for transmission;

–*network bandwidth* (β). It is defined as the maximum amount of data, which can be transmitted in a certain unit of time through a data communication channel.

Among the data communication methods, implemented in ParaLab, there are the following two well-known communication methods [3]. The first method is aimed at *passing messages* as indivisible information blocks (*store-and-forward routing* or *SFR*). The second communication method (*cut-through routing* or *CTR*) is based on representing the transmitted messages as a set of information blocks of smaller sizes (*packets*).

2.2 Selecting the Problem and the Parallel Method

The following widely used parallel algorithms applied to solving complicated computational problems in various scientific and technical applications are implemented in the system ParaLab: the algorithms for data sorting, the algorithms for matrix operations, the algorithms for solving the systems of linear equations, graph processing, the algorithms for solving differential equations in partial derivatives and the algorithms for global multiextremal optimization.

As a rule, for every task there are several parallel solving methods implemented. For the matrix-vector multiplication task we implemented algorithms based on block, rowwise and columnwise matrix decomposition. For the matrix multiplication problem there are parallel Fox's and Cannon's algorithms and the algorithm based on striped matrix decomposition. For the problem of solving the system of linear equations we present the parallel variants of Gauss method and conjugate gradient method. For the sorting problem we implemented parallel variants of bubble sort, Shell sort and quick sort. For the graph processing task there are parallel algorithm for building minimal spanning tree, Dijkstra's and Floyd's algorithms for shortest paths problem. For the problem of solving the differential equation in partial derivatives we have parallel Gauss-Seidel algorithm. Parallel index method is implemented for the problem of multiextremal optimization.

The main problem parameter in ParaLab is the amount of the initial data. User can set additional parameters for some types of problems. For example there is a possibility to choose the boundary conditions for the problem of solving the differential equation in partial derivatives, to choose the type of function for the problem of multiextremal optimization, to create a graph with the help of built-in graph editor for the graph processing problem.

2.3 Carrying Out the Computational Experiment

ParaLab provides various forms of graphical demonstration of parallel computation results in order to observe the process of carrying out a computational experiment of solving complicated time consuming computational problems. Before the parallel algorithm execution user can set the visualization parameters for demonstration speed, the mode of communication operation visualization, the required level of details to be shown.

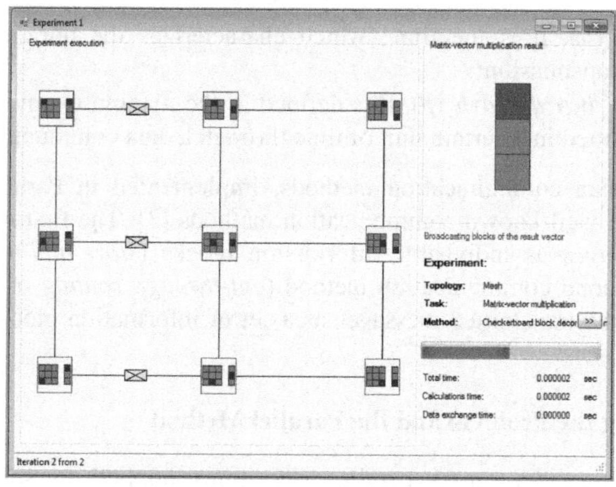

Fig. 2. The window of the computational experiment while solving the problem of matrix-vector multiplication

The system ParaLab provides different schemes of carrying out experiments to give convenient possibilities for studying parallel algorithms of solving complicated computational problems. Problems may be solved in the sequential execution mode, in the time sharing mode with the possibility to simultaneously observe the algorithm iterations in all the computational experiment windows. Carrying out series experiments, which requires long-continued computations, may take place in the automatic mode. Experiments may be also carried out in the step-by-step mode.

2.4 Accumulating and Analyzing the Experiment Results

To accumulate the results of the executed experiments, ParaLab provides a special memory, which is hereinafter referred to as the *experiment log*. The results are stored in the experiment log automatically. Accumulated results can be used for observing and analyzing.

For the experiments saved in the experiment log, we build the graph that shows how the execution time and the speedup depend on problem and computational system parameters. These graphs are built in accordance with the theoretical models we use to estimate the execution time of the parallel algorithm.

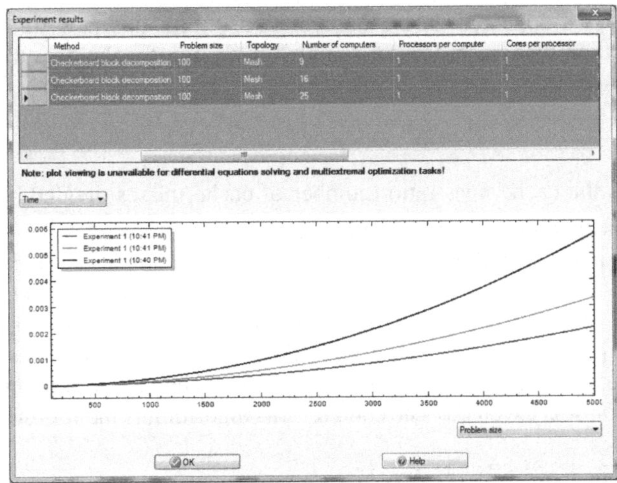

Fig. 3. The experiment log window

3 Modeling Parallel Computations

3.1 Model for the Local Computations

While creating a model to estimate the time of local computations we assume that this time is the sum of the calculation time and the memory access time:

$$T_1 = T_{calc} + T_{mem} \quad (1)$$

Here the calculation time is the result of multiplication of the executed operations number N by the time of one operation execution τ. The memory access time is the result of division of the maximum amount of data M by the memory bandwidth β. To make the estimation more precise we should consider that the data comes from memory not in byte-by-byte mode but in full cache lines, the length of one cache line is equal to L bytes. The worst case is when every data element should be downloaded from the memory and it falls in the separate cache line. We should also consider the RAM latency α that can significantly influence the time of computations. Thus, the model for the local computations execution time can be the following:

$$T_1 = N \cdot \tau + M \cdot (\alpha + L / \beta) \quad (2)$$

This model doesn't reflect the modern processor architecture, where the processor has small but fast local memory, which is called cache memory. In order to get the fast access to the necessary data this data is downloaded from RAM to cache before the computations with the use of different prediction algorithms. This download can be performed simultaneously with computations and doesn't affect the time of computation execution. The situation when the necessary data is not in the cache and the processor should wait for them to be downloaded from RAM is called *cache miss*. To make the model of computational time more precise we need to know the number of cache misses appeared during computations. With this new information we can correct the time that the processor spends on waiting for the data to be downloaded from the RAM:

$$T_1 = N \cdot \tau + \gamma M \left(L / \beta + \alpha \right) \tag{3}$$

where γ is the cache miss ratio (number of cache misses divided by the number of cache access operations), which can be theoretically estimated.

To make a decision about the model accuracy the computational experiments were carried out on the computer with the Intel core 2 quad Q6600 processor. The architecture of this processor includes first-level caches with the bandwidth of 153 Gb/sec and latency of 1,22 nsec. The RAM of the target system has a bandwidth of 12,4 Gb/sec and latency of 8,31-80 nsec. The algorithm of matrix-vector multiplication was executed. The code for this algorithm is the following:

```
for (i=0; i<Size; i++) {
  pResult[i] = 0;
  for (j=0; j<Size; j++)
    pResult[i] += pMatrix[i*Size+j]*pVector[j];
}
```

To calculate the time of one operation execution τ we measured the time spent on performing the algorithm for small object size, when matrix and vectors can fit in cache L1. We divide this time by the number of performed operations and get the time of one operation execution $\tau = 3{,}78$ nsec.

Table 1. Comparison of the experimental and theoretical execution time of the matrix-vector multiplication algorithm

Matrix Size	Experimental Time	Theoretical Time	Relative Error
2000	0,0303	0,0304	0,0021
4000	0,1222	0,1217	0,0036
6000	0,2748	0,2740	0,0029
8000	0,4894	0,4872	0,0044
10000	0,7637	0,7611	0,0034

In current version of ParaLab the simplier model for estimating the time of local computation is realized. This model only uses the number of operations and the time of one operation execution τ. We plan to implement the described approach to local computations time estimation in the next version of ParaLab.

3.2 Model for Data Passing Operations Execution

The time necessary for transmitting data between the processors defines *the communication overhead* of the parallel algorithm execution in a multiprocessor system. The basic set of parameters, which can help to evaluate the data transmission time, consists of the following values:

- **initializing time** (α) characterizes the duration of preparing the message for transmission, the search of the route in the network etc.;
- **control data transmission time** (t_c) between two neighboring processors (i.e. the processors, connected by a physical data transmission channel); to control data we may refer the message header, the error detection data block etc.;
- **transmission time of one data byte along a data transmission channel** ($1/\beta$); the duration of this transmission is defined by the communication channel bandwidth.

Let's consider *store-and-forward routing (SFR)*. In case of this approach the processor, which contains a message for transmission, gets all the amount of data ready for transmission, defines the processor, which should receive the data, and initializes the operation of data transmission. The processor, to which the message has been sent, first receives all the transmitted data and only then begins to send the received message further along the route. The time of data transmission t_{comm} for the method of transmitting the message of m bytes along the route of length l is defined by the expression:

$$t_{comm} = \alpha + (t_c + \frac{m}{\beta})l \tag{4}$$

If the messages are long enough, the control data transmission time may be neglected, and the expression for data transmission time may be written in a simplified way:

$$t_{comm} = \alpha + \frac{m}{\beta}l \tag{5}$$

Let's consider *cut-through routing (CTR)*, when the receiving processor may send the data further along the route immediately after receiving the current packet without waiting for the termination of the whole message data transmission. The data transmission time in case of packet communication method will be defined by the following expression:

$$t_{comm} = \alpha + \frac{m}{\beta} + t_c l \tag{6}$$

If we compare the obtained expressions, it is possible to notice that in the majority of cases the packet communication leads to faster data transmission. Besides, this approach decreases the need for memory for storing the transmitted data.

3.3 The Data Passing Operations in Multiprocessor and Multicore Architectures

As it was previously mentioned, in ParaLab the computational system consists of computational nodes, the network links between them are determined by the topology (farm, ring, etc.). Every node has one or more processors, every processor consists of one or more cores. We assume that the internal links between cores (busses) in frame of one node form the full graph topology.

To make the time estimation model easier we assume that the computations and data passing operations cannot overlap, which means that the computations stop when the cores are performing the data transmission, and vice versa.

Every collective data passing operation between cores can be divided into 3 stages:

1. Data transmission between cores in frames of one computational node and sending the data into the external network (via network adapters),
2. Data transmission between different computational nodes through the local network,
3. Receiving the data from the network adapter by the different cores in frames of one computational node.

To calculate the final time of the communication operation we only take into account the time of the second stage (passing the data through local network). The time spent on transmitting the data through the bus is 3 to 4 degrees less than that.

3.4 An Example of Computational Experiment Time Estimation

Let's consider the complexity of the parallel algorithm for matrix-vector multiplication based on rowwise matrix decomposition. Every core performs the multiplication of the matrix stripe by the vector, each stripe has *n/p* rows, where *n* is the size of the matrix and *p* is number of cores. One scalar product of the matrix row and a vector involves *n* multiplications and *(n-1)* additions. Let's assume that the multiplication and addition have the same duration τ. Besides, let us assume that the computer system is homogeneous, i.e. all the processors of the system have the same performance. With regard to the introduced assumptions, the computation time of the parallel algorithm is:

$$T_p(calc) = \lceil n/p \rceil \cdot (2n-1) \cdot \tau \tag{7}$$

The 'all gather' operation is used to put the result vector on all the processes of the parallel program. This operation can be performed in $\lceil log_2 p \rceil$ iterations. At the first iteration the interacting pairs of processors exchange messages of size $w\lceil n/p \rceil$ byte (w is the size of one element of the vector in bytes). At the second iteration the size becomes doubled and is equal to $2w\lceil n/p \rceil$ etc. As a result, the all gather operation execution time when the Hockney [2] model is used can be represented as:

$$T_p(comm) = \sum_{i=1}^{\lceil log_2 p \rceil} \left(\alpha + \frac{2^{i-1} w \lceil ^n/_p \rceil}{\beta} \right) = \alpha \lceil log_2 p \rceil + \frac{w \left\lceil \frac{n}{p} \right\rceil \left(2^{\lceil log_2 p \rceil} - 1 \right)}{\beta} \tag{8}$$

where α is the latency of data communication network, β is the network bandwidth. Thus, the total time of parallel algorithm execution is

$$T_p = \left(\frac{n}{p} \right) \cdot (2n-1) \cdot \tau + \alpha \cdot log_2 p + w \cdot \left(\frac{n}{p} \right) \cdot (p-1)/\beta \tag{9}$$

(to simplify the expression (8) it was assumed that the values n/p and $log_2 p$ are whole numbers).

Let us analyze the results of the computational experiments carried out in order to estimate the efficiency of the discussed parallel algorithm. The obtained results will be used for the comparison of the theoretical estimations and experimental values of the computation time. The experiments were carried out on the computational cluster on the basis of the processors Intel XEON 4 EM64T, 3000 Mhz and the network Gigabit Ethernet under OS Microsoft Windows Server 2003 Standard x64 Edition. The value of τ was equal to 1.93 nsec. The value of latency α and bandwidth β are correspondingly 47 msec and 53.29 Mbyte/sec. All the computations were performed over the numerical values of the double type, i.e. the value w is equal to 8 bytes.

The comparison of the experiment execution time T_p^* and the theoretical time T_p calculated in accordance with the expression (9) is shown in Table 2.

Table 2. The comparison of the experimental and theoretical execution time for parallel algorithm of matrix-vector multiplication based on rowwise matrix decomposition

Matrix Size	2 processors		4 processors		8 processors	
	T_p	T_p^*	T_p	T_p^*	T_p	T_p^*
1000	0,0069	0,0021	0,0108	0,0017	0,0152	0,0175
2000	0,0132	0,0084	0,014	0,0047	0,0169	0,0032
3000	0,0235	0,0185	0,0193	0,0097	0,0196	0,0059
4000	0,0379	0,0381	0,0265	0,0188	0,0233	0,0244
5000	0,0565	0,0574	0,0359	0,0314	0,028	0,015

4 Conclusion

The Parallel Laboratory software system (ParaLab) provides the possibility of carrying out computational experiments for studying and investigating the parallel algorithms of solving complicated computational problems. The system may be used for organizing a set of laboratory works on various courses in the area of parallel programming. This laboratory works will allow the learners to do the following:

- *To model multiprocessor systems* with various data communication network topologies,
- *To obtain the visual presentations of the computational processes and data communication operations* which take place in case of parallel solving various problems,
- *To construct the efficiency estimations* of the parallel methods to be studied.

In general, ParaLab is the integrated environment for studying and investigating the parallel algorithms of solving complicated computational problems. A wide set of available means to visualize the process of carrying out an experiment and to analyze the obtained results allows to study the parallel method efficiency on various computer systems, to make conclusions concerning the scalability of the algorithms and to determine the possible parallel computation speedup.

The processes of study and research realized by ParaLab are aimed at mastering the fundamentals of parallel computation theory. They allow the leaners to form the basic concepts of the models and methods of parallel computations through observation, comparison and analysis of various visual graphic forms demonstrated in the course of the experiment execution.

For those who only start to study the problem of parallel computations, ParaLab is very useful, as it allows them to master the parallel programming methods. Experienced users may use the system in order to estimate the efficiency of new parallel algorithms, which are being developed.

References

1. Foster, I.: Designing and Building Parallel Programs: Concepts and Tools for Software Engineering. Addison-Wesley, Reading (1995)
2. Hockney, R.W., Jesshope, C.R.: Parallel Computers 2. Architecture, Programming and Algorithms. Adam Hilger, Bristol (1988)
3. Kumar, V., Grama, A., Gupta, A., Karypis, G.: Introduction to Parallel Computing. The Benjamin/Cummings Publishing Company, Inc. (1994)
4. Quinn, M.J.: Parallel Programming in C with MPI and OpenMP. McGraw-Hill, New York (2004)
5. Buyya, R.: High Performance Cluster Computing, vol.1: Architectures and Systems, vol. 2: Programming and Applications. Prentice Hall PTR, Prentice-Hall Inc., Englewood Cliffs (1999)
6. Xu, Z., Hwang, K.: Scalable Parallel Computing Technology, Architecture, Programming. McGraw-Hill, Boston (1998)
7. Voevodin, V.V., Voevodin, V. V.: Parallel Computations, BHV, Saint-Petersburg (2002)
8. Gergel, V.P.: Theory and Practice of Parallel Computations. BINOM (2007)
9. Korneev, V.V.: Parallel Computational Systems, Knowledge, Moscow (1999)
10. Tanenbaum, E.: Computer Architecture, Piter, Saint-Petersburg (2002)

A Message Forward Tool for Integration of Clusters of Clusters Based on MPI Architecture

Francisco Isidro Massetto[1], Augusto Mendes Gomes Junior[2], Fernando Ryoji Kakugawa[2], Calebe de Paula Bianchini[3], Liria Matsumoto Sato[2], Ching-Hsien Hsu[4], and Kuan Ching Li[5]

[1] Federal University of ABC Santo Andre, SP Brazil
francisco.massetto@ufabc.edu.br
[2] University of São Paulo São Paulo, SP Brazil
{augusto.gomes,fernando.kakugawa,liria.sato}@poli.usp.br
[3] Instituto Presbiteriano Mackenzie São Paulo, SP Brazil
calebe.bianchini@mackenzie.br
[4] Chung Hua University Hsinchu, Taiwan
chh@chu.edu.tw
[5] Providence University Shalu, Taichung Taiwan
kuancli@pu.edu.tw

Abstract. Advances in microprocessor technology, power management and network communication have altered the course of development of multiprocessor architectures in order to bring higher level of processing. The introduction of multi-core technology has boosted computing power provided by high-speed network of workstations and SMPs, providing large computational power at an affordable cost using solely commodity components. In this paper, it is presented a tool for integration of several clusters in a single High-Performance System based on MPI standard. The Gateway Process is responsible for MPI process communication channels control and message forwarding, through the use of a protocol that guarantees message ordering and sender/receiver synchronization. It is implemented to support system scalability, offering resources for point to point and collective operations. Results of experimental tests show that the proposed tool is practical and efficient.

1 Introduction

Advances in microprocessor and network technologies have influenced the development of high performance systems based on fast processor nodes interconnected via high speed network, such as Gigabit Ethernet, Myrinet or InfiniBand. Examples of such systems are Clusters of PCs, Network of Workstations, among others. A number of these systems are found either on industry or academia, and corresponding researches are improving traditional ways of computing. Other initiatives are also emerging researches for High Performance Computing. High Performance Grid Computing [1] is currently the focus of researches and Grid Middlewares (e.g., Globus [2], OurGrid[3]) offering resources for parallel application's execution and process communication.

C.H. Hsu and V. Malyshkin (Eds.): MTPP 2010, LNCS 6083, pp. 105–114, 2010.

Programming libraries are needed for the development of parallel and distributed applications in cluster and grid environments, and crucial to obtain the high performance inherent in these systems. MPI[4], PVM[5], Orca[6], and CLR[7] offer a number of resources for data transmission and exchange among processes in parallel applications. Among these libraries, MPI is the most used programming interface, becoming a standard among application developers, offering a variety of operations that includes point to point communication, collective operations and processes management. However, MPI has some drawbacks due to infrastructure features. High-performance systems are more commonly built using a local area network, with nodes using private network addresses and a front-end node with two network interfaces: one with a public network address to be available using internet and another with a private address to reach compute nodes. Even MPICH-G2, a Grid-enabled implementation MPI, it does not cope with this cluster common feature. To find an alternative way to make possible communication among nodes inside and outside of the cluster, and consequently improve the performance over the original design is a very interesting project. The flexibility on the design may assist us to quickly adapt the structure in a few steps, although we need an effectively resources allocation method and a robust communication mechanism.

Therefore, using current MPI implementations is not possible to develop a scalable high-performance system aggregating several private clusters, since nodes inside one cluster are not able to communicate to other nodes outside of the cluster. By this way, this paper presents a tool for message forwarding based on MPI standard in order to integrate several clusters of PCs with MPI into a single system, based on the scalability properties of Cluster of PCs. The system provides great extendable style, when you need and you can have an expanding system, as also may reduce unnecessary costs on computing resources allocation and infrastructure.

The remaining of this paper is organized as follows. Section 2 shows related work where this project is based, while Section 3 explains the system architecture of proposed solution. Section 4 shows results of preliminary tests and finally, conclusion remarks and future work are discussed in Section 5.

2 Related Work

A number of research initiatives have been developed and proposed to integrate cluster systems into a scalable and portable high-performance system. Projects as HyMPI, PACX-MPI, and MPICH-G2 have proposed solutions for such integration process of cluster platforms and they will be discussed next.

MPICH-G2 [8] is a Grid-Enabled MPI implementation and developed by Argonne National Laboratory which makes use of Globus Toolkit as underlying software infrastructure. All processes have a global numbering and each process in a cluster or a supercomputer can addresses messages to any other processor. In terms of functionality, it is able to be used either on a cluster of computers or a cluster of clusters. In both situations, MPICH-G2 creates a mesh of connections among all processors, as depicted in Figure 1, which we observe that is possible to send a message directly from a processor to any other interconnected processor in this environment.

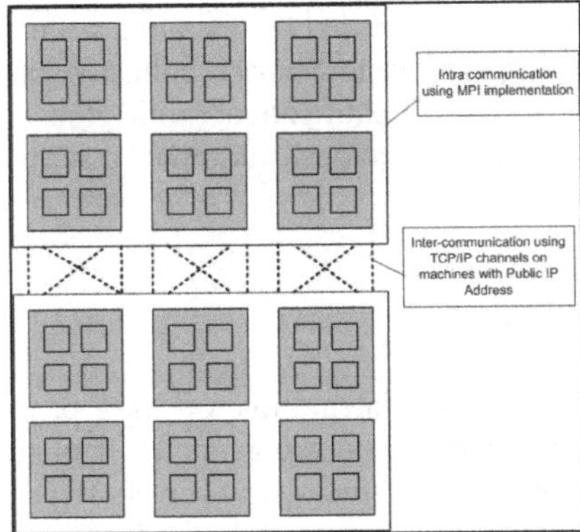

Fig. 1. MPICH-G2 Communication Model

Unfortunately, MPICH-G2 can be used only on infrastructures composed by nodes within public IP address, and therefore, it restricts the use on nodes with private addresses in implementations of clusters. In fact, this limits a broader utilization, and researchers urge to improve it, and turning it flexible and efficient is highly needed.

PACX-MPI [9] proposes a different solution based on pseudo-MPI processes. Inside this implementation, a cluster might have computing nodes with private IP address and a front-end node, and each front-end node executes two additional MPI processes (numbered as processes 0 and 1). These processes are responsible for forwarding messages to/from other processes inside the cluster, as illustrated in Figure 2. Programmers and application developers must adapt their parallel application and deal with the process model, and their functionality is restricted to solely forward messages.

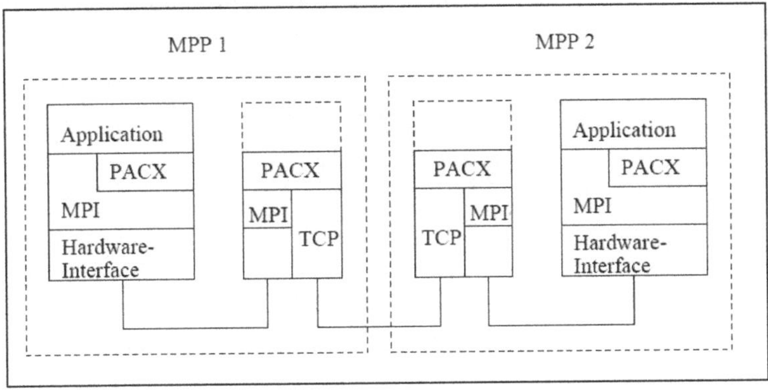

Fig. 2. PACX-MPI Communication model

HyMPI - Hybrid MPI [10, 11] is proposed as an alternative to integrate several clusters with private IP address without making use of MPI processes. Instead, it utilizes a process named *Gateway Process*, which must be executed on the front-end node of the cluster as a service provided by operating system. Not only responsible for forwarding messages, process management and collective operations orchestration, and controlling all the participants identification (both global and local), but also port number control and collective operations. Additional discussion of features of Gateway Process will be done next.

3 Gateway Architecture

The entire system is based on integration of MPI processes being executed on different administrative domain clusters, e.g., a cluster of clusters. The most common cluster implementation is based on nodes with private IP address and a front-end node acting as the interface between compute nodes and internet. Therefore, the front-end node must execute a process which permits process communication and system control. The proposed architecture is new and different from those existing solutions of communication, which users could easily visit other nodes through common and traditional communication ways, from node to node.

Every parallel application instrumented with MPI must make use of Gateway process not only to send and receive messages, but also to manipulate MPI_COMM_WORLD process group, performing operations such as retrieve size and number of processes.

The *Gateway* process is the core of entire system, providing the following features:

- Communication channels establishment and closing for processes during the initiation and finalization phases, such as port number assignment and management,
- Message forwarding for point to point and collective operations,
- Control mechanisms for barrier operations.

Figure 3 depicts how Gateway executes waiting for MPI process messages. Even processes running on nodes with public IP address, though different administrative domains, the Gateway process is needed to control processes numbering and port assignment.

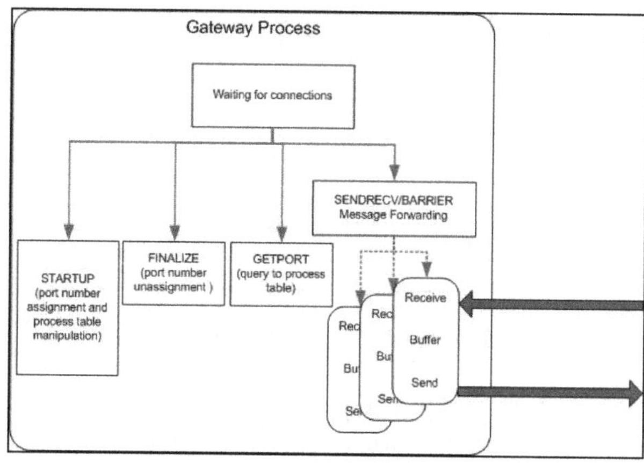

Fig. 3. Gateway Process architecture

3.1 Gateway Internals

The Gateway process is executed as a service on the front-end node receiving data and control messages from processes. A message envelope describing message type and contains an extra field for response from gateway. We describe next the message exchange process between MPI processes and the Gateway.

- STARTUP: message sent during process startup (MPI_Init or MPI_Init_thread operation). MPI process sends a startup message to Gateway asking which port number will be assigned based on its rank. Gateway keeps a range of port numbers which selects one of available ports and responds to MPI process. Next, Gateway populates the process table with the number of each process for current application. This table will be used during the execution of MPI application.
- GETPORT: once all processes performed STARTUP phase, they are all able to communicate among each another, although only Gateway process knows all ports used by all processes. If any process must send a message to another one, it must ask for the first time the port number of destination process. Gateway then receives GETPORT message, its process table, creates the response message with port number assigned to destination process and sends it back to requesting process.
- FINALIZE: message sent during the MPI_Finalize operation by MPI process. When gateway receives this message, it marks on the process table that the process performed this operation and makes the assigned port number availxstart a communication with a process outside the cluster. It means that the Gateway must create a new thread to forward the message either to destination process or to another gateway, in case of receiver located inside another cluster. Gateway receives the data message, buffers it and forwards the message to the destination (process or gateway). Additional details about communication protocol will be discussed on section 3.2.
- BARRIER: This is a particular message exchanged among MPI process that depends on gateway behavior. Unlike the point to point or other collective operations, the MPI_Barrier is a two-way control message among processes. In the first phase all process notifies a coordinator that they must enter a barrier and must synchronize. Next, all processes receive from the coordinator a message notifying the end of barrier.

3.2 Communication Strategy and System Architecture

The Gateway process is developed to be able to control all the communication as also process identification for multiple applications running simultaneously. For communication, the Gateway process handles multiple threads for each communication channel between two processes.

Each process inside a cluster that must send a message to other processes outside the cluster requests to the gateway a communication thread, and sending a message envelope, specifying some information such as sender id, receiver id, message type, element count, data type (MPI_INT, MPI_FLOAT, MPI_CHAR, etc), and message tag. Next, the Gateway process creates a thread to receive the message from the sender, buffers it and forwards to the destination process (if receiver process runs inside a cluster) or to another gateway (in cases where receiver is running inside another cluster).

In order to control all the threads, a signal array control is handled by the Gateway to assure the termination of thread. It means that, if the Gateway must send two messages to the same destination, it will occur serially, to guarantee message ordering and sender/receiver synchronization. Depending on operation, the Gateway performs different activities as described next.

3.2.1 Standard Point to Point Communication

The standard point to point operation (MPI_Send) is implemented as depicted in Figure 4. The sender process sends the message envelope to Gateway process, which allocates memory to buffer the message, receiving it and then forwards to the destination.

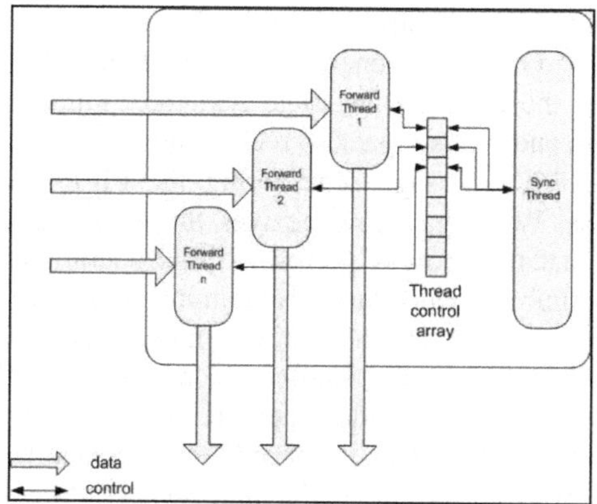

Fig. 4. Communication mechanism for point to point operations

When the gateway receives the message, it notifies the sender process that the transmission is concluded. By this way, the forwarding message control is coordinated by the Gateway process to the destination.

3.2.2 Non blocking Point to Point Communication

In non-blocking communication (MPI_Isend), the behavior of Gateway is the same as the standard communication. There exists one main difference, in which the sender process creates a thread for message transmission, and as soon asthe message is received, the Gateway process notifies the sending thread.

3.2.3 Collective Operations

In order to optimize Gateway features, collective operations are implemented in a two-phase protocol. In the first phase, all process inside a cluster elects a local root that control a local phase. Later in the global phase, all local roots sends/receives message to/from the global root to conclude the entire operation.

For the MPI_Bcast operation, the global root sends the message to all local root processes using standard point to point communication. Later, all local root processes proceed to the local phase to all processes inside the same cluster. For the MPI_Reduce operation, all processes send messages to the local root, which collects and performs the related operation (MPI_SUM, MPI_MAX, among others) in the local phase. Finally, all local root processes proceed to the global phase, forwarding result data to the global root process.

3.2.4 Barrier Operation

The MPI_Barrier operation is a particular operation that must use a two-way communication channel on each thread that is handled by the Gateway process, as depicted in Figure 5. This is necessary since MPI_Barrier is implemented as a two-way protocol, where a coordinator process receives a message from all processes in entire system, starting the synchronization operation. As soon as the coordinator receives all messages, it sends back a notification, unlocking all barrier participants.

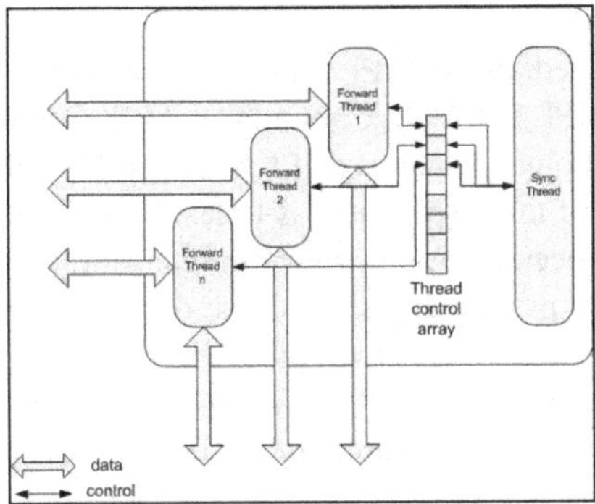

Fig. 5. Communication mechanism for MPI_Barrier operation

The main difference between MPI_Barrier, other collective operations and Point-to-Point communications (Figures 5 and 4, respectively) is related to maintenance of communication channel. As for Point-to-Point operation, a one way communication thread is created while for MPI_Barrier the Gateway process creates a two-way communication thread.

4 Tests and Discussions

The main purpose of this section is to assure the entire system functionalities and stability. All tests were done gradually to guarantee correctness on MPI primitives according to application behavior and gateway functional features. Figure 6 depicts infrastructure used in the set of tests discussed in this section.

This figure shows three clusters in different administrative domains, and all of them are implemented using private IP address for compute nodes and a front-end node running Gateway process. We will discuss tests performed in this computing environment.

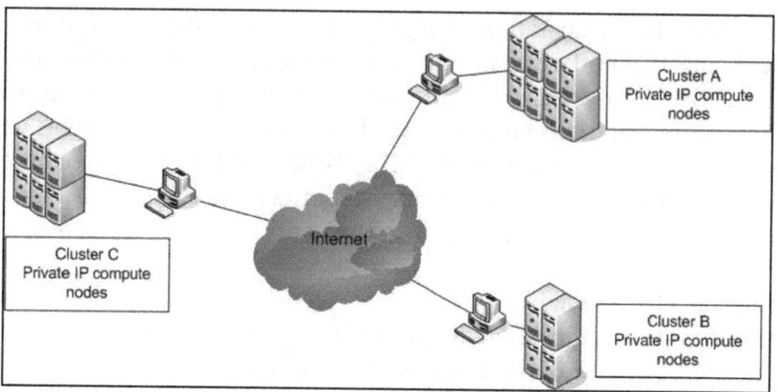

Fig. 6. Computational environment for tests with Gateway process

4.1 Test #1 – Single MPI Application Control, Startup, Finalization and Point to Point Communication on the Same Domain

This test was performed using MPI primitives for application startup, environment control, finalization (MPI_Init, MPI_Comm_Size, MPI_Comm_Rank and MPI_Finalize) and standard Point-to-Point operations (MPI_Send and MPI_Recv).

The main purpose of this experiment is to test port number control and assignment for communicating processes. As for the communication, the classical parallel matrix multiplication algorithm was applied, based on considerable amount of message transmission between master and slaves processes.

We checked log files generated by Gateway and MPI processes to verify port number assignment and data transmitted. All MPI processes received a unique and valid port number and startup and finalization protocol between MPI process and Gateway working correctly.

Generated log files containing control information about sender and receiver synchronization and message ordering. In all tests, checked files generated the same sender/receiver behavior, indicating that communication protocol was correctly developed and the entire system was stable during execution.

4.2 Test #2 – Multiple MPI Applications with Communication among Nodes in the Same Domain

This test explored the capability of Gateway on dealing multiple MPI applications in the same administrative domain. According to the Gateway specification, for each different MPI application, a different port number range is used to be assigned to the process. Several applications were launched and all generated logs covered the specification issues. Three instances of parallel matrix multiplication algorithm were launched in the cluster and port all port numbers were correctly assigned to each MPI process in each application. This assured that the Gateway implementation is according to the specification and runs correctly.

4.3 Test #3 – Process Communication of a Single and Multiple MPI Applications Running in Different Clusters

In this phase, all performed tests include communication using the Gateway message forwarding capability. In this test, we executed a single MPI application using Point-to-Point communication, collective and barrier operations to check communication features and thread handling. Point to point (using MPI_Send and MPI_Recv) and collective (using MPI_Bcast, MPI_Gather, MPI_Scatter and MPI_Barrier) operations were instrumented in the parallel matrix multiplication application.

We launched simultaneous applications to assure Gateway stability, thread handling features and generated log files. In this test we could assure all message transmission and ordering, added to thread manipulation and synchronization between sender and receiver processes. All log files generated were checked and all data transmitted were saved into external data to assure that the results were made as expected.

Some communication times were obtained in order to test overall communication performance system, as depicted in Table 1.

Table 1. Average communication times (in seconds) for point to point operations on a single Ping-Pong application, from a process on Cluster A to different destinations

Message Size	To Cluster B	To Cluster C
10 Kbytes	0,887	1,891
100 Kbytes	6,918	7,120
1 Mbyte	54,122	60,110
10 Mbytes	433,833	559,872

In this test we calculate average tranmission time of Ping Pong application, in wich process inside cluster A sends different sizes messages (from 10 Kbytes to 10 Mbytes) to process on custers B and C, according to system environment depicted in Figure 6. Differences between receivers on cluster B and C are due to geographic location and message routing.

Gateway process receives message, buffers it and forwards to destination process. Further communication tests using other applications will be performed to get more precise results.

5 Conclusions and Future Work

In this paper we present a tool for integration of clusters into a single High Performance Distributed System. Using the Gateway it is possible to allow processes running inside a cluster to communicate with other processes outside the cluster transparently and using a global numbering.

For this solution we have developed functionalities such as communication channel control and mechanism for message forwarding, including Point-to-Point and collective operations for MPI bindings. This solution showed to be viable for this integration since applications do not have to be changed and the entire system is presented to the user (or developer) as a single high performance system.

However, due to cluster implementations, the Gateway process is a bottleneck for processes communication. This feature can have effects on message transmission performance if the communication channel turns to be busy or overloaded.

As for future work, we will continue the development and improvement of Gateway and additional and precise tests will be performed, including performance execution and transmission times. Additionally, application launch, user identification and credential generation will be integrated on the features of Gateway.

References

[1] Foster, I., et al.: The Physiology of The Grid – An Open Grid Services Architecture for Distributed Systems Integration, `http://www.globus.org/research/papers/ogsa.pdf`

[2] The Globus Project, `http://www.globus.org`

[3] Andrade, N., et al.: OurGrid: An Approach to Easily Assemble Grids with Equitable Resource Sharing. In: Feitelson, D.G., Rudolph, L., Schwiegelshohn, U. (eds.) JSSPP 2003. LNCS, vol. 2862, pp. 61–86. Springer, Heidelberg (2003)

[4] Snir, M., Gropp, W.: MPI the Complete Reference. MIT Press, Cambridge (1998)

[5] Geist, A., et al.: PVM: Parallel Virtual Machine. A User's Guide and Tutorial for Networked Parallel Computing. The MIT Press, Cambridge (1994)

[6] Bal, H.E., et al.: Orca: A Language for Parallel Programming of Distributed Systems. IEEE Transactions on Software Engineering 18(3), 190–205 (1992)

[7] Johnson, K.L., et al.: CRL: high-performance all-software distributed shared memory. ACM SIGOPS Operating Systems 29(5) (1995)

[8] Karonis, N., Toonen, B., Foster, I.: MPICH-G2: A Grid-Enabled Implementation of the Message Passing Interface. Journal of Parallel and Distributed Computing (JPDC) 63(5), 551–563 (2003)

[9] Gabriel, E., Resch, M., Ruhle, R.: Implementing MPI with optimized algorithms for metacomputing. In: Message Passing Interface Developer's and Users Conference (1999)

[10] Massetto, F.I., Sato, L.M., Li, K.C.: A Novel Strategy for Building Interoperable MPI Environment in Heterogeneous High Performance Systems. Journal of Supercomputing (in press, available via Online First) (2010)

[11] Massetto, F.I., Gomes, A.M., Sato, L.M.: HyMPI – A MPI Implementation for Heterogeneous High Performance Systems. In: Chung, Y.-C., Moreira, J.E. (eds.) GPC 2006. LNCS, vol. 3947, pp. 314–323. Springer, Heidelberg (2006)

Dynamic Resource Provisioning for Interactive Workflow Applications on Cloud Computing Platform

Hui-Zhen Zhou[1], Kuo-Chan Huang[2], and Feng-Jian Wang[1]

[1] Department of Computer Science
National Chiao Tung University
1001, University Road, Hsinchu, Taiwan
hzzhou.cs96g@g2.nctu.edu.tw, fjwang@cs.nctu.edu.tw
[2] Department of Computer and Information Science
National Taichung University
140, Min-Shen Road, Taichung, Taiwan
kchuang@mail.ntcu.edu.tw

Abstract. Cloud computing opens new opportunities for application providers because with the policy "add as needed and pay as used" they can economize the cost for computing resources. In cloud environments, issues such as resource allocation and dynamic resource provisioning based on users' QoS constraints are yet to be addressed for interactive workflow applications. This paper develops an effective load metric, remaining tasks, for interactive workflow applications. Based on this metric load dispatching and dynamic resource provisioning approaches are proposed which outperform existing methods under a series of simulation evaluations. Experimental results show that the proposed approaches offer application providers better maintenance of QoS-satisfied response time under time-varying workload, at the minimum cost of resource usage.

Keywords: cloud, interactive workflow, load dispatching, resource provisioning.

1 Introduction

Cloud computing [5] has become the most-mentioned computing environment and some cloud computing service providers have began to provide commercial services, such as Amazon's EC2 [7], where users are charged according to the amount of computing resources they actually use. With the power and flexibility of cloud computing, companies around the world may realize their objectives efficiently, especially in both technical and economic aspects.

A typical cloud platform may consist of multiple clusters located at different places worldwide for providing different users with resources near them. Users need rent, instead of buying the computing resources. Applications are plugged into the cloud and acquire computing services as they want without knowing where the resources are located. Users pay for resources they actually use without any huge hardware/software investment in advance. The cloud computing service providers make the profits by

C.H. Hsu and V. Malyshkin (Eds.): MTPP 2010, LNCS 6083, pp. 115–125, 2010.

providing high-quality services through efficiently allocating the resources on demand. In this paper, we present a framework handling the execution of interactive workflow applications on a cloud computing platform.

An interactive workflow [11] is used for controlling user navigation, performing view play, and interacting with the user for clicking buttons and hyperlinks. Unlike scientific workflows [12], [13], which can be applied with a complex static scheduling to minimize the makespan [14], [15], interactive workflows are involved with human interactions and mainly consider the factors such as response time, stability and security, etc. Thus, the computational behaviors in interactive workflows are more similar to those in web-based applications than those in scientific workflows.

In such a scenario, application developers are asked to accomplish two goals simultaneously: minimized user response time and minimized resource usage cost. The activities to accomplish both goals may conflict with each other. For example, user response time can be shortened by using more computing resources while the cost may be reduced by using fewer resources. Since the workload of an application service usually varies with time, a dynamic resource provisioning mechanism may help to achieve the goal instead. To deal with the issue, this paper proposes a dynamic resource provisioning manager REM_DRP (REMaining tasks based Dynamic Resource Provisioning). REM_DRP provides scalable processing power with dynamic resource provisioning mechanisms, where the number of servers used is dynamically adapted to the time-varying incoming request workload. To evaluate our framework and mechanisms, we applied GridSim [9] to simulate the cloud environment. In the simulation, the workload estimation for interactive workflows is investigated comprehensively. To evaluate the performance of REM_DRP, we compare it with QuID [1], [4], a dynamic resource provisioning approach proposed recently.

The remainder of the paper is organized as follows. Section 2 presents literature survey related to our work. Section 3 describes our scalable framework for interactive workflow applications on the cloud. Section 4 presents dispatching methods for workload balancing, our simulation environment, and experimental results. Section 5 presents our dynamic resource provisioning algorithm and its performance evaluation. Section 6 concludes the paper and points out some future research directions.

2 Related Work

The appearance of cloud computing revolutionizes how organizations operate and people work. However, new challenges are introduced while companies benefit from the planning flexibility in technical and economic aspects. Harold *et al.* [5] address some challenges and opportunities of automated control in cloud computing. In accordance with the cloud computing context, they present a proportional thresholding mechanism to enhance stability for feedback controllers. In utility computing, a similar feedback control policy for adaptive resource provisioning is discussed in [6]. They both dynamically adjust the resource shares in individual tiers in order to meet the QoS requirement for multi-tier web applications, whereas our approach aims for interactive workflow applications.

In the context of the dynamic resource provisioning, S. Ranjan et al. [4] introduce three mechanisms for web clusters. The first mechanism, QuID [1], optimizes the performance within a cluster by dynamically allocating servers on-demand. The second, WARD [2], is a request redirection mechanism across the clusters. The third one is a cluster decision algorithm that selects QuID or WARD under different workload conditions.

For multi-tier internet applications, Bhuvan *et al.* [3] propose a provisioning technique which employs two methods that operate at two different time scales: predictive provisioning at the time-scale of hours or days, and reactive provisioning at time scales of minutes to respond to a peak load. They model a multi-tier application as a network of queues where each queue at a tier represents a server, and the queues from a tier feed into the next tier. Given the request arrival rate and per-tier response time, the number of servers needed at each tier is computed individually by the proposed algorithm. While the above techniques are aimed for multi-tier web applications, our work in this paper targets at interactive workflow applications.

3 A Scalable Computing Framework

This section presents a scalable framework for interactive workflow applications on the cloud computing platform. The framework deals with the scenario that an interactive workflow application, hosted on a cloud computing platform, runs many workflow instances simultaneously according to the incoming user requests. Since the amount of incoming requests changes with time and the cloud platform is a pay-per-use service, the application has to dynamically manage the resources it uses to maintain acceptable response time and reduce the total cost of resource consumption under various workloads.

In the framework, each resource, representing a distinct computing server, is capable of processing multiple interactive workflow requests. Prior to ready for service, the required data and workflow definitions have to be deployed to the resources in some way.

To efficiently utilize resources, there are two key issues considered in the framework. The first is finding the least loaded resource for dispatching incoming requests. The second issue deals with dynamic resource provisioning (DRP) for adaptively handling dynamic workloads. With resource state monitoring, each workflow enactment request will be sent to the least loaded resource for service. The effectiveness of least load dispatching largely depends on how to accurately capture the computing load on each resource. To find the most effective load metric, several candidate load metrics are proposed and evaluated with simulation studies, which will be presented in section IV. Our strategy for dynamic resource provisioning is presented in section V. The policy will be compared with the one in [1]. In the dynamic resource provisioning strategy, the most effective load metric evaluated for request dispatching is used to represent the resources' load status.

Fig. 1 shows an overview of the framework in handling user requests for an interactive workflow application running on a resource cloud. The architecture consists of four main components Dispatcher, Resource Allocator, Dynamic Resource Provisioning Manager (DRP), and Resource Manager. When a user sends a workflow enactment request to the system, Dispatcher serves as the front-end guard of the entire system. According to the request, Dispatcher initializes and puts a workflow instance on the resource with the least load, returned by Resource Allocator. We apply a session affinity based scheme, that all subsequent requests at a workflow are handled by the same application server. Tasks on a resource are executed in a non-preemptive FCFS (First-Come, First-Served) order.

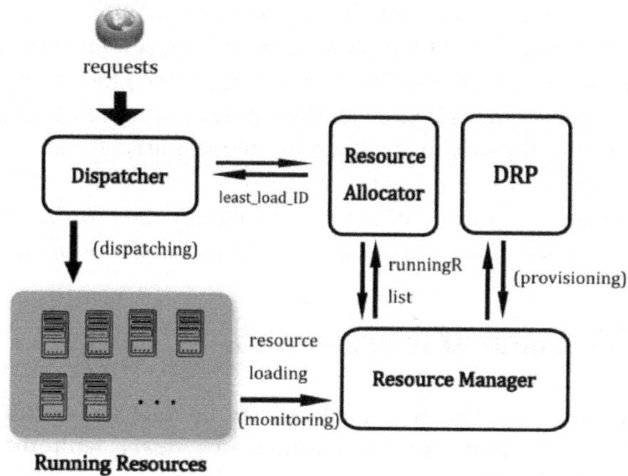

Fig. 1. A scalable computing framework

Resource Monitor on each resource is responsible for monitoring and recording the resource's load status, such as waiting queue length, response time, arrival rate, waiting time, resource utilization, etc.. Resource Manager maintains the status information of all resources. Based on the load information, Resource Allocator can identify the least loaded resource and passes the ID of that resource to Dispatcher. Additionally, Resource Manager also complies with the decision made by DRP. For each schedule interval, DRP fetches the load information from Resource Manager, diagnoses the state of running resources, and initiates requests to Resource Manager to acquire or release resources whenever needed.

4 Least Load Dispatching

In this section, several candidate load metrics for interactive workflow applications are proposed and compared for their effectiveness. The following describes the candidate metrics to be evaluated:

- **Arrival rate.** It counts the number of new tasks arriving at the resource's waiting queue within a time-interval.
- **Average response time.** It calculates the average response time of all tasks finishing execution within a specific time interval. The response time for a task accounts for both the waiting time in queue and the processing time on the resource.

- **Remaining tasks.** A task in an interactive workflow is executed only when all its preceding tasks complete and the instantiation event is triggered. The number of remaining tasks of a resource can be defined as the number of all unexecuted tasks for the running workflows. We might use the number of remaining tasks to predict the future workload.
- **WF counts.** It measures the number of workflows currently being served at a resource.

Our simulation environment for the following experiments is based on GridSim [9]. GridSim is a discrete event simulator built on top of the simulation package SimJava [10] and can be used to model and simulate various entities in parallel and distributed computing environments. GridSim controls all the entities, delivers the events, and advances the simulation time. As Fig. 2 shows, our simulation environment is constructed by instantiating entities from the classes in GridSim and SimJava.

Workflow tasks are modeled as *Gridlet* objects and executed on GridSim resources. Task execution time and user thinking time are generated from the negative exponential distributions with the mean values of 3 seconds and 7 seconds respectively based on the TPC benchmark [8]. The task length is expressed in terms of the time it takes to run on a standard PE (Processing Element) with a MIPS rate of 100. Therefore, in our simulation environment, the processing capability of a resource is expressed in MIPS (Millions of Instructions Per Second). The workflow related parameters in our simulations are summarized in Table 1.

Other parameters used to configure the simulation environment are listed in Table 2. The resources are interconnected through a network of 100Mbps. The measurement intervals of arrival rate and response time are determined by running a series of simulations of 900 workflows using various arrival rate and response time measurement intervals. The values in Table 2 deliver the shortest average response time in the simulations. We also include random selection and the Round-Robin load balancing algorithm in the experiments for performance comparison.

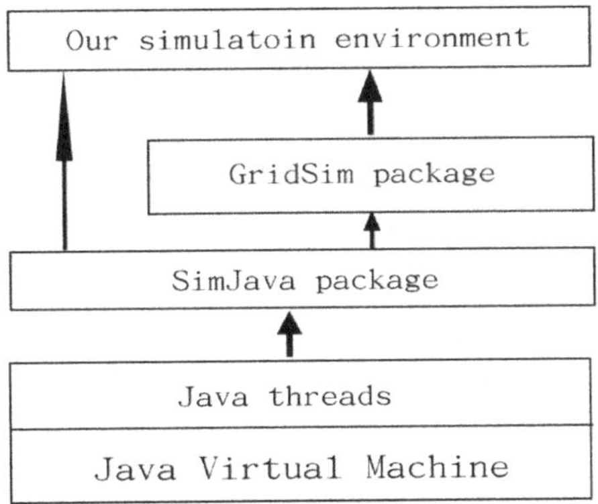

Fig. 2. Simulation environment

Table 1. Workflow related parameters

Number of tasks	4 ~ 15 (random generation)
Mean task execution time	3 sec.
Mean user thinking time	7 sec.
Maximum degree of a task	3
Input file size	100 bytes
Output file size	100bytes

Table 2. Simulation parameters

Request arrival interval	2.2 (Poisson distribution)
Arrival rate measurement interval	35 sec.
Response time measurement interval	15 sec.
Computing speed	100 MIPS * 16

Fig. 3 shows the performance comparisons of the candidate load metrics. Obviously, the remaining tasks metric outperforms the others in terms of task response time. Arrival rate and response time are the most frequently used load metrics in web applications. The former performs better than the latter in the experiment. The two metrics are based on information collected in the past. Sometimes, past information cannot accurately predict the future workload. Moreover, these time-interval based metrics have a potential problem that it's hard to find a perfect interval for collecting an appropriate amount of load information. Once an interval is decided, any load variation outside the interval is ignored. The remaining tasks metric not only gives a more detailed load information than the WF counts do but also seizes the counteraction between request arrivals and completed tasks.

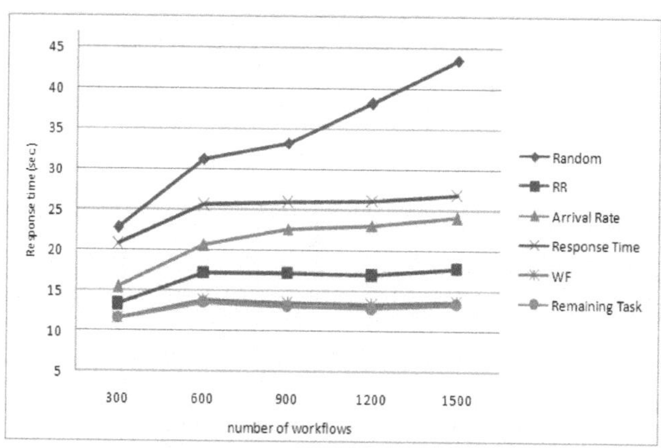

Fig. 3. Comparison of load metrics

5 Dynamic Resource Provisioning

In this section, we propose an auto-scaling algorithm denoted as REM_DRP (REMaining tasks based DRP) to dynamically provide an adequate amount of resources to an interactive workflow application. To maintain acceptable response time, in REM_DRP, each resource is configured with a workload limit Rw and a threshold value for the entire system is Rw*|R|, where |R| is the number of running resources. The workload limit on each resource is based on the load metric of remaining tasks. When the total system workload exceeds the threshold value, the system would deploy additional resources to share the workload. On the other hand, if the system workload is below the threshold value the system would remove some resources to reduce the costs of resource usage. We also define equations 1 and 2 to compute the number of resource(s) to be added or removed in each decision respectively, where T is the workload limit on each resource represented by the number of tasks.

$$R_{more} = \lceil (RemainingTasks - Threshold) / T \rceil \quad (1)$$

$$R_{less} = \lfloor (Threshold - RemainingTasks) / T \rfloor \quad (2)$$

Fig. 4 is an example to describe how REM_DRP work. We mark three different conditions of workload in Figure 4. Zones A and B are examples whose number of remaining tasks of the running resources exceed the implicit threshold, while zone C is the contrary. In REM_DRP, the system is asked to add resources in Zone A where the arrivals exceed the completions but not in zone B since the workload is declining. Zone C characterizes that the running resources are underutilized and the unnecessary resources are removed regardless of the number of arrivals or completions. A decision making algorithm in REM_DRP is defined as follows to dynamically adjust the number of resources used.

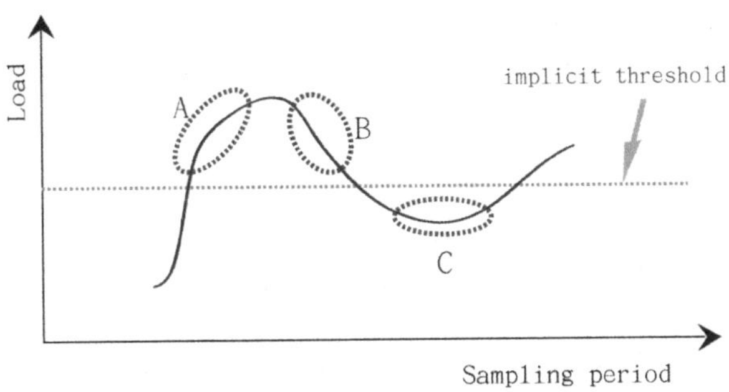

Fig. 4. Illustration of dynamic workloads

```
Decision Making Algorithm

Input: |R|=current number of running resources;
       |rem|: remaining tasks of all running resources;
       |ari| = number of task arrivals for all resources
               in the previous interval;
       |done| = number of all completed tasks in the
                previous measurement interval;
       T = workload limit on each resource (number of
           tasks).
Output: |R'|: the required number of resources for the
              next interval

if |rem| > ( |R| * T)
    if |done| < |ari|
        get R_more by equation (5.1)
        |R'| = |R| + R_more
    else
        |R'| = |R|
else if |rem| < ( |R| * T )
    get R_less by equation (5.2)
    |R'| = |R| − R_less
else
    |R'| = |R|
```

The following experiments compare the effectiveness of REM_DRP and QuID [1], [4]. Two experiments are conducted. There are at most 16 resources available in the first experiment and 50 resources in the second. In both experiments, three workloads of different arrival rates are run. Remaining tasks metric is used for dispatching. Moreover, the warm-up time of idling resources is 20 seconds and 4 resources are assumed to be ready for accepting requests at the beginning of experiments. Simulation parameters are summarized in Table 3. Parameters of REM_DRP and QuID are listed in Table 4.

Table 3. Simulation parameters

Simulation parameters	Experiment 1	Experiment 2
Total resources	16	50
3 rounds (workflows)	{ 300, 400, 200 }	
Request arrival interval	{ 3.5, 2.5, 4.5 }	
Initial/minimum running resources	4	
Warm-up time	20 sec.	
Load metric for dispatching	Remaining tasks metric	

Table 4. Parameters of REM_DRP and QuID

	REM_DRP	QuID
Measurement interval	3 sec.	60 sec.
Utilization rate		75%

Experimental results are summarized in Table 5. For REM_DRP, the workload limit on each resource is set to 9 tasks in the very beginning. The value 9 was determined by simulation studies on the performance of various values for the workload limit. The results indicate that REM_DRP can outperform the others. For comparison with QuID, in experiment 1, where at most 16 resources are available, REM_DRP has shorter response time than QuID by 31%, but 0.63 more in resource usage. In experiment 2, REM_DRP outperforms QuID in both response time and resource usage. Comparing with the static provisioning approach, let the target response time be 8.2 seconds, REM_DRP requires roughly 11.96 resources in average while static provisioning requires 16 resources. From another aspect, let the resources usage be 12 for economic reasons, REM_DRP provides an average response time of 8.3 seconds and static provisioning provides 22.36 seconds. REM_DRP is two times quicker.

Utilization rate is used in QuID for measuring the workload on each resource. One potential drawback of utilization rate is that when utilization rate reaches 100%, it cannot effectively calculate the amount of resources to increase. Moreover, utilization rate is a time-interval based measurement, such as arrival rate and average response time. Therefore, it is a crucial issue to determine an appropriate measurement interval. However, this is difficult since response time and resource usage are not monotonically increasing or decreasing with the measurement intervals.

Table 5. Simulation results

<table>
<tr><td rowspan="3"></td><td colspan="2" rowspan="2">Static provisioning</td><td colspan="4">Dynamic resource provisioning</td></tr>
<tr><td colspan="2">Experiment 1 :
4 ~ 16 resources</td><td colspan="2">Experiment 2 :
4 ~ 50 resources</td></tr>
<tr><td>Fixed 16 Resources</td><td>Fixed 12 Resources</td><td>REM_DRP</td><td>QuID</td><td>REM_DRP</td><td>QuID</td></tr>
<tr><td>Avg. response time (sec.)</td><td>8.2</td><td>22.36</td><td>9.03</td><td>13.1</td><td>8.3</td><td>13.31</td></tr>
<tr><td>Avg. resource usage</td><td>16</td><td>12</td><td>12.2</td><td>11.57</td><td>11.96</td><td>12.63</td></tr>
</table>

6 Conclusions

Applications require the capability of adjusting the amount of deployed resources dynamically in order to take benefits of the pay-only-what-you-consume policy on the cloud platform. This paper proposes a framework for developing interactive workflow applications on the cloud platform. Currently many server applications adjust the amount of resources at runtime manually. The framework in this paper allows applications to automatically manage the amount of resources according to the system workload. It offers application providers the benefits of maintaining QoS-satisfied response time under time-varying workload at the minimum cost of resource usage.

The framework mainly deals with two issues: dynamic request dispatching and resource provisioning. For dynamic request dispatching, an improved least load dispatching approach is proposed which adopts a load metric of remaining tasks based on the characteristics of interactive workflow applications. Experimental results based on simulations indicate that remaining tasks can achieve better dispatching performance than arrival rate and response time which are commonly used in existing dispatching methods.

For dynamic resource provisioning, REM_DRP is proposed as a feedback controller to automate resource provision by taking advantage of the characteristics of interactive workflow applications. Experimental results show that REM_DRP outperforms static provisioning and the utilization rate based QuID approach in both average response time and resource usage.

References

1. Ranjan, S., Rolia, J., Fu, H., Knightly, R.: QoS-Driven Server Migration for Internet Data Centers. In: The Tenth International Workshop on Quality of Service, Miami, FL (2002)
2. Ranjan, S., Karrer, R., Knightly, E.: Wide Area Redirection of Dynamic Content in Internet Data Centers. In: The IEEE INFOCOM, HongKong (2004)
3. Urgaonkar, B., Shenoy, P., Chandra, A., Goyal, P., Agile, T.W.: Dynamic Provisioning of Multi-Tier Internet Applications. ACM Transactions on Autonomous and Adaptive Systems 3(1) (2008)
4. Ranjan, S., Knightly, E.: High-Performance Resource Allocation and Request Redirection Algorithms for Web Clusters. IEEE Transactions on Parallel And Distributed Systems 19(9) (2008)
5. Lim, H., Babu, S., Chase, J., Parekh, S.: Automated Control in Cloud Computing: Challenges and Opportunities. In: 1st Workshop on Automated Control for Datacenters and Clouds (2009)
6. Padala, P., Shin, K.G., Zhu, X., Uysal, M., Wang, Z., Singhal, S., Merchant, A., Salem, K.: Adaptive Control of Virtualized Resources in Utility Computing Environments. In: 2nd ACM SIGOPS/EuroSys European Conference on Computer Systems (2007)
7. Amazon Elastic Compute Cloud (EC2), `http://aws.amazon.com/ec2/`
8. TPC-W: Transaction Processing Council, `http://www.tpc.org/tpcw/default.asp` Standard Specification Version 1.8, `http://www.tpc.org/tpcw/spec/tpcw_V1.8.pdf`
9. Sulistio, A., Cibej, U., Venugopal, S., Robic, B., Buyya, R.: A Toolkit for Modelling and Simulating Data Grids: An Extension to GridSim. Concurrency and Computation: Practice and Experience (CCPE) 20(13), 1591–1609 (2008)
10. Howell, F., McNab, R.: Simjava: a Discrete Event Simulation Package for Java with Applications in Computer Systems Modelling. In: First International Conference on Web-based Modelling and Simulation, Society for Computer Simulation, San Diego CA (1998)
11. Siebel Business Process Framework, `http://download.oracle.com/docs/cd/B40099_02/books/BPFWorkflow/booktitle.html`

12. Ludtke, S., Baldwin, P., Chiu, W.: EMAN: Semiautomated Software for High Resolution Single-Particle Reconstructions. J. Struct. Biol. (128), 82–97 (1999)
13. Singh, G., Deelman, E., Berriman, G.B., et al.: Montage: a Grid Enabled Image Mosaic Service for the National Virtual Observatory. Astronomical Data Analysis Software and Systems (ADASS) (13) (2003)
14. Zhang, Y., Koelbel, C., Kennedy, K.: Relative Performance of Scheduling Algorithms in Grid Environments. In: 7th IEEE International Symposium on Cluster Computing and the Grid (2007)
15. Mandal, A., Kennedy, K., Koelbel, C., Marin, G., Crummey, J.M., Liu, B., Johnsson, L.: Scheduling Strategies for Mapping Application Workflows onto the Gird. In: The IEEE Symposium on High Performance Distributed Computing (HPDC 14), pp. 125–134 (2005)

A Xen-Based Paravirtualization System toward Efficient High Performance Computing Environments

Chao-Tung Yang[1], Chien-Hsiang Tseng[1,2], Keng-Yi Chou[1], Shyh-Chang Tsaur[3], Ching-Hsien Hsu[4], and Shih-Chang Chen[5]

[1] High-Performance Computing Laboratory Department of Computer Science, Tunghai University, Taichung City, 40704, Taiwan, R.O.C.
{ctyang,g97350007}@thu.edu.tw
[2] ARAVision Incorporated, Sindian City, Taipei Count, 23141, Taiwan, R.O.C.
warp@aracity.com
[3] Department of Electronic Engineering, National Chin-Yi University of Technology, Taiping City, Taichung County, Taiwan
sctsaur@gmail.com
[4] Department of Computer Science and Information Engineering,
[5] College of Engineering,
Chung Hua University, Hsinchu, Taiwan 300, R.O.C.
{robert,scc}@grid.chu.edu.tw

Abstract. A virtual machine provides platforms to install an OS within another OS which provides resources. It can be accomplished to construct a computational cluster system on a single machine. The real cluster with machines provides full utilization of its resource for users while a virtual machine assigns the resources of the host to residing OSs. Xen is such kind of virtual machine to construct the virtualization system. It is chosen to be our system's virtual machine monitor because it provides better efficiency, supports different operating system work simultaneously, and gives each operating system an independent system environment. The performance of the virtualization system is examined by comparing with a non-virtualization system which is a real cluster system. The experiments show less power consumption and better computing efficiency by executing programs such as matrix multiplication, LINPACK, lower-upper triangular and Primes test sets. The results show better choices of constructing a large-scaled computing system using a virtual machine.

1 Introduction

The Millennium Ecosystem Assessment (MA) was called for by the United Nations Secretary-General, Kofi Annan, in 2000, and initiated in 2001. The objective of the MA was to assess the consequences of ecosystem changes for human well-being, and to focus on the action by the scientific basis to enhance the conservation and sustainable use of the ecosystems, and to make contribution to human well-being. The MA has engaged the work with more than 1,360 experts worldwide. It mainly estimates the world ecosystems of the past, now and future so to formulate

C.H. Hsu and V. Malyshkin (Eds.): MTPP 2010, LNCS 6083, pp. 126–135, 2010.

corresponding policies. With the improved technique of a virtual machine (VM) and development of hardware, it is the best solution to save energy and the consequences by using a VM to construct the virtual cluster computing system

In this paper, implementing and integrating a virtualization [1-7] system with a cluster computing system on Xen [8-11] is well introduced. The categories of Xen are listed and the reasons of choosing different types of Xen are given to judge which one is better for research. The details of user interfaces, resource brokers and system architectures are given to illustrate each part of the work flow.

In Section 2, we will discuss the virtual cluster computing system in details. In Section 3, we will mention our system architecture, web interfaces, resource brokers, and experiments by using efficient virtualization. Finally, conclusions and the future work are stated in Section 4.

2 Preliminary

In order to evaluate the viability of the difference between virtualization and non-virtualization, the virtualization software we used in this paper is Xen. Xen is a virtual machine monitor (hypervisor) that allows you to use one physical computer to run many virtual computers. Xen has a combination of features that make it uniquely well suited for many important applications. Xen is chosen to be our system's virtual machine monitor because it provides better efficiency, supports different operating systems simultaneously, and gives each operating system an independent system environment.

This free software is mainly divided into two kinds of simulate types, paravirtualization and Full virtualization. Paravirtualization implements virtualization technology, mostly via the modified kernel of Linux. In general, most virtualization strategies are classified into two major categories:

- **Full virtualization** (also called *native virtualization*) is similar to emulation. As in emulation, unmodified operating systems and applications run inside a virtual machine. Full virtualization differs from emulation in that operating systems and applications are designed to run on the same architecture as the underlying physical machine. This allows a full virtualization system to run many instructions directly on the raw hardware. The hypervisor in this case monitors access to the underlying hardware, and gives each guest operating system the illusion of having its own copy. It must no longer use software to simulate a different basic architecture.
- **Paravirtualization:** In some instances, this technique is also referred to as enlightenment. In paravirtualization, the hypervisor exports a modified version of the underlying physical hardware. The exported virtual machine is of the same architecture, which is not necessarily the case in emulation. Instead, targeted modifications are introduced to make it simpler and faster to support multiple guest operating systems.

The characteristic of paravirtualization is as follows:

- A virtual machine quite like a real machine on operating efficacy
- Supporting more than 32 cores of computer structures at most

- Supporting x86/32, with PAE technique and x86/64 hardware platform
- Good hardware driver support, almost for any Linux device driver

There are restricts with full virtualization, and it can be only executed when the hardware satisfies these conditions in the following:

- Intel VT technique (Virtualization Technology, Intel-VT)
- AMD SVM technique (Secure Virtual Machine, AMD-SVM or, AMD-V)

As a result of the widespread of virtual machine softwares in recently years, two best x86 CPU manufacturers Intel/AMD, with efficiency of x86 computer and increasing of CPU computing, both have published the new integrated virtualization on CPU, one for Intel Vander pool and another for AMD Pacifica. These technologies also support Xen, and make efficient step-up more than initial stages.

3 System Implementation

This section introduces the system architecture including the Beowulf cluster and the computing nodes in Section 3.1. The web page of user interfaces and the work flow of resource brokers are illustrated in Section 3.2. The experiment focuses on power consumption and computing efficiency, and both results are given in Section 3.3.

3.1 System Architecture

Beowulf is a design for high-performance parallel computing Clusters on the inexpensive personal computer hardware. The name comes from the main character in the Old English epic poem Beowulf. A Beowulf Cluster is a group of usually identical PC computers running a Free and Open Source Software (FOSS) Unix-like operating system, such as BSD, Linux or Solaris. They are networked into a small TCP/IP LAN, and have libraries and programs installed, which allow processing to be shared among them.

The Beowulf cluster consists of one or more head nodes. The head node acts as a server for Network File System (NFS), and as a gateway to the outside world. As an NFS server, the master node provides user file spaces and other common system software to the compute nodes via NFS. As a gateway, the head node allows users to gain access through it to the tail nodes. Usually, the head node is the only machine that is also connected to the outside world using a second network interface card (NIC). The sole task of the tail nodes is to execute parallel jobs.

As a traditional cluster framework mentioned before, we use an ideas of virtualization in the cluster system to economize power. Therefore, there are some distinct on the framework of clusters, all physical compute hosts are below massage passing interfaces, but in the virtualization computing cluster, full/ paravirtualized machines are used instead of physical machines, as shown in Fig. 1.

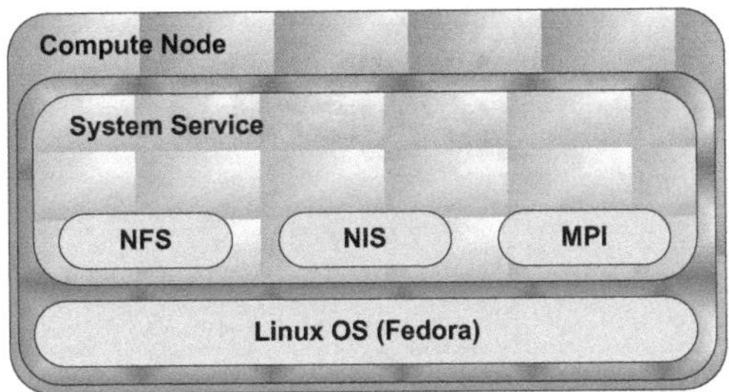

Fig. 1. Software stack without Xen

In this paper, the cluster system is established by using four homogeneous computers, all with Intel Xeon E5410 CPU, 8 GB physical memory, 500 GB HD and the fedora 8 OS installed. A gigabit switch is used to construct the network environment in the cluster system.

While the architecture of compute node in Fig. 1 represents the non-virtualization system, the architecture with Xen in Fig. 2 represents the virtualization system. In the experiment section, virtualization system and non-virtualization system are compared to show the effect on computing efficiency and economizing on power.

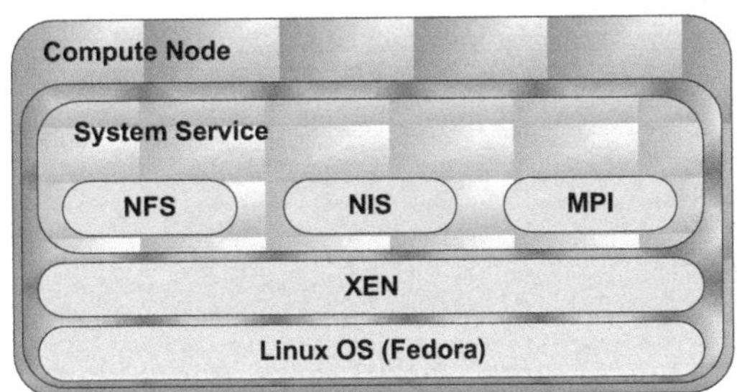

Fig. 2. Software stack of master node with Xen

3.2 User Interface and Resource Broker

3.2.1 End User's Web Page

We design a useful web interface for end users to submit their jobs, and for administrators to manage their VM templates and host machines. Easy to submit and easy to get result are very important for end users. We design a single page for all of the steps when they want to submit their jobs. In a single page, they can choose VM types, number of VM processors, the size of memory, and how many virtual machines they want to use. At the same page, they can also upload their job files (include Makefile, source code and etc.). After they upload their job files, this interface will compile their job files automatically. More details for the end users' web pages are shown on Fig. 3, 4, 5, 6, and 7.

Fig. 3. Login Page

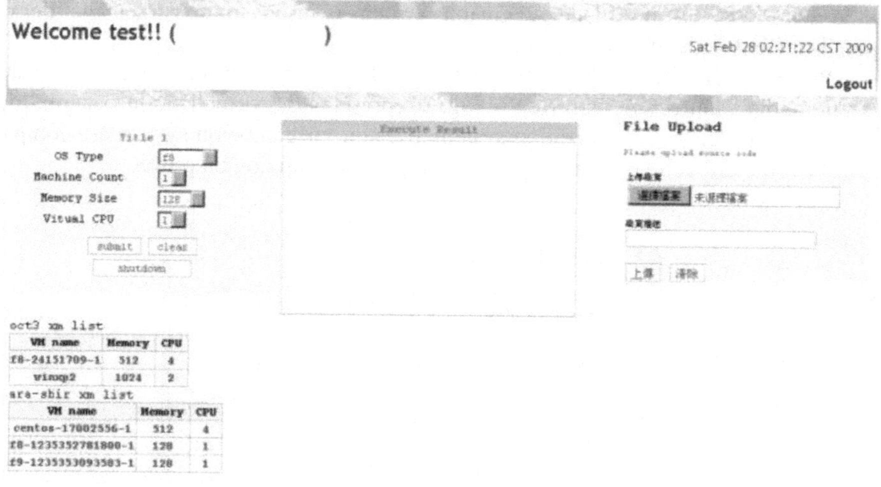

Fig. 4. End user's single page

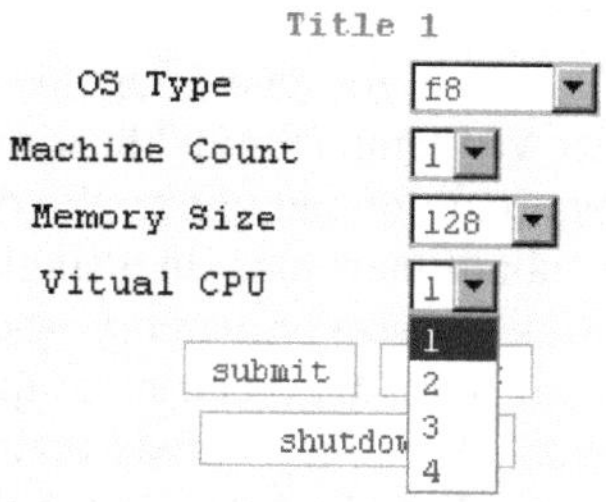

Title 1
OS Type f8
Machine Count 1
Memory Size 128
Vitual CPU 1
1
2
3
4
submit
shutdow

Fig. 5. To choose OS, machine count, virtual memory, virtual CPU

oct3 xm list

VM name	Memory	CPU
f8-24151709-1	512	4
winxp2	1024	2

ara-sbir xm list

VM name	Memory	CPU
centos-17002556-1	512	4
f8-1235352781800-1	128	1
f9-1235353093583-1	128	1

Fig. 6. Virtual machine lists

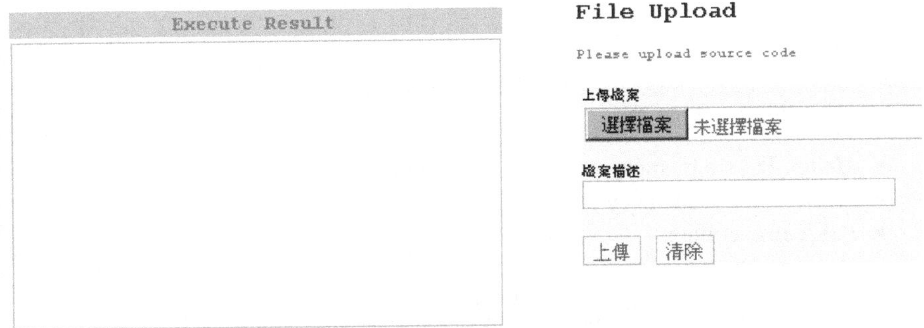

Fig. 7. File upload and result

3.2.2 Resource Broker

A resource broker is the most important part of the whole system because it manages resources, jobs and licenses. To combine the resource broker with a web portal, it lets end users easily use the system to complete their jobs. The work flows among end users, user interfaces and resource brokers are given in Fig. 8 and listed as follows:

- Step 1. End users use UI to upload source code.
- Step 2. The system call GetAvailableResource() to gather available resources for end users.
- Step 3. The resource broker assigns and utilizes resources for end users.
- Step 4. Start the computation progress of jobs for end users.
- Step 5. The computational system returns results.
- Step 6. End users receive results through UI.

In order to count the number of virtual machines, it avoids physical swap and control licenses. Variables are defined in the following for the equation (1) of total available virtual machines:

- *NodeTotalvnum*: Total available virtual machines.
- *Cn*: Total available virtual machines per physic node.

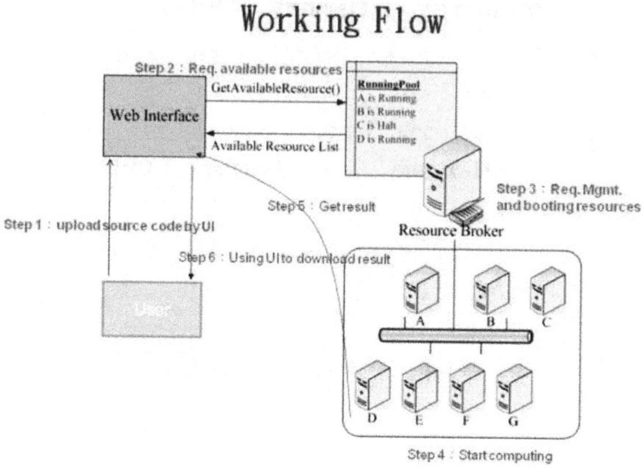

Fig. 8. Flow of resource broker

- *Memp*: Physic memory on one physic node.
- *Memv*: Virtual memory on virtual node.
- *L*: License counts.

$$\mathrm{Node}_{\mathrm{TotalVnum}} = \sum_{n=1}^{n=L} C_n \,, C_n = \left(\frac{\mathrm{Mem}_p}{\mathrm{Mem}_v}\right) \text{ and } \mathrm{Node}_{\mathrm{Totalvnum}} > \mathrm{Node}_{\mathrm{Realvnum}} \quad (1)$$

3.2.3 Admin's Web Page

The management interface for admin is using the same login page shown in Fig. 3. The login page will check the user to grant his (her) permission. If the user has permission, jobs are submitted, boot VM, upload source code, and download his (her) results.

3.3 Experiment

This section uses matrix multiplication, LINKPACK, LU and Primes test sets to show the results of a traditional cluster and a virtual cluster. It also shows the improvement of proposed architecture on power consumption and computing efficiency.

The experiment programs are listed below:

- **Matrix Multiplication** in mathematics, is the operation of multiplying a matrix with either a scalar or another matrix. Besides this matrix multiplication, the matrix size could be varied into different values.
- **LINPACK** is a collection of FORTRAN subroutines that analyze and solve linear equations and linear least-squares problems. The package solves linear systems whose matrices are general, banded, symmetric indefinite, symmetric positive definite, triangular, and tridiagonal square. In addition, the package

computes the QR and singular value decompositions of rectangular matrices, and applies them to least-squares problems. LINPACK uses column-oriented algorithms to increase efficiency by preserving locality of reference.

- **LU** (Lower-Upper Triangular), demands that the number of CPU must be power of 2, and majors in calculating for degree of particle of non-linear communication.
- **Primes** is a test program give the parameter that an integer any number than itself will find out how many primes below this number.

Since an Intel Xeon E5410 cpu needs 80 watts to work, it requires 160 watts to work for each machine. Therefore it requires 640 watts for the traditional cluster with four machines, but only 160 watts for the virtualized system with only one machine. Table 1 provides the information of thermal power of cluster on power consumption by performing matrix multiplication. The detail information shows that the virtualization system has great improvement compared with physical machines.

Table 1. Thermal energy of physical and virtualization (unit: watt hour)

Array Size	Physical	Virtualization
32	0.000475	0.000092828
64	0.00091	0.000184
128	0.001878	0.000633
256	0.006573	0.00185
512	0.038035	0.012013
1024	0.477288	0.353829
2048	32.99902	7.764905
4096	252.4356	58.53319

About efficiency, LU, LINPACK and Primes test sets are used to prove our presumption that virtualization is better than traditional cluster. The result of LINPACK sets in Fig. 9 has shown that while array size is large than 645, the efficiency of physical cluster is worse than Virtualizations.

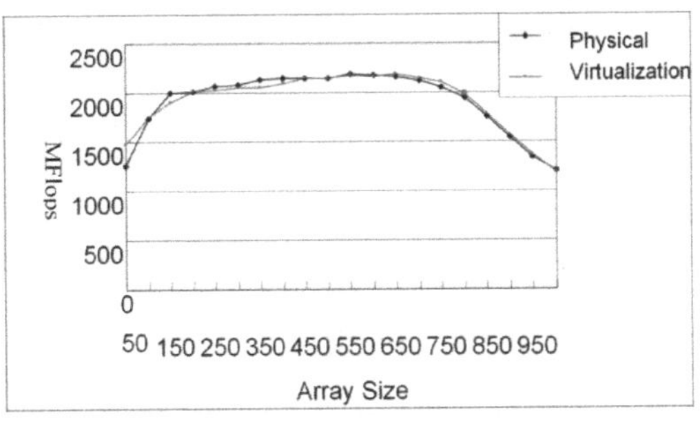

Fig. 9. LINPACK sets

In Fig. 10, the test sets, LU, is the second efficiency of comparison between non-virtualization and virtualization. Obviously, the efficiency of virtualization is much better than traditional cluster. Fig. 11 shows that the Xen based virtualization system provides comparable performance compared with physical machines. The execution time of both platforms is very close in each set of results.

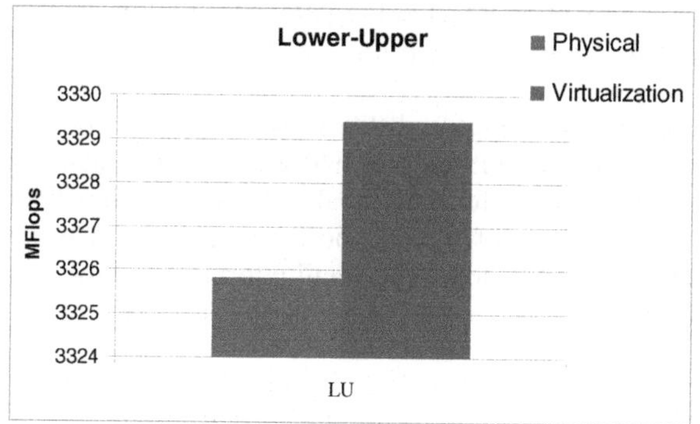

Fig. 10. LU sets

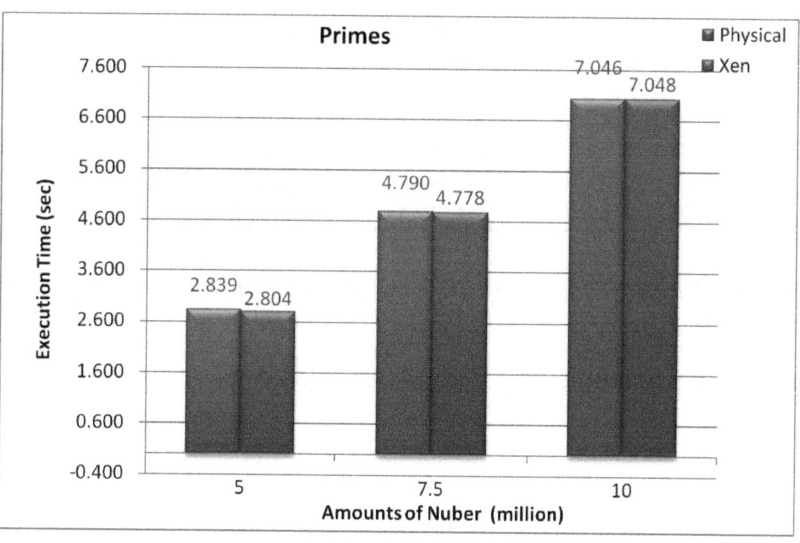

Fig. 11. Calculate primes in Any number

4 Conclusions and Future Work

This paper compares the power consumption and computing efficiency on a virtualization system, which is a node with Xen installed, with a non-virtualization system which is a four-node traditional cluster. The web page is easy for end users to submit their jobs on the system, and is established with a resource broker. The experiments show that the virtualization system consumes less power and provides comparable or better computing efficiency than a non-virtualization system. The future work of this system is to improve the resource broker and build more functions for end users.

Acknowledgement

The work of this paper was supported by National Science Council of Taiwan under Grant number NSC-98-2220-E-029-004.

References

1. Cherkasova, L., Gardner, R.: Measuring CPU Overhead for I/O Processing in the Xen Virtual Machine Monitor. In: Proceedings of the annual conference on USENIX Annual Technical Conference, pp. 387–390. USENIX Press, California (2005)
2. Dong, Y., Li, S., Mallick, A., Nakajima, J., Tian, K., Xu, X., Yang, F., Yu, W.: Extending Xen with Intel Virtualization Technology. Intel Technology Journal 10(3), 1–14 (2006)
3. Huang, W., Liu, J., Abali, B., Panda, D.K.: A Case for High Performance Computing with Virtual Machines. In: Proceedings of the 20th annual international conference on Supercomputing, pp. 125–134. ACM Press, New York (2006)
4. Liu, J., Huang, W., Abali, B., Panda, D.K.: High Performance VMM-Bypass I/O in Virtual Machines. In: Proceedings of the annual conference on USENIX '06 Annual Technical Conference, p. 3. USENIX Press, California (2006)
5. Menon, A., Santos, J.R., Turner, Y., Janakiraman, G., Zwaenepoel, W.: Diagnosing Performance Overheads in the Xen Virtual Machine Environment. In: VEE '05: Proceedings of the 1st ACM/USENIX international conference on Virtual execution environments, pp. 13–23. ACM Press, New York (2005)
6. Sharifi, M., Hassani, M., Mousavi, S.L.M., Mirtaheri, S.L.: VCE: A New Personated Virtual Cluster Engine for Cluster Computing. In: 3rd IEEE International Conference on Information and Communication Technologies: from Theory to Applications, pp. 7–11 (2008)
7. Smith, J.E., Nair, R.: The Architecture of Virtual Machines. Computer 38(5), 32–38 (2005)
8. Barham, P., Dragovic, B., Fraser, K., Hand, S., Harris, T., Ho, A., Neugebauer, R., Pratt, I., Warfield, A.: Xen and the Art of Virtualization. In: Proceedings of the nineteenth ACM symposium on Operating systems principles, pp. 164–177. ACM Press, New York (2003)
9. Emeneker, W., Jackson, D., Butikofer, J., Stanzione, D.: Dynamic Virtual Clustering with Xen and Moab. In: Min, G., Di Martino, B., Yang, L.T., Guo, M., Rünger, G. (eds.) ISPA Workshops 2006. LNCS, vol. 4331, pp. 440–451. Springer, Heidelberg (2006)
10. Emeneker, W., Stanzione, D.: HPC Cluster Readiness of Xen and User Mode Linux. In: 2006 IEEE International Conference on Cluster Computing, pp. 1–8. IEEE Press, New York (2006)
11. Vallee, G., Scott, S.L.: OSCAR Testing with Xen. In: 20th International Symposium on High-Performance Computing in an Advanced Collaborative Environment, p. 43. IEEE Press, New York (2006)

A Scalable Multi-attribute Range Query Approach on Cluster-Based Hybrid Overlays

You-Fu Yu, Po-Jung Huang, Quan-Jie Chen, Tian-Liang Huang, and Kuan-Chou Lai

Department of Computer and Information Science, National Taichung University
Taichung, Taiwan, R.O.C.
kclai@mail.ntcu.edu.tw

Abstract. Resource discovery in distributed computing systems is a critical issue to find and retrieve distributed resources rapidly. In general, most of previous proposed strategies focus on developing keyword searching approaches with preserving system scalability. In this paper, we propose a cluster-based hybrid overlay, which supports efficient keyword searching with the highly churn rate. The cluster-based hybrid overlay groups the nodes with the same attributes to form unstructured attribute-groups, and then clusters these attribute-groups with similar attributes to form attribute-clusters. Our proposed hybrid overlay could provide efficient multi-attribute and range-query searches with load balancing in large-scale P2P networks. Experimental results show that the proposed overlay performs well.

Keywords: P2P; Range query; Multi-attribute; Hybrid; Overlay.

1 Introduction

The Peer-to-peer (P2P) system is built as an overlay on the existing Internet infrastructure to provide resource sharing services. Recently, due to the development of P2P technology, efficient resource sharing approaches are applied in large-scale computing systems with various distributed resources. In general, a decentralized P2P system distributes all searching and locating loads across all participating peers; and the efficiency of the resource discovery mechanism becomes one of the most important issues.

In general, P2P overlays could be categorized into unstructured and structured. In an unstructured P2P overlay, there is no control and no accurate information on the network topologies, message broadcasting becomes essential. Most of the unstructured P2P overlays employ the flooding approach to discover resources, in which flooding is conducted in a hop-by-hop fashion counted by a time-to-live (TTL) count. Although the flooding approach can facilitate various resource-searching modes, flooding can seriously limit system scalability due to the number of redundant query messages grows exponentially during the message propagation. Therefore, unstructured approaches have to address the scalability problem.

On the contrary, the structured P2P overlay adopts accurate information to construct the network topology for message forwarding. Most of the structured approaches hash

C.H. Hsu and V. Malyshkin (Eds.): MTPP 2010, LNCS 6083, pp. 136–145, 2010.

the key attributes of distributed resources to form distributed hash tables (DHTs) and then construct the structured overlay for resource searching. Therefore, the hashing mechanism allows the searching operation to be performed with a bounded complexity. However, these structured P2P overlays are difficult to offer varied resource-searching modes because the hashed indices cannot capture the original resource characteristic and then the hashing-directory decouples the semantics of the information to be looked up from their indices. So, it is difficult to support complex queries such as range queries in DHT-based overlays.

In this paper, we propose a cluster-based hybrid P2P overlay by integrating the advantages of structured and unstructured approaches. The cluster-based hybrid overlay groups the nodes with the same attributes to form attribute-groups according to the unstructured approach; then, these attribute-groups with similar attributes are clustered to form attribute-clusters according the structured approach. Finally, these attribute-clusters are connected to form a hybrid overlay. Based on the cluster-based hybrid overlay, we also propose an efficient multi-attribute range query strategy which could maintain the load balancing in the unstructured attribute-groups. The proposed strategy could efficiently discover suitable resources in large-scale P2P networks.

The rest of the paper is organized as follows: Section 2 discusses related works. Section 3 presents the proposed cluster-based hybrid P2P overlay. Section 4 describes the multi-attribute range query strategy. Experimental results are given in Section 5. We conclude the paper in Section 6.

2 Related Works

The problem of resource discovery in P2P systems is well known. In order to improve the effectiveness, flexibility and scalability of the P2P system, many resource discovery strategies have been proposed. In general, P2P overlays could be categorized into unstructured and structured. The unstructured approaches (e.g., LightFlood [1] and pFusion [2]) try to reduce the amount of redundant messages when broadcasting the queries; and the structured approaches (e.g., Chord, CAN and Tapestry [3]) construct the structured overlay to make the keyword search effectively.

Some proposed approaches adopt the hierarchical architecture to reduce the redundant communications. For example, S-Club [4] and Gang et al. [5] employ the concept of the service group and user's interests separately to construct the overlay-groups for reducing the searching scope. Although these approaches could reduce the searching scope, their searching performance is still ineffective due to their unstructured architectures; therefore, there may be satisfied resources which couldn't be found.

SORD [6] approach constructs a two-tier structured architecture to provide the multi-attribute search, but the proposed strategy generates redundant searching messages to find the required resources. James et al. [7] also construct a structured two-tier overlay which supports contextualized keyword searching with additional cost for maintaining the connections. Therefore, their approach is limited due to the lack of scalability. Marzolla et. al. [8] also propose a two-level hierarchical topology to improve the single tree architecture, in which the lower level is constructed by many groups of trees and the upper level connects these groups by the unstructured approach.

Although their network topology reduces the maintenance cost in the single tree architecture, the unstructured upper level causes redundant searching cost. GTap [9] organizes nodes into groups and improves Tapestry by additional group links and backup links for flexible routing. Even though additional links enlarge the size of routing table, the latency of discovering groups is still a constant. However, the maintenance cost of these DHT-based overlays confounds the system scalability. Zhang et. al. [10] propose an assisted P2P overlay in which nodes participate in both the structured index overlay and the unstructured search overlay. The assisted P2P overlay records popular and unpopular properties for the index overlay and the search overlay respectively. Their approach could reduce the cost of maintaining unpopular data in the index overlay; however, the search delay and bandwidth overhead also increase.

In general, the DHT-based P2P networks need additional discovery mechanisms to support the effective multi-attribute or range query searching. PIRD [11] constructs a structured multi-ring overlay according the node ID which is hashed by node's attributes. Ronaldo et al. [12] hash the attributes to generate the attribute-matrixes and provide the multi-attribute search by the attribute-matrixes to find out the closest resources. However, their approach wastes a lot of time for computing these matrixes. Simon et al. [13] propose a hierarchically structured P2P system to provide the range query. Due to the hierarchical architecture, their system needs to travel more hops in order to route the search messages between different tiers. SARIDS [14] exploits the double-link architecture to improve the performance of the range query. Because SARIDS needs to forward search messages to different attribute rings, a lot of volume of traffic is arose from these forwarding operations and a lot of computing time is wasted for filtering out the correct resource.

This paper proposes a cluster-based hybrid P2P overlay to support multi-attribute and range-query searches. The proposed overlay is evaluated by OMNeT++ [15]. OMNeT++ is an event-driven P2P simulation environment with GUI. OMNeT++ supports many components, modules and communication protocols for simulating P2P systems.

3 Cluster-Based Hybrid P2P Overlay

In this section, the cluster-based hybrid P2P overlay is introduced. The proposed overlay hashes the fixed attributes of distributed resources to form attribute-vectors, and then constructs attribute-groups and attribute-clusters.

The attribute-vector is hashed from the attribute values of distributed resources to form a binary string. Because different attributes have different ranges of attribute values, the attribute value has to be normalized. The hash function is defined as follows:

$$f(V_a) = \left\lfloor \frac{2^m \times (V_a - MIN_a)}{MAX_a - MIN_a} \right\rfloor, \tag{1}$$

where MAX_a and MIN_a are the maximum and minimum values of attribute a respectively, V_a is the value of attribute a, and the attribute value is represented by m

bits. Assume that there are n kinds of static attributes, each attribute hashes its attribute value and then arranges in order to form the attribute-vector by $n \times m$ bits.

After generating the attribute-vector, each attribute-group groups the distributed resources with the same attribute-vector by the unstructured approach. The similarity function between two attribute vectors is computed as the following formula.

$$sim(X_{av}, Y_{av}) = \text{the number of attribute value which is zero in } (X_{av} \oplus Y_{av}), \quad (2)$$

where X_{av} and Y_{av} are two attribute vectors. When the value of $sim(X_{av}, Y_{av})$ approaches n, it means that these two attribute vector, X_{av} and Y_{av}, are more similar.

After generating attribute-groups, the attribute-groups in which the value of $sim(X_{av}, Y_{av})$ of any two attribute-groups is n-1, are clustered into the same attribute-cluster. Therefore, these attribute-clusters become a cluster-based hybrid overlay. In such an overlay, the attribute-group with the smallest attribute vector becomes the central attribute-group. Therefore, there are at most $2^{n \times m}$ attribute- groups and $n \times 2^{m(n-1)}$ attribute-clusters in the proposed overlay. Because an attribute-group may be placed in different attribute-clusters, these attribute-clusters are connected to form the cluster-based hybrid overlay. Fig. 1 shows a cluster-based hybrid overlay, when n=2 and m=2.

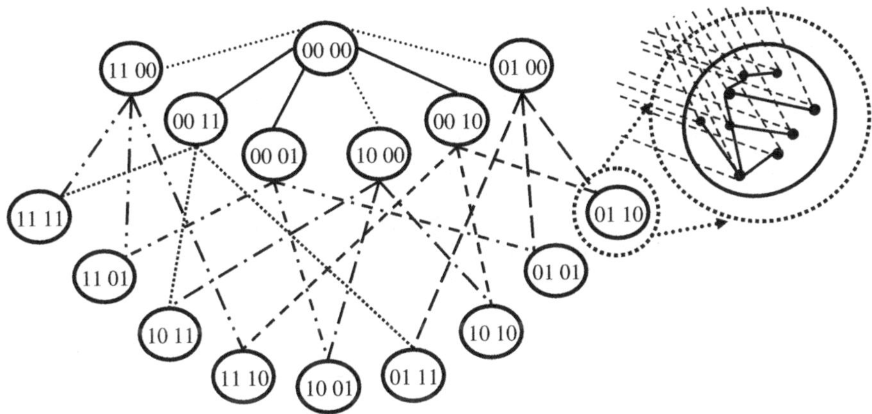

Fig. 1. Cluster-based hybrid overlay with four attribute-groups, when n=2, m=2

When a search query is submitted, the query is hashed into an attribute vector by Eq.(1) and then is routed to the attribute-group with the same attribute vector. Fig. 2 illustrates the query forwarding algorithm. G_{av}, Q_{av} and GC_{xav} are the attribute vectors of G, Q and GC_x respectively. In order to reduce the traffic and to avoid the circular routing, the query is dropped when the group G receives the same query repeatedly (line 1-3). If the value of $sim(G_{av}, Q_{av})$ equals to n, it means the attribute vectors of G and Q are the same. Therefore, the group G returns "*HIT*" (line 4-5). If the group G is not the destination of Q, it chooses the child or parent with the maximum value of $sim(G_{av}, Q_{av})$ to forward the query (line 6-15). However, if there are no children or no parents suitable to forward the query, it means the destination of the query doesn't exist in the P2P network. Therefore, the group G returns "*Miss*"(line 17).

In general, the time complexity of resource discovery in most structured P2P networks is $O(N)$, where N is the number of nodes. However, because the proposed cluster-based hybrid overlay could be viewed as a hierarchical tree-like architecture with n layers, in which each central attribute-group becomes the parent node, as shown in Fig. 1, the time required for forwarding queries to the specified attribute-group is $O(2n)$, where n is the value of different attribute types. Notice that the forwarding time of our proposed overlay is unconcerned with the number of resources. Due to that there are $2^{n \times m}$ different attribute-groups, so the time complexity of resource discovery in most structured P2P networks is $O(\log 2^{n \times m}) = O(n \times m)$. Thus the proposed overlay performs better in the resource discovery when m is greater than 2.

```
Algorithm : Query Forwarding (query)
Input : a query Q
Output : "MISS" or "HIT" or the next attribute groups to forward
Begin
when group G receives an query Q from group GF
        if G has received Q before
                drop Q ;
        else if sim(G_av, Q_av) = n
                return HIT ;
        else if G has children GC_1,GC_2,...GC_x
                for i := 1 ~ x
                        if sim(GC_xav, Q_av) > sim(G_av, Q_av)
                                return GC_x ;
                if G does not has any parents
                        return MISS ;
                else
                        return GP which has the maximum sim(G_av, Q_av) ;
        else if G has parents GP
                return GP which has the maximum sim(G_av, Q_av) ;
        else
                return MISS ;
End
```

Fig. 2. Query forwarding algorithm

Because the population of the P2P networks changes frequently, the dynamic operations in the proposed overlay are discussed as follows. After the first resource is chosen as the central attribute-group of all the attribute-clusters, the following resources join the overlay by connecting with the first one, and develop their own attribute-groups and attribute-clusters.

When a resource joins the cluster-based hybrid overlay, it contacts with the bootstrap server to get the addresses of other existing resources in the cluster-based hybrid overlay. Then the new joining resource generates its attribute vector and tries to discover the attribute-group with the same attribute vector in order to join it. After joining the attribute-group, the new joining resource starts to connect the central attribute-groups in other attribute-clusters.

After joining the attribute-group, each resource exchanges the update messages with its neighbors periodically. The update messages include a list of the addresses of its neighbors in the same attribute-group. If the resource doesn't receive any update message form its neighbors in a fixed-period of time, it tries to build a new connection between one of neighbors in the address list and breaks the original one's connection.

In the cluster-based hybrid overlay, we also propose a load balancing mechanism, as shown in Fig. 3, to improve the performance of load balancing. Because the attribute-groups are constructed by the unstructured approach, each resource is regarded as the neighbor of other resources in the same attribute-group. In addition, each resource maintains a list of records for the current attribute values of dynamic resources (e.g., communication bandwidth) and exchanges the list with its neighbors. We also simply classify the resources as either high loading or low loading according the load. When the resource is in high loading, it will transfer the redundant connections to the neighbors whom are in low-loading (line 3). Therefore, the high-loading resources process less query forwarding. In addition, the high-loading resource doesn't reply and forwards the query to neighbors whom are in low-loading (line 4). Only the low-loading resources reply the query and forward the query to other low-loading neighbors (line 5-7) until finding out enough resources. Thus, the proposed mechanism makes use of transferring the redundant connections and forwarding queries to the low-loading neighbors for load balancing.

```
Algorithms : Load Balancing (Q)
Input: a query Q
Output: satisfied resource
Begin
1: When resource R receives a query Q
2:        if R is in high loading;
3:               transferring the connections of parents and children to the low-loading
   neighbors until the number of the connections of each parent and child is equal to 1;
4:               forwarding the query Q to the low-loading neighbors;
5:        else // R is in low loading
6:               return R;
7:               forwarding the query Q to the low-loading neighbors ;
End
```

Fig. 3. Load balancing algorithm

4 Multi-attribute Range Query Strategy

The proposed cluster-based hybrid overlay could maintain efficient keyword searches and provide range-query searches without additional cost. In this study, when the query arrives at the attribute-group with the same attribute vector, it means the query is successful because all of the resources in the same attribute-group satisfying the query.

Because the proposed overlay uses the attribute vector to preserve the resource's attributes, a user can discover the satisfied resources according to these attribute vectors. Users could define the value of each attribute, and then hash these values into the attribute vector by Eq.(1). Therefore, when the multiple-attribute search is issued, the query can be forwarded to the satisfied attribute-groups according to the query forwarding algorithm.

To support the range query, this study also proposes a range query searching algorithm. For the partial query, users can issue the query with partial attributes without the remaining attributes. The randomly attributes is denoted by "*" (e.g. If the values of the first and third attribute are "01", and the value of the second attribute is random, the attribute vector is represented as "01 * * 01"), and the results of exclusiveOR with any other values are zero (i.e. only the same attribute value). In addition, users can define the attribute values within a range (e.g., 00 01 ~ 11 10).

```
Algorithms : Range query search (range query)
Input: a range query (an attribute vector list along with a target attribute vector)
Output: satisfied resource
Begin
When G receives a range query RQ from GF
    if G has received RQ before
        drop RQ ;
    else
        TG := Query Forwarding (RQ_t) ;
        if TG = "MISS"
            return "There is no satisfied resource" ;
        else if TG = "HIT"  // it means TG = G and sim(G_av, RQ_tav) = n
            if G has a parent GP which sim(GP_av, RQ_tav) = n and GP!=GF
                forward RQ to GP ;
                if G has children GC_1, GC_2, ..., GC_x
                    for i := 1 ~ x
                        if sim(GC_xav, RQ_lav) = n and GC_x != GF
                        forward RQ to GC_x ;

                if sim(G_av, RQ_lav) = n
                    Load Balancing (Q);
            else
                if G has children GC_1, GC_2, ..., GC_x
                    for i := 1 ~ x
                        if sim(GC_xav, RQ_lav) = n and GC_x != GF
                            forward PQ to GC_x ;

                if sim(G_av, RQ_lav) = n
                    Load Balancing (Q);
        else
            forward RQ to TG ;
End
```

Fig. 4. Range query searching algorithm

Fig. 4 illustrates the range query searching algorithm. Let *RQ* denote a range query and the attribute group *G* receive *RQ* from *GF*. *TG* is the results of forwarding query *RQ*. *GP* and GC_x denote *G*'s parents and children respectively. GP_{av} and RQ_{av} are the attribute vectors of *GP* and *RQ* respectively. The range query *RQ* will be packed into an attribute list RQ_l along with a target attribute vector RQ_t. RQ_l lists all of the combinations of *RQ*. RQ_t is similar to the partial query which set the attribute values within a range to "$*$".

If the group *G* doesn't receive the range query *RQ* repeatedly and the result of forwarding query RQ_t is not "*MISS*" (line 1-7), the query will try to forward to *TG*. If *TG* is "*HIT*", the group *G* forwards the query to its parent and children with the similarity = *n* (line8-14). However, if there is no parent with the similarity = *n*, it means the group *G* is the root of the range query. Therefore, the range query is forwarded to the children with the similarity = *n* (line 17-21).

In general, most structured DHT-based P2P overlays perform the range query searching by two phases. In the first phase, the upper and lower bound nodes are found out. In the second phase, the upper and lower bound nodes employ the "one-hop" search which routes the queries hop by hop to find out all the nodes between them. Assume there are *x* nodes between the upper and lower bound nodes, the time complexity is $O(\log N + x/2)$. Therefore, when *x* is greater than $2 \times \log N$, it would spend more time for one-hop search.

The procedure to perform the range query is illustrated in Fig. 4. The range query searching algorithm firstly finds out the center attribute-group in which the attribute vector matches with the attribute vector of the query. This step needs at most $O(n)$ hops. And then, the algorithm employs the "radial" search which broadcasts the query cluster-by-cluster to find out all the resources which satisfy the query. This step also needs at most $O(n)$ hops. Therefore, the time complexity of the range query searching algorithm is still $O(n)$.

5 Experimental Results

In the section, we employ OMNeT++ 4.0 to evaluate the performance of the proposed cluster-based hybrid overlay. Fig. 5(a) demonstrates a cluster-based hybrid overlay generated by OMNeT++ 4.0. We also implement Chord for performance comparison.

We evaluate the performance of range query searches as follows. When there are 4096 nodes and the query range varies from 10 to 100, Fig. 5(b) shows the comparison results of range query searches. Fig. 5(b) shows that when the scope of the range query gradually increases, Chord wastes more time on one-hop search. However, the proposed cluster-based hybrid overlay keeps a stable performance in different query ranges.

Fig. 5(c) shows the average search time for the keyword search when *m* is fixed to 2 and the number of nodes varies from 256 to 16384. Table 1 shows the combinations of different attribute vectors. Fig. 5(c) shows that the cluster-based hybrid overlay preserves the efficiency of the keyword search. Thus the cluster-based hybrid overlay performs better in resource discovery when *m* is greater than 2.

Fig. 5(d) shows that the average search time in different numbers of nodes when *n* is fixed to 2. Fig. 5(d) shows that the average search time is independent of the number of nodes. Therefore, the proposed cluster-based hybrid overlay indeed supports better scalability.

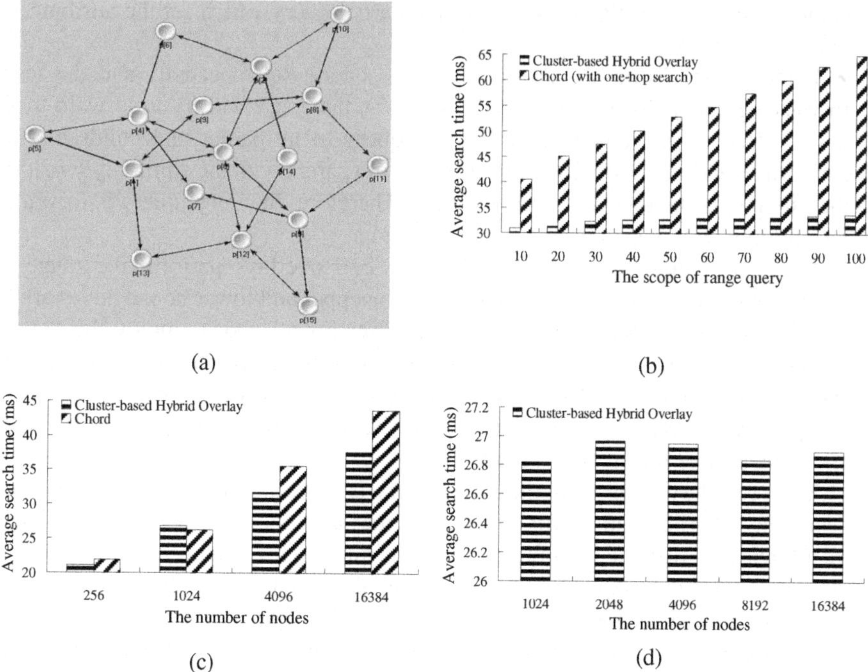

Fig. 5. (a) Example of a cluster-based hybrid overlay generated by OMNeT++ 4.0. (b) Average searching time in different query ranges. (c) Average search time in different nodes when m = 2. (d) Average search time in different nodes when n = 2.

Table 1. Combinations of different attribute vectors

Number of attribute-groups	256	1024	4096	16384
Attribute vector combination	n=4, m=2	n=5, m=2	n=6, m=2	n=7, m=2

6 Conclusions

This study proposes a cluster-based hybrid overlay by integrating the advantages of structured and unstructured P2P overlays. The attribute-vector preserves the original characteristics of distributed resources for the multi-attribute searching. The attribute-groups are constructed by the unstructured approach to achieve load balancing. The attribute-clusters connect each other to form the cluster-based hybrid overlay to support efficient multi-attribute range query searches. Experimental results show that the proposed overlay performs well and the system performance is independent of the number of resources.

Acknowledgements

This study was sponsored by the National Science Council, Taiwan, Republic of China under contract numbers: NSC 97-2221-E-142-001-MY3, NSC 97-3114-E-007-001- and NSC 98-2218-E-007 -005.

References

1. Jiang, S., Guo, L., Zhang, X., et al.: LightFlood: Minimizing redundant messages and maximizing scope of peer-to-peer search. IEEE Transactions on Parallel and Distributed Systems 19(5), 601–614 (2008)
2. Zeinalipour-Yazti, D., Kalogeraki, V., Gunopulos, D.: pFusion: A P2P Architecture for Internet-Scale Con-tent-Based Search and Retrieval. IEEE Transactions on Parallel and Distributed Systems 18(6), 804–817 (2007)
3. Stephanos, A.T., Diomidis, S.: A Survey of Peer-to-Peer Content Distribution Technologies. ACM Comm. 36(4), 335–371 (2004)
4. Hu, C., Zhu, Y., Huai, J., et al.: S-Club: An Overlay-based Efficient Ser-vice Discovery Mechanism in CROWN Grid. In: Proceedings of 2005 IEEE International Conference on e-Business Engineering, pp. 441–448 (2005)
5. Chen, G., Low, C.P., Yang, Z.: Enhancing Search Performance in Unstructured P2P Networks Based on Users' Common Interest. IEEE Transactions on Parallel and Distributed Systems 19(6), 821–835 (2008)
6. Al-Oqily, I., Karmouch, A.: SORD: A Fault-Resilient Service Overlay for MediaPort Resource Discovery. IEEE Transactions on Parallel and Distributed Systems 20(8), 1112–1124 (2008)
7. Salter, J., Antonopoulos, N.: An optimized two-tier P2P architecture for contextualized keyword searches. Future Generation Computer Systems Journal 23(2), 241–251 (2007)
8. Marzolla, M., Mordacchini, M., Orlando, S.: A P2P resource discovery system based on a forest of trees. In: Proc. GLOBE '06 - 2nd Int. Workshop on Grid and Peer-to-Peer Computing (to appear, 2006)
9. Zhang, Y., Li, D., Chen, L., et al.: Flexible Routing in Grouped DHTs. In: Eighth International Conference on Peer-to-Peer Computing, pp. 109–118 (2008)
10. Zhang, R., Hu, Y.C.: Assisted Peer-to-Peer Search with Partial Indexing. In: IEEE INFOCOM 2005, pp. 1146–1158 (2007)
11. Shen, H., Li, Z., Li, T., et al.: PIRD: P2P-based Intelligent Resource Discovery in Internet-based Distributed Systems. In: Proc. of ICDCS 2008, pp. 858–865 (2008)
12. Ferreira, R.A., Koyuturk, M., Jagannathan, S., et al.: Semantic indexing in structured peer-to-peer networks. J. Parallel Distrib. Journal 68, 64–77 (2008)
13. Rieche, S., Vinh, B.T., Wehrle, K.: Range Queries and Load Balancing in a Hierarchically Structured P2P System. In: 33rd annual IEEE conference on local computer networks - LCN 2008, pp. 28–35 (2008)
14. Lin, Y.H.: SARIDS: A Self-Adaptive Resource Index and Discovery System. Master thesis, National Tsing-Hua University (2009)
15. OMNeT++, OMNeT++ Discrete Event Simulation System, `http://www.OMNeTpp.org`

Satellite Image Structure Analysis with the GRID Technologies

A.I. Alexanin, M.G. Alexanina, P.V. Babyak, S.E. Diyakov, and G.V. Tarasov

Institute of Automation and Control Processes, Far Eastern Branch of Russian Academy of Sciences (FEB RAS), Vladivostok, 5 Radio st., 690041, Russia
george@dvo.ru
http://www.satellite.dvo.ru/en.html

Abstract. A problem considered is to increase computational capacity of Satellite Data Processing Center on the base on external computational resources. As the problem solution, the facilities of the Satellite Center and Supercomputing Center were integrated by means of the GRID technologies. The developed GRID-services network infrastructure make the possibility to considerably increase the speed of data processing and provide feasible time interval for result reception.

1 Introduction

The Satellite Center of FEB RAS is a research department for monitoring and analysis of the ocean and atmosphere state in the Far East of Russia. The main activity of the Center includes the following areas: new algorithms and technologies development for satellite data processing; design and development satellite data processing system [1]; delivery information to users both directly and through the global information system of Europe Space Agency (EoPortal).

Diversity of satellite information and available processing packages designed for different operating systems, requires a development of the new algorithms and correspondent processing technologies. It is necessary to integrate computation and technology facilities of different holders for every day processing of the receiving satellite data. There are two reasons to use GRID technologies for computation arrangement. Control the data processing flow, which consists of many kind of computational tasks, and to increase the data processing time. The software packages used such as SeaDAS (USA), AAPP, RTTOV, MetOffice-1Dvar (Europe) are complex and under development permanently, and as the result, it is difficult to upgrade the software packages for parallel computation.

As the result we use two types of parallelism in our practice: task parallelism and data parallelism. The task parallelism is the result of the necessity to process a data file with the different algorithms for different goal realizations. For example, the output products number of one file of the MODIS radiometer processed with SeaDAS package can be reached up to 200 units. The data parallelism is the main approach for the processing acceleration. There are two kind of the

C.H. Hsu and V. Malyshkin (Eds.): MTPP 2010, LNCS 6083, pp. 146–151, 2010.

last one – object and image-stripe parallelism. Such products as temperature and humidity profiles of the atmosphere or sea eddy shapes (1) are made with the object parallelism – each eddy shape is computed simultaneously in this case.

Most of all the image-stripe parallelism is used. A source image is cut into sections, and all sections are processed simultaneously. Such tasks as sea eddy or typhoon detection automatically are based on an analysis of the structure charts. Last ones are computed as dominant orientations of thermal contrasts (DOTC) on a set of infrared (IR) satellite images (NOAA and MTSAT-1R data used). DOTC is the statistical significant contrast orientation in the vicinity of the point selected. Due to the shear nature of the atmosphere and ocean flows the DOTC may be considered as velocity flow orientation (2). This property is used for typhoon and sea eddy automatic detection and their geometry computation [2].

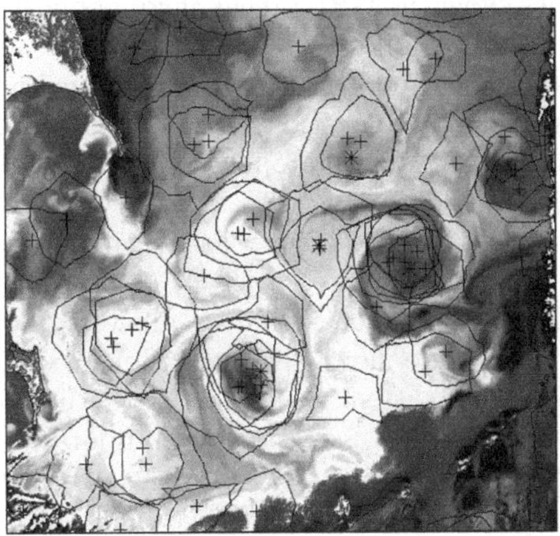

Fig. 1. The shapes of the sea synoptic scale eddies without the false object filtration, IR-image of the Okhotsk sea, September 6, 2006: + - the object centers, * - the real eddy centers

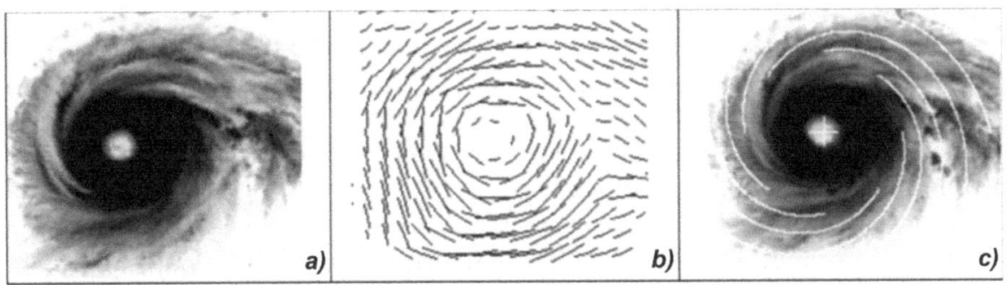

Fig. 2. Typhoon cloudy IR-image (a); its structure chart (b); the center, eye size and cycloid computed (c)

2 GRID Environment for Satellite Image Processing

GRID is a common technology for computational infrastructure creation and at present the GRID is widely used in many scientific projects. The most famous grid environment in Europe Union is EGEE [4], consolidated near a hundred institutes and other organizations to improve quality of computations in such areas as earth sciences, high energy physics, bioinformatics and astrophysics. Within EGEE the such projects as GENESI-DR [4] and SEE-GRID-SCI [4], linked to processing and allocation of the satellite information to various consumers intensively is developed. Leading world space agencies are engaged in implantation and usage GRID technologies: project G-POD (Grid Processing on Demand) in the European Space Agency (ESA) [5], project IPG (Information Power Grid Infrastructure) in National Aeronautics and Space Administration (NASA) [6], and also many other projects and the organizations. Mentioned successful researches show the GRID-technology application allows to solve two main tasks. At first, GRID allows to integrated different types of resources by means of standard services and methods. Common ones are file exchange service and computational resource allocation service. At second, the GRID allows to control the computation task flow automatically by means of grid-schedulers and other broker services.

In the given article the task of increasing a computational capabilities of Satellite Center FEB RAS by integration of external computing resources by the GRID-technologies use is considered. It allows to have automatic processing of the full satellite data flow. The extension of computational capability of the satellite processing system was made on the base of the Supercomputer Center facilities (The Center "Far-Eastern Computation Resource", Institute of Automation and Control Processes, FEB RAS). That Centers are connected through the Vladivostok Scientific and Educational Metropolitan Area Network (MAN). The most computation capacity procedures are allocated in the Supercomputer Center which peak performance is 1.4 TFLOPS now. The scheme of the data processing is presented on figure 3 [3].

The distributed computation system is the core of Satellite Center information system. It makes the launch a set of the processing algorithms after an event appearance in the information system. The GRID network infrastructure integrates the computation facilities of the Supercomputer Center (84-nodes cluster each 2xPowerPC CPU) and the 8-cores computation server of the Satellite Center. The Globus Toolkit software was used for deploying basic GRID-services. The software allows to create common GRID infrastructure, and its opened, widespread, and reliable toolbox.

The developing GRID-net includes two main resources – GRID Processing Unit and Computing Element. The GRID Processing Unit belongs to the Satellite Center information system. The Computing Element – to the Supercomputer Center. They are connected by the up to 1Gbit/s channel. Both sides have the same configuration of the Globus Toolkit and use conformance certificates for authentication and authorization. Resource bidirectional communications are based on two main services – GRAM and RFT. RTF is used for reliable data and

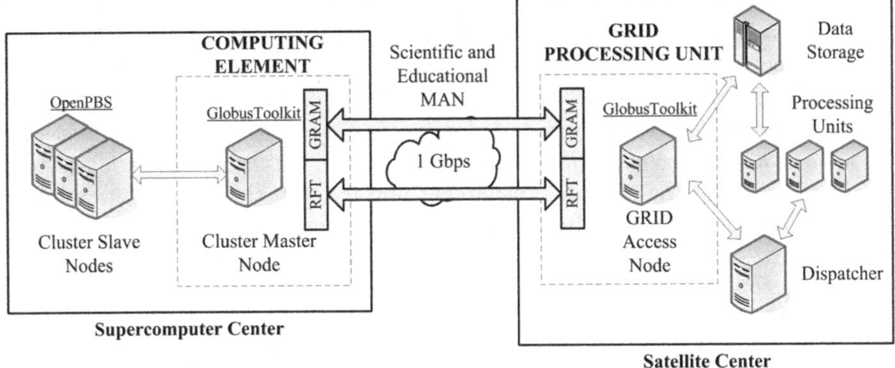

Fig. 3. GRID environment for satellite data processing

file exchange, GRAM — for command and task execution control. The GRID Processing Unit forms a job passport in XML format for every task received due to the GRAM service. The Computing Element makes a schedule of data processing, solves the task, and returns the results and logs to GRID Processing Unit.

3 Example: DOTC Image Processing

As a sample of of external computing resources involvement we consider the processing task of infra-red images for construction the charts [7]. Input data for the processing system is a set of images. The processing system fulfills the following set of operations:

1. For each image processing system forms the template of the job. Job parameters are: the template of the processing command, a filename which should be processed, the name of the processing unit, and additional parameters of a processing package. In the example considered the processing unit is fixed and corresponded to unit integrated into GRID-environment of two centers.
2. On the basis of the template the processing system forms the executed job which represents a script in BASH command language interpreter. There are three main executed blocks in this script: generation of adjusting and auxiliary files for the job starting with GRAM-service (creation of the passport of the job), execution the client command of GRAM-service and waiting the job finishing.
3. The system's dispatcher executes by SSH protocol a script in a GRID Processing Unit and waits it is finished. All further operations of the processing algorithm are fulfilled on this unit. There are three stages of the script.
   (a) The first stage forms the grid job passport in XML format. The name of an executable file of processing package, URLs of a source and destination files, number of processors and other processing package parameters are specified in appropriate sections of the passport.

   (b) The second block executes 'globusrun-ws' command with specified parameters: grid job passport file name and resource address name (Computing Element) on which this job will be processed.
   (c) the third stage waits the completion of 'globusrun-ws' command, and forms an appropriate return code for successful or unsuccessful 'globusrun-ws' command completion when script is completed. Return code is recognized by the system's dispatcher and fixed in the list of the handled or unhandled jobs. The operator of processing system can initiate processing once more on the base of list of unhandled jobs.
4. Grid job passport are handled automatically by the Computing Element GRAM-service. After job passport handling the GRAM-service transfers control to a local resource management system (OpenPBS) which executes processing package on the cluster nodes.

The following application package has been developed for parallel processing of one image and DOTC field construction. There are follow tools of the package:

- bldstr2 — sequential algorithm to process one image [7]. Parameters for the given algorithm are input and output file names, and also necessary processing parameters including description a range of the window sizes. The processing window - is a rectangular subregion of the image in which it is calculated a contrast gradient. Algorithm it is iterated passes by all possible variants of subregions within the image and from all of variant are selected dominant which are written in an output file.
- pr_vsplit — algorithm for a source image partitioning into vertical bars (image-striping) . One of the parameters is the quantity of the bars.
- pr_vmerge — merging algorithm all bars into complete output image.
- tasker — parallel processing control algorithm, written using MPI. 'tasker' is a parallel extension over bldstr2. Input parameter is the file containing commands to execute 'bldstr2' algorithm. Number of command equals to a number of bars. In a parallel mode one instance of 'tasker'-process is declared as a master-process (MPI rank = 0). Master defines quantity of accessible salve-processes, reads command list and iteratively transfers commands to appropriate free slave-processes. Each slave-process executes received command until master-process sends special 'end' command.
- mpbldstr2.sh — script for parallel processing of one image. The given script consistently executes pr_vsplit, tasker and pr_vmerge.

The approach effectiveness was estimated on a task of high resolution DOTC chart calculations for Okhotsk sea region for some day. There were 16 images in Mercator projection and of 50 MB of total volume. The cutting of an image into 10 sections gave us the 160 tasks. A desktop computer with 3 GFLOPS performance rate has solved all tasks for 7 days. 10-CPU parallel configuration has completed all tasks for 8 hours. 160-CPU parallel configuration has completed all tasks only for 4 hours. The reason of two-times speed-up only was in the GRID scheduler settings used – limited count of simultaneously tasks per one user was

25 ones only. Settings with unlimited count of task per user allowed to solve the task for a half hour. The time lost due to the communication procedures were less 1%.

4 Conclusion

The task of integration of external computing resources in Satellite Center processing system using GRID-technologies is solved. Experimental executions of a task of DOTC field computing have shown high performance of applying external computing resources. The received positive experience in applying GRID-technologies in processing system allows to transfer in GRID environment different image processing algorithms.

Acknowledgment. The work was supported by Russian Foundation for Basic Research (grant No. 08-07-00227), Far-Eastern Branch of Russian Academy of Sciences (grant No. 09-III-A-03-065, 09-I-P2-05, 09-I-P17-03, 09-I-PI-02).

References

1. Shokin, Y.I., Pestunov, I.A., Smirnov, V.V., Sinjavsky, Y.N., Skachkova, A.P., Dubrov, I.S., Levin, V.A., Alexanin, A.I., Alexanina, M.G., Babjak, P.V., Gromov, A.V., Nedoluzhko, I.V.: The distributed system of delivery, acquisition, and data processing Siberia and the Far East areas monitoring. Journal of the Siberian Federal University. Engineering and technologies 4(1), 291–314 (2008) (in Russian)
2. Alexanin, A.I., Eryomenko, A.S.: Automatic Computation of Tropical Cyclone Tracks on Data the Geostationary Meteorological Satellites. Issledovanie Zemli iz Kosmosa 5, 22–31 (2009) (in Russian)
3. Babyak, P.V., Tarasov, G.V.: An experience of GRID-technology use in a processing system of FEB RAS Satellite Center. In: Proc. of the Sixth Russian Open Conference Modern Problems of Remote Sensing of the Earth from Space: Physical bases, methods and technologies of monitoring of the environment, potentially dangerous phenomena, and objects. Release 6, vol. I, pp. 71–80. Azbuka-2000, Moscow (2009) (in Russian)
4. EGEE Collaborating Projects Liaison Office (CPLO) and CERN NA2 Team. Success and Sustainability: EGEE Collaborating Projects Achievements in 2009 and Future Plans (September 2009), `http://press.eu-egee.org/fileadmin/documents/publications/CP-booklet-final-for-print.pdf`
5. Plaza, A.J., Chang, C.-I. (eds.): High performance computing in remote sensing. Computer & Information Science Series, p. 466. Chapman&Hall/CRC, Boca Raton (2008)
6. Foster, I., Kesselman, C., Tuecke, S.: The Anatomy of the Grid: Enabling Scalable Virtual Organizations. International Journal of High Performance Computing Applications 15(3), 200–222 (2001)
7. Aleksanin, A.I., Aleksanina, M.G.: Detection of Stable Synoptical Features of Sea Surface from a Series of Infrared Satellite Images. Pattern Recognition and Image Analysis 17(4), 480–486 (2007)

A Dynamic File Maintenance Scheme with Bayesian Network for Data Grids

Chao-Tung Yang[1], Chien-Jung Huang[1],Tai-Lung Chen[2,*], and Ching-Hsien Hsu[3]

[1] Department of Computer Science,
Tunghai University, Taichung, 40704, Taiwan
ctyang@thu.edu.tw, comgodzore@gmail.com
[2] Institute of Engineering and Science,
Chung Hua University, Hsinchu 30012, Taiwan
ctl@sclab.csie.chu.edu.tw
[3] Department of Computer Science and Information Engineering,
Chung Hua University, Hsinchu 30012, Taiwan
chh@chu.edu.tw

Abstract. With the large amount of data which is produced by scientific experiments and simulations, the data replication has become an important issue in data grid. In order to shorten the file accessing time, multiple copies of file and stores can be created in the appropriate location. In this paper, we present a dynamical maintenance service of replication to maintain the data in grid environments. Based on the Bayesian Networks (BN), the optimization strategy was proposed in this study; named Implicit Dynamic Maintenance Service with Bayesian Network (IDMSBN).

1 Introduction

In the past years, more executing projects about high energy physics in Europe, climate predict, earthquake simulation, research of bioinformatics and so on deal with large-scale files in Terabyte or even in Petabyte all over the world. So that more and more computing power, network bandwidth and storage devices are required for better performance. Those necessary of devices can't enough supported by a computer center with supercomputers. Grid technology is the best approach developed to solve this kind problem. The main motive of Grid Project supply people to share their devices including computing and storage, such as CPU, memory, and hard disk...etc., geographically distributed around the world work together as a massive virtual computer. For the experiments or simulations, Grid needs to integrate those equipments and communicate each other by some middleware. Globus Toolkit is a middleware of open source software for building grid environment. It contains many components such as security, information infrastructure, resource management, data management, communication, fault detection and portability.

[*] Corresponding author.

C.H. Hsu and V. Malyshkin (Eds.): MTPP 2010, LNCS 6083, pp. 152–162, 2010.

Data replication [1, 2], which operates base on the Globus Toolkit, can make the same data scatter in distributed sites. A superior replication strategy can reduce the file access time, access latency, and accelerate download speed of desired file to reduce the executing time. On account of this reason, if one storage element is crashed, client nodes can fetch their desired data from other storage elements. It improves the fault tolerance and makes the whole grid environment become more stable and reliable. Another advantage of data grid is that users can download data parallel from the chosen sites or some applications chose automatically after estimating by some evaluated model in grid environment. The bandwidth utilization is the most important factor to affect the downloading speed and executing time. Because of network environment unstable, hence the replica sites are not always the best choices for downloading data to reduce the transmission time. Replicas should be distributed to the suitable storage elements of the Grid environment for computing device reduce the time users get the data and execute with the higher performance.

This paper presents a different replica maintenance strategy called Implicit Dynamic Maintenance Service with Bayesian Network (IDMSBN) differ from Dynamic Maintenance service (DMS) and others. DMS framework can automatically maintain the data by the relative status and information metadata. Such as access frequency of the data, the free space of storage element that the data will be replicated or migrated to, and the condition of network of the file site to other sites. Both two above methods need to catch metadata by collecting metadata of server. DMS need more storage space to record metadata about site to site and when amount of sites and files increase, it perforce can enlarge system IO and reduce performance.

The contributions of this paper provide another aspect of replica in Bayesian Network, and reduce metadata frequently read & write times. Also IDMSBN can add some effective factor that can describe with condition probability for replica prediction.

2 Related Work

The Bayesian networks (BN) can represent the joint probabilistic distribution tightly between variables and clearly show the conditionally independence. A BN have two parts: a set of parameters and a structure. The structure contains nodes, which represent any kind of variables, and arcs, which connect nodes that are related probabilistically. The BN graph is directed acyclic graph (DAG), meaning that it can't start at some node and end at same node. The set of parent nodes of a node Xi is denoted by parents (Xi).

$$P(X_1, X_2, \ldots, X_n) = \prod_{i=1}^{n} P(X_i \mid parents(X_i)) \tag{1}$$

There are many vital elements about the data maintenance in data grids. In [4], Foster introduced six dynamic replication strategies which compared those six strategies by using a simulator called PARSEC to measure average response time and the total bandwidth consumed of each strategy. In [5, 6], Rashedur M. Rahman *et al.* proposed a static replica placement algorithm to place replicas in p best candidate nodes that can minimize the total response of each node by using Lagrangean Relaxation which is a heuristic approach to measure the response time of each client node to its nearest server node. They use a simulator called OptorSim [7] which is developed by EU Data Grid project to compare their method called dynamic p-median with static p-median and best client. In [8], Sang-Min Park *et al.* proposed a dynamic replica maintenance algorithm called Bandwidth Hierarchy based Replication (BHR). In [9], Ruay-Shiung Chang *et al.* indicated that the BHR algorithm will be better performance than other strategies only when the capacity of storage element is small.

There are some issues about predictive technique for replica selection in grid environment. R. M. Rahman *et al.* [10] used a neural network (NN) for transfer time prediction of different sites that have replicas. NN is like neuron perceptions of human brain, which trained to learn behavior. They used the predictor trace data to train the NN with back-propagation algorithm and predict current data transfer performance in the grid, throughput. Bayesian Methods often are developed to use for diagnosing system [11] and can detection of outbreaks of disease, whether natural or bioterrorist induced. Several algorithms provided for the priori and posterior probabilities of Bayesian Network [12].

The rest parts of this paper are organized as follows. In Section 3, we describe the details of design and implement of data grid framework, system component, and the algorithm of IDMSBN, also we will introduce our parameters and evaluation model to determine when data should be adjusted to the appropriate locations. In Section 4, we simulate some scenarios by using the IDMSBN algorithm compare with other replication strategies and show the experimental results. In the last part, Section 5, we describe the conclusion and the future works of this paper.

3 System Description

3.1 System Architecture

The software stack diagrams of our data grid system are shown in Figure 1 and Figure 2. There are three layers in the data grid system, such as bottom layer, middle layer, and top layer. The functionalities of each layer are described below.

Bottom Layer: It shows the software that each node has been installed in the grid environment. The two major components of Bottom Layer are Information Provider and Grid Middleware. The Information Provider is consists of two blocks, Ganglia and Network Weather Service (NWS). Ganglia can gather the machine information such as the number of processor and how many cores do the processor have, the load of processor, the free and total size of memory, and the usage of disk. NWS is used to gather the network bandwidth between each node and also gather the latency of each link. The Grid Middleware is consists of Globus Toolkit. Globus Toolkit is used to join a node into the grid environment.

Middle Layer: The main component in this layer is Site. Each Site is constructed by several nodes, usually located in the same place or connected with same switch or hub, each node in Site should connect to each other by internet. Moreover, each Site usually builds up as a cluster but has real IP of each node and the first node of this Site called the head node of this Site.

Top Layer: The components in this layer are Applications, Services, Monitoring Service, and Records. By using the data gird system, users can easily control the grid environment by the Services downward the Applications. Services are consists of Anticipative Recursively-Adjusting mechanism, Replica Selection Service, One-way Replica Consistency Service, and Dynamic Maintenance Service which provide by the portal. Services can operate by the information that Monitoring Service and Records gathered. Records can provides the machine or file information before downloading file or adjusting the file location. Monitoring Service provides a web front-end page for users to observe the variation during the process of jobs.

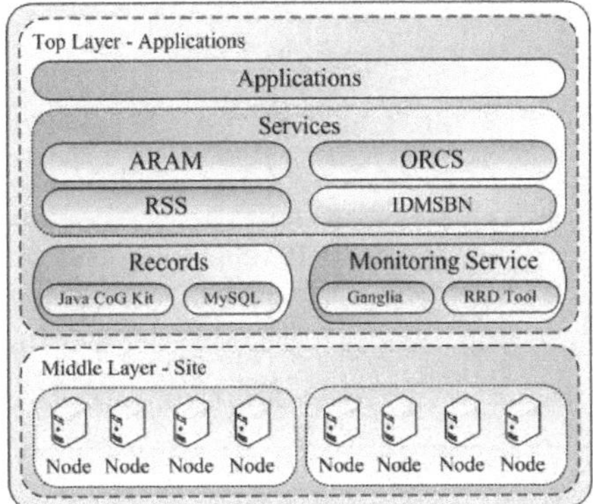

Fig. 1. The software stack diagram of all nodes

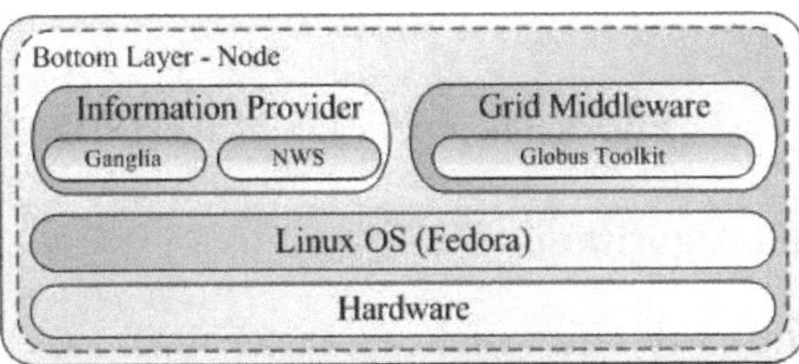

Fig. 2. The software stack diagram of all sites, services, and portal

3.2 The Operation of IDMSBN

The major contribution of this paper is constructing a dynamic replica maintenance service for the data grid system. IDMSBN is constructed and used to maintain the replicas in the grid environment. The operation of IDMSBN is shown in Figure 3. Before the IDMSN DMS start to maintain the file dynamically, Information Service and Replica Location Service will record the related information into the database for IDMSBN to measure by using the cost model which we will describe latter in Section 3.3.

The functionality of Information Service and Replica Location Service are describing below:

Information Service: The Information Service will periodically gather the idle ratio of changeable factors such as CPU, Memory, free space of storage devices, and the network bandwidth…etc. It will record real-time information of those dynamic factors into the Information Database (Info. DB) for IDMSBN to use.

Replica Location Service: The Replica Location Service (RLS) will record the information of each file into the File Information Database (File Info. DB) such as logical file name, file size, the physical location of the file, the time that the file has been created or updated, and the access frequency of the file… etc. It is convenience for users to search the desired file in the appropriate location which is closest to the user in the grid environment by using Replica Location Service to download.

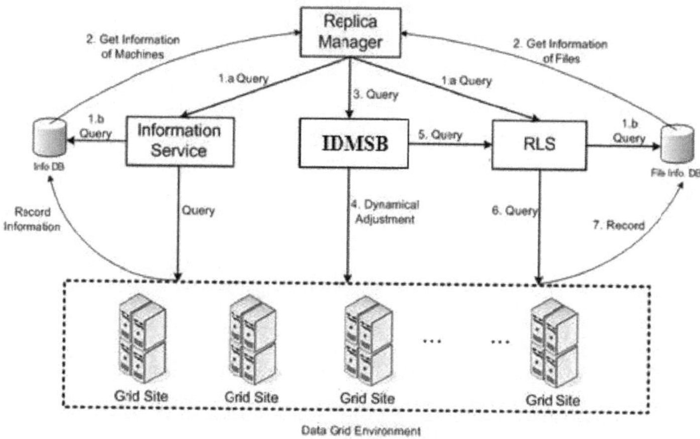

Fig. 3. The operation of IDMSBN

3.3 Parameters and Algorithm

There are many factors will affect replica transfer time and execution performance in grid environment. By the reason, lots of parameters will be figured out many static, dynamic other factors that will affect the overall performance. On the following are those factors that we can choose few for the model.

Static Parameters: These factors are not changeable whatever the grid environment is changed or not. As Xuanhua Shi et al. mentioned in [12], these factors are system attributes of each site, such as the type and frequency of CPU, the storage capacity of hard disk of each storage elements, the capacity of memory, and the transferring rate of network card.

Dynamic Parameters: These factors are changeable when the grid environment is changed. When a job is executed, it will consume the computing power, use part of memory space, download or upload data, and store the computational results into storage elements. The usage rate of CPU, memory space, bandwidth, and free space of each node is changed. The most important factor affects the performance is the network bandwidth.

Other Parameters: The parameters mentioned above can be measure for quantification easily. But the parameters have influence in the data replica operation, for example that sites are located in same region or the date of replica created. Maybe best parameters are finding to substitute for that past used.

Generally, we often are interested in knowing the probability about P(X|Y). Let X denotes hypothesis and Let Y denote evidence. Then, Bayesian method have be used to know P(Y|X) for getting P(X|Y), because it is easier to model hypothesis lead to the evidence.

$$P(X \mid Y) = \frac{P(Y \mid X) \cdot P(X)}{P(Y)} \tag{2}$$

$$P(X \mid Y) = \frac{P(Y \mid X) \cdot P(X)}{\sum P(Y \mid X') \cdot P(X')} \tag{3}$$

The terms in Bayesian rules are referred to as follows: P(X) is the prior probability of X, P(X|Y) is the posterior probability of X given E, P(Y|X') is the sample probability and, P(Y|X) is the likelihood of Y of given H.

The selected parameters will be used to form the factor of BN that be defined as fallows.

AF: File access frequency is more or less.
Region: The site are in the same location or not.
CP: Computing performance is high or low
Bandwidth: Network bandwidth of WAN is high or low
Replica: Probability the replication move to other sites or not.

This paper has the assumption that the many parameter is simplify with domain experience in Figure 4. At initial, give the some probability to the connection of parameter or do the some sample training to get it. The value of threshold by user can be used to decide that do replica or not.

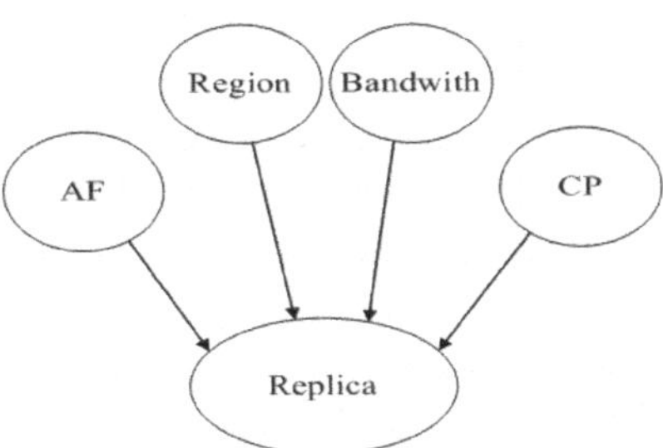

Fig. 4. The Bayesian network structure

The IDMSBN Algorithm as follow. First, we get the weight and probability of the using parameters according to domain experience or training sample. Then, it uses the probability and weight of parameters to inference the probability of replica happen when the file occur operation. When passed a period of time, Algorithm can catch the latest metadata (the probability of parameters and the CPT) and calculate replica probability value of file. The metadata in DB is changeable by passed time, so the threshold value of file is different when calculate probability by BN learning from history value. As Figure 5 shows, if the probability value excesses the threshold the user defined, IDMSBN can duplicate same file to other best site without replication. And every storage elements have kept the disk spaces maintain free space to save the more large files by deleting other files. When the site have no space, files will be deleted replication on the site hasn't used for a long time to liberate the hard disk. No matter how file be deleted, you can find a replication exists on the one of sites at the least.

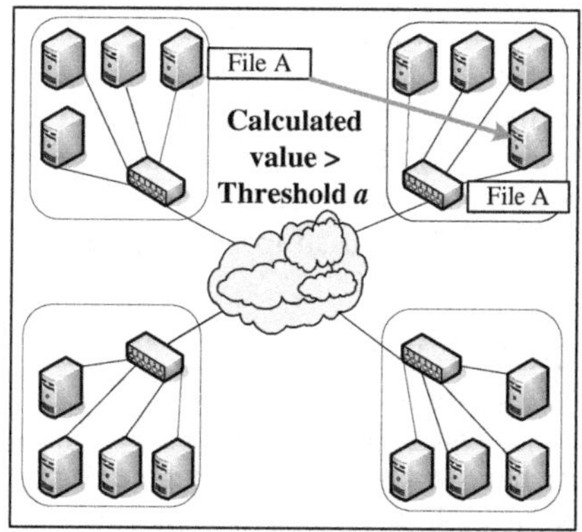

Fig. 5. Example of replication

4 Experimental

4.1 Experimental Environment

Because of the possible existence of cycle in a general undirected graph, a pre-scheduling stratend data while v_j is receiving the same message from another source node v_k, then v_i will discard the transmission $<v_i, v_j>$ and select next target to send data.

This section, we compare the performance of IDMSBN algorithm with other four replication strategies, Least Frequency Used (LFU), Least Recently Used (LRU), and simple Accessed Remotely (AR).Strategy AR mean site have no replicas, files are accessed remotely entirely. The LFU and LRU strategies are always replicates when the request is occurred, but if the free space of storage element is not enough, files be chosen to delete is different. LRU choose the oldest file to delete when there is no enough free space for storing the replicated data, and LFU choose files with least frequency of request to delete if there is no enough space.

Optorsim simulation is used to be measure the performance of replica strategies. In the BN's Algorithm use NETICA API write in java for BN. Figure 6 is the environment of experiment that contains 16 nodes. The computer element and storage element can be in same node.

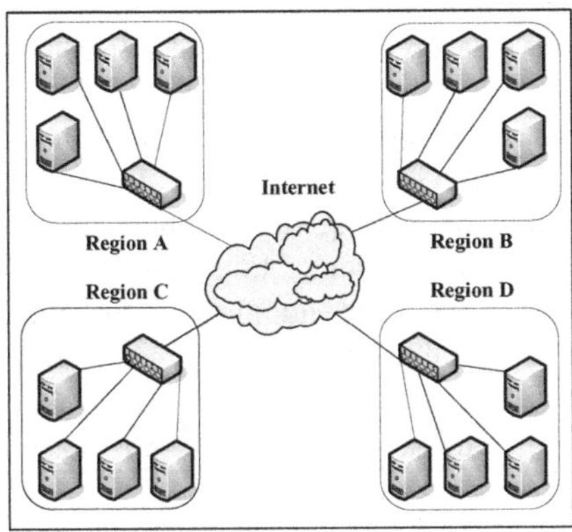

Fig. 6. The environment of experiment

In general, there are a lot of important parameters which can be used for Optorsim in Table 1 In the node that have storage element, the storage disk has be set 200GB capacity for initial status. The intranet bandwidth in the region is 100Mbps and internet bandwidth is 1000 Mbps. Other parameters relates to job scheduling contain number of jobs, number of job type, file number of each job, size of each file, and job delay. Number of job mean to the number of jobs submitted during the simulation run. There are 500 jobs having 30 different kind type of job which contains 15 size of file (250MB-4000MB). Job delay is time interval for submitting job in executing simulation.

Table 1. The parameters are used in the simulation

Parameters	Values
Node storage disk	200GB
Intranet bandwidth	1000Mbps
Internet bandwidth	100Mbps
Number of jobs	500
Number of job types	30
File number of each job	15
Size of each file	250-4000MB
Job delay	2500ms

4.2 Evaluation Metrics and Results

There are tree evaluation metrics using for the simulation, as shown below.

- Total job time
- ENU
- CE usage

The total job time is the simulation time taken for job to run, form beginning to completion. It is used to measure performance of different strategies in seconds. ENU (The effective network usage) mean that ratio between file requests which use network resources and total number of file requests. It is effectively the ratio of files transferred to files requested. Generally speaking, the lower ENU value indicates that the used strategy is better tot putting files in the right places. As follow equation, Nremote_file_accesses is the number of times the local site reads a file from other site, Nfile_replications is the total number of file replications that occur, and Nall_file_accesses is the total number of file that be requested in a local site.

$$ENU = \frac{N_{\text{remote\_file\_accesses}} + N_{\text{file\_replications}}}{N_{\text{all\_file\_accesses}}} \quad (4)$$

The CE usage means the percentage of time that Computing Elements have been running jobs or otherwise active (the average of the CE usage for each site).

The experimental results are shown in Figure 7, Figure 8, and Figure 9. The threshold value of IDMSBN is set up 0.3.

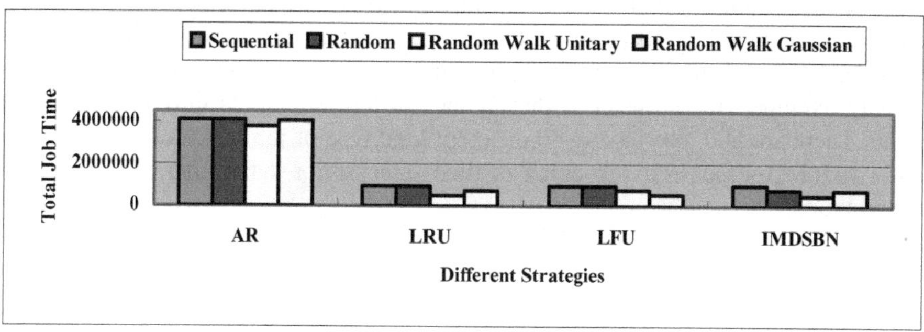

Fig. 7. The performance comparison of four strategies

The total job time of IMDSBN sometimes have better result in the different file access pattern. Because, some files are popular for the user on the other site. If there are more replicas can reduce the total job wait time. But if IDMSBF add the file replication you can find the ENU value increase a little with number of replication. The higher effective network usage is not good in the busy network environment, because network bandwidth be shared by other sites.

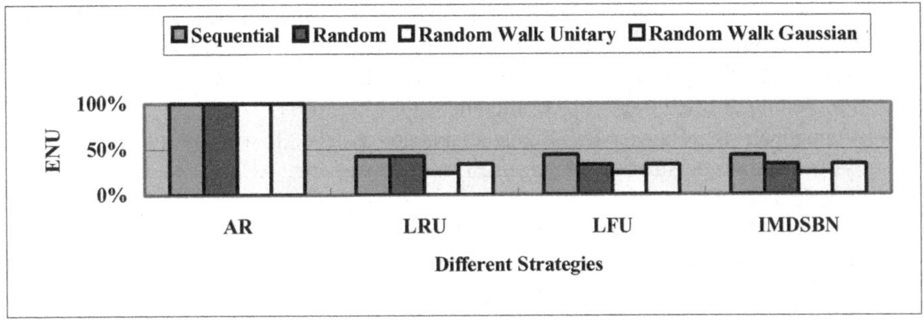

Fig. 8. The ENU comparison of four strategies

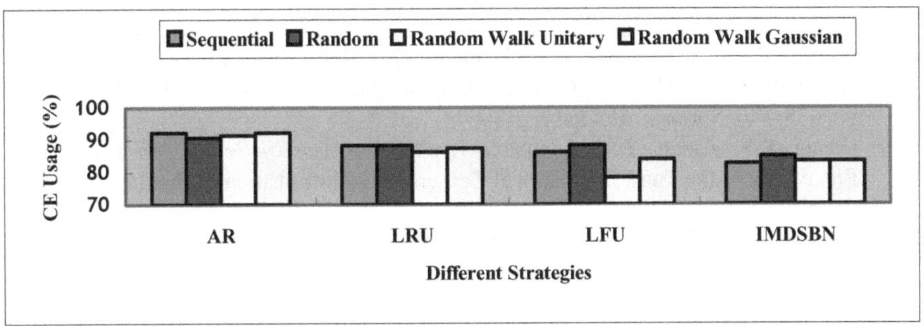

Fig. 9. The CE Usage comparison of four strategies

5 Conclusions and Future Work

The dynamical maintenance service of IDMSBN is needed to be designed to provide the services of efficient resource management in data grid. The problem of DMS metadata overload has been main challenges in data grid environment. In this paper, we have presented a dynamical maintenance service of IDMSBN for data grid. In general, IDMSBN provided a different view of replication maintenance parameters. Our future work will improve the parameters for the hierarchical structure and try to find the other approach parameters.

Acknowledgement

The work of this paper was supported by National Science Council of Taiwan under Grant number NSC-98-2220-E-029-004.

References

1. Chervenak, A., Foster, I., Kesselman, C., Salisbury, C., Tuecke, S.: The Data Grid: Towards an Architecture for the Distributed Management and Analysis of Large Scientific Datasets. Journal of Network and Computer Applications, 187–200 (2001)

2. Stockinger, H., Samar, A., Allcock, B., Foster, I., Holtman, K., Tierney, B.: File and Object Replication in Data Grids. Journal of Cluster Computing, 305–314 (2002)
3. Yang, C.T., Fu, C.P., Huang, C.J.: A Dynamic File Replication Strategy in Data Grids. In: Proceeding of IEEE Region 10 Conference, pp. 1–5 (2007)
4. Venugopal, S., Buyya, R., Ramamohanarao, K.: A Taxonomy of Data Grids for Distributed Data Sharing, Management and Processing. ACM Computing Surveys 38, Article 3 (2006)
5. Rahman, R.M., Barker, K., Alhajj, R.: Replica Placement Design with Static Optimality and Dynamic Maintainability. In: Proceedings of the Sixth IEEE International Symposium on Cluster Computing and the Grid, pp. 434–437 (2006)
6. Rahman, R.M., Barker, K., Alhajj, R.: Effective Dynamic Replica Maintenance Algorithm for the Grid Environment. In: Chung, Y.-C., Moreira, J.E. (eds.) GPC 2006. LNCS, vol. 3947, pp. 336–345. Springer, Heidelberg (2006)
7. OptorSim - A Replica Optimizer Simulation, `http://edg-wp2.web.cern.ch/edg-wp2/optimization/optorsim.html`
8. Park, S.M., Kim, J.H., Ko, Y.B.: Dynamic Grid Replication Strategy Based on Internet Hierarchy. In: Li, M., Sun, X.-H., Deng, Q.-n., Ni, J. (eds.) GCC 2003. LNCS, vol. 3033, pp. 838–846. Springer, Heidelberg (2004)
9. Chang, R.S., Chang, J.S.: Adaptable Replica Consistency Service for Data Grids. In: Proceeding of the Third International Conference of Information Technology, pp. 646–651 (2006)
10. Rahman, R.M., Barker, K., Alhajj, R.: A Predictive Technique for Replica Selection in Grid Environment. In: The Seventh IEEE International Symposium on Cluster Computing and the Grid, pp. 163–170 (2007)
11. Wagner, M.M., Moore, A.W., Aryel, R.M.: Handbook of Biosurveillance. In: Bayesian Methods for Diagnosing outbreaks, ch. 18, pp. 273–287 (2006)
12. Shi, X.H., Jin, H., Qiang, W.Z., Zou, D.Q.: An Adaptive Meta-scheduler for Data-Intensive Applications. In: Li, M., Sun, X.-H., Deng, Q.-n., Ni, J. (eds.) GCC 2003. LNCS, vol. 3033, pp. 830–837. Springer, Heidelberg (2004)

High Performance Computation for Large Eddy Simulation

Alexander Starchenko and Evgeniy Danilkin

Tomsk State University, 36 Lenin Prospekt, 634050, Tomsk, Russia
{starch,ugin}@math.tsu.ru

Abstract. Parallel implementation of the algorithm of numerical solution of the Navier-Stokes equations for large eddy simulation (LES) of turbulence is presented in this research. The Smagorinsky model has been applied for sub-grid simulation of turbulence. This model with numerical Van Leer's scheme for advection terms is in good agreement with experimental data. Various ways of geometrical decomposition for the parallel numerical solution of the transport equations have been investigated. A theoretical analysis of the parallel algorithms effectiveness was performed and the recommendations for their application were developed. The parallel realization of the iteration methods for the numerical solution of the Poisson equation for velocity and pressure coupling are discussed. Special techniques such as the order of nested loops and the effect of cash memory have been applied to increase the performance of the developed parallel programs.

Keywords: domain decomposition, parallel computation, turbulence modelling, LES approach.

1 Introduction

It is assumed that the turbulent motion of incompressible flow is described by the Navier-Stokes equations, and convective heat exchange – by the energy equation:

$$\frac{\partial u_i}{\partial t} + u_j \frac{\partial u_i}{\partial x_j} = -\frac{1}{\rho}\frac{\partial p}{\partial x_i} + \vartheta \nabla^2 u_i, i = 1,2,3, \tag{1}$$

$$\frac{\partial u_i}{\partial x_i} = 0 \tag{2}$$

$$\frac{\partial T}{\partial t} + u_j \frac{\partial T}{\partial x_j} = \alpha \nabla^2 T \tag{3}$$

where $u_i = u_i(\vec{x}, t)$ are the components of the instant velocity field, T is the temperature, ϑ is the coefficient of the kinematic viscosity, α is the thermal diffusivity coefficient, ρ is the density, p is the instant pressure value, $\nabla^2 \equiv \partial^2/(\partial x_j \partial x_j)$ is the Laplace operator. The summation is performed over the repeated indices.

C.H. Hsu and V. Malyshkin (Eds.): MTPP 2010, LNCS 6083, pp. 163–172, 2010.

At present there are three main and frequently applied techniques to simulate turbulent flows. The first approach is the direct numerical simulation (DNS). By the direct numerical simulation we mean the numerical solution of the system of differential equations (1)–(3) approximated by high-order finite differences. The second approach is the large eddy simulation (LES) which applies a larger grid as compared with the previous method. It is based on the concept of "filtering" the turbulence, i.e. averaging the Navier-Stokes equations and the energy over space with some weighing factor – a "filter". As it is impossible to simulate the smallest scales of turbulence by means of a large grid, the subgrid-scale models are used to take account of small-scale dissipation of the turbulence energy and the backscattering of the flow energy from the small to large scales. Along with this, the large-scale turbulent motions are considered explicitly and do not require simulation. The third approach is based on the Reynolds-averaged Navier-Stokes (RANS) equations. Here the turbulent flow is considered as a summation of the two components: the averaged and fluctuative parts.

2 Problem Definition

In this paper the simulation of the turbulent flow and the heat exchange is realized by means of large eddy simulation. The large eddies are computed explicitly by means of numerical simulation and the small-scale turbulence is parametrized, i.e., it is defined by the characteristics of the large-scale motions. The way to solve these problems is opened by the Kolmogorov universal equilibrium theory which asserts: if there is no possibility to resolve all the scales of the turbulent motion by means of the numerical simulation, one should simulate small-scale eddies as coarsely isotropic structures as compared with anisotropic large-scale eddies [1].

Most often to separate large- and small-scale motion structures one uses the filtering operation according to which the variables of the cell scale are determined by the equation

$$\bar{f}(x,t) = \int_{R^3} G(x-y) f(y) dy, \quad \int_{R^3} G(x) dx = 1, \tag{4}$$

Here G is the filter function with a characteristic length scale Δ. By filtering the Navier-Stokes (1)–(2) and energy (3) equations one gets

$$\frac{\partial \bar{u}_i}{\partial x_i} = 0, \tag{5}$$

$$\frac{\partial \bar{u}_i}{\partial t} + \frac{\partial \bar{u}_i \bar{u}_j}{\partial x_j} + \frac{\partial \tau_{ij}}{\partial x_j} = -\frac{1}{\rho} \frac{\partial \bar{p}}{\partial x_i} + \vartheta \frac{\partial^2 \bar{u}_i}{\partial x_j \partial x_j}, i = 1,2,3, \tag{6}$$

$$\tau_{ij} = \overline{u_i u_j} - \bar{u}_i \bar{u}_j, \tag{7}$$

$$\frac{\partial \bar{T}}{\partial t} + \frac{\partial \bar{u}_j \bar{T}}{\partial x_j} + \frac{\partial q_j}{\partial x_j} = \alpha \frac{\partial^2 \bar{T}}{\partial x_j \partial x_j}, \tag{8}$$

$$q_j = \overline{Tu_j} - \bar{T}\bar{u}_j, \tag{9}$$

Equation system (5)–(9) remains not closed because (7) and (9) contain the non-filtered components of velocity and temperature. τ_{ij} and q_j characterise the influence of small-scale eddies on the evolution of the large-scale eddies, and they should be simulated by establishing their correlations with the average velocities $\bar{u}_i$ and the temperature $\bar{T}$. It is the latter that is the essence of the subgrid-scale simulation procedure.

2.1 Choice of Subgrid-Scale Model

There exist a great number of approaches to the subgrid-scale simulation, but the eddy viscosity models (EVM) for explicitly resolved turbulent motions are in wide use to date. In these models the stress tensor is calculated according to the formula: $\tau_{ij} = -2K_T \bar{S}_{ij}$, where $\bar{S}_{ij} = \frac{1}{2}\left(\frac{\partial \bar{u}_i}{\partial x_j} + \frac{\partial \bar{u}_j}{\partial x_i}\right)$, and heat flux with $q_j = -\frac{K_T}{Pr_T}\frac{\partial \bar{T}}{\partial x_j}$.

Here $\bar{S}_{ij}$ is the strain-rate tensor constructed according to the filtered velocity field $\bar{u}_i$, $K_T = K_T(\bar{u}, \vec{x}, t) \geq 0$ is the turbulent eddy viscosity depending on the solution, Pr_T is the turbulent Prandtl number. The choice of the relationship $K_T = K_T(\bar{u}, \vec{x}, t)$ is very diverse. Smagorinsky [2] proposed to approximate the effects of subgrid eddies on the evolution of the large-scale eddies by using the expression $\tau_{ij}^{Smog} = -2C_S^2 \Delta_g^2 |\bar{S}| \bar{S}_{ij}$, where $\Delta_g = h$ is the filter width of the model, $|\bar{S}| = \sqrt{\bar{S}_{ij}\bar{S}_{ij}}$ is the norm of strain-rate tensor, C_S is Smagorinsky's constant.

3 Approximation and Solution Method

The approximation of the differential problem is carried out on the basis of the finite-volume method. The main idea of this method consists in separation of the computation domain into non-overlapping finite volumes bordering on one another so that one nodal point of a domain is contained only in its own finite volume. The values of the velocity components are determined on the faces of the finite volumes, and the scalar values – in the centre. By separating the domain in this way, one integrates every equation with respect to every finite volume. While computing the integrals one is to apply a piecewise-polynomial interpolation for the dependent values. The approximation of the advective terms of the equation is performed with the use of the upwind scheme MLU Van Leer [3]. To solve the equation of motion the explicit scheme is applied, therefore, the result of approximate integration of the motion equation with respect to one finite volume reduces to the formula:

$$\begin{aligned} ap^0 \Phi_{i,j,k}^{n+1} = ap_{i,j,k}\Phi_{i,j,k}^n + ae_{i,j,k}\Phi_{i+1,j,k}^n + an_{i,j,k}\Phi_{i,j+1,k}^n + at_{i,j,k}\Phi_{i,j,k+1}^n \\ + aw_{i,j,k}\Phi_{i-1,j,k}^n + as_{i,j,k}\Phi_{i,j-1,k}^n + ab_{i,j,k}\Phi_{i,j,k-1}^n + b_{i,j,k} \end{aligned} \tag{10}$$

Here $\Phi_{i,j,k}^{n}$ is the generalized scalar value which may represent any component of the velocity or temperature. The obtained finite-difference scheme has the first-order accurate approximation in time and the second-order accurate in space and it is conditionally stable and monotone.

The adjustment of the velocity and pressure fields in hydrodynamic part of the model is carried out by means of the correction of the pressure field.

4 Parallel Realization

A geometric decomposition of the grid domain is chosen as a principal means of algorithm parallelization. In the case considered here the three different ways of decomposition of the grid function values with respect to the compute nodes are possible – the one-dimensional scheme, the two-dimensional or block decomposition and three-dimensional decomposition of the computational grid nodes (Fig. 1).

Every processor element together with a separate grid sub-domain acquires all the values of the grid function $\Phi_{i,j,k}^{n}$ belonging to this sub-domain [4]. After the decomposition stage when the distribution of data into blocks in order to construct a parallel algorithm has been realized, one should pass to the stage of establishing communications between the blocks, where the computations will be fulfilled in parallel, i.e. to planning the communications. Due to the stencil of the explicit difference scheme used in order to compute the next approximation in the neighbouring nodes every sub-domain should possess the values of the grid function from the neighbouring boundary processor element. For this purpose halo cells are used to store the data from the neighbouring computational node, and the transfer of these boundary values necessary to afford the computation uniformity is organized.

A comparison was made of different ways of parallelizing the above mentioned explicit algorithm to solve the unsteady Navier-Stokes equations in order to determine the optimum (from the view point of minimization) communication costs by using an advective-diffusive equation as an example.

We shall turn to the tentative theoretical analysis of the efficiency of different ways of decomposing the computational domain for the case under consideration. We shall evaluate the work time of the parallel program as the work time of the sequential program T_{calc} divided by the number of the processors applied plus the time spent on communication: $T_p = T_{calc}/p + T_{com}$. The communication time for different ways of decomposition may be approximately expressed by means of the amount of the data being transferred at each step [5]:

$$T_{com}^{1D} = t_{send} \cdot 2N^2 \times 2, \qquad T_{com}^{2D} = t_{send} \cdot 2N^2 \times 4/p^{1/2}, \\ T_{com}^{3D} = t_{send} \cdot 2N^2 \times 6/p^{2/3}, \tag{11}$$

where N^3 is the size of finite-difference grid, p is the number of the compute nodes, t_{send} is the communication time of one number.

From formulas (11) one can get that at $p > 5$ the 2D and 3D decompositions will be more effective, and at $p > 11$ in the case of the 3D decomposition a smaller amount of the grid values of the function will be required, and one can expect that in this case the communication time will be minimum.

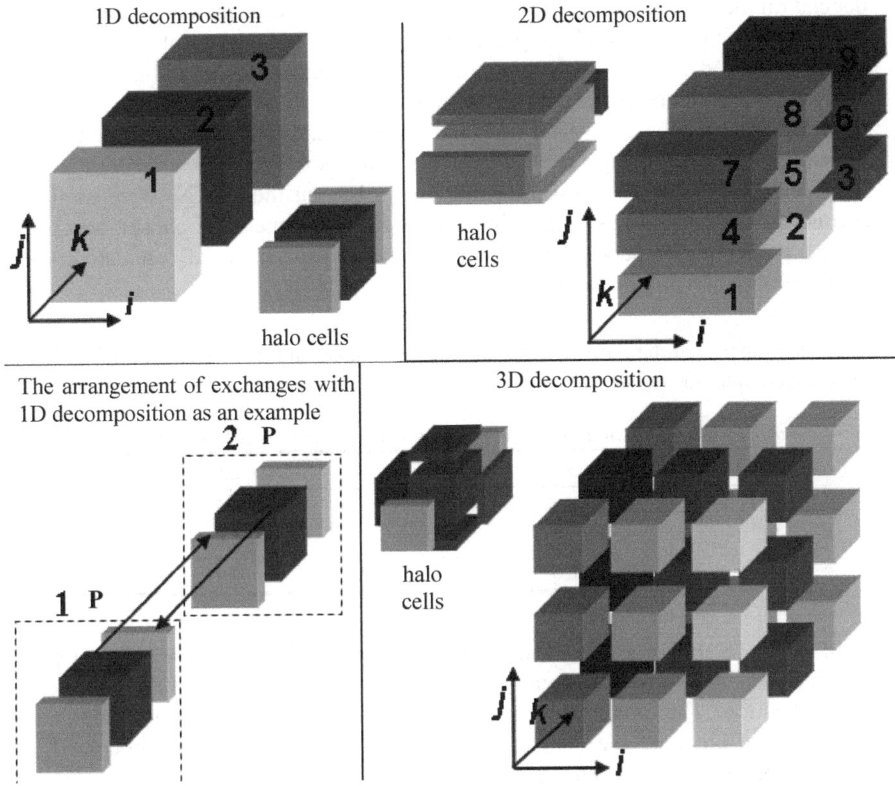

Fig. 1. Different ways of decomposition. The scheme to provide an exchange for 1D decomposition.

The computations were performed on the cluster system TSU SKIF Cyberia applying the grid $360 \times 360 \times 360$, and up to 144 processors were used. The results of the computational experiment demonstrated high speed-up in solving the problems of this class. The main attention was paid to comparison of the communication time and the computation time with different methods of decomposition.

At the first stage one general program was applied, the array sizes were not changed from run to run, at each processor unit the numeration of the array elements started from zero. Despite the fact that according to the theoretical analysis the 3D decomposition is an optimal variant of paralleling, the computation experiments showed that one can reach the best results by applying the 2D decomposition by using from 25 to 144 processors.

In order to explain the obtained results one should pay attention to the assumptions which were made in the preliminary theoretical analysis of the formulated problem. Firstly, it was supposed that in spite of the way of distributing the data one processor unit performs the same amount of the computation work which is to result in the identical time expense. Secondly, it was assumed that the time spent on the interprocessor communication of the same amount of data in any sequence does not depend on the way of their retrieval from the random-access memory (RAM).

To understand what is taking place in reality the following set of test computations has been carried out. To estimate the reliability of the first assumption a problem was analyzed when the program was run by applying one processor, and at the same time different ways of the geometric decomposition of data were imitated with the same amount of computations performed by every processor. The two versions of programs were realized: in the first one the static arrays having the size N^3 were used, in the second – "dynamic" ones, when the array sizes were specially chosen according to the decomposition type in order to minimize the time expense connected with the retrieval of data from the RAM.

Table 1. Influence of the grid domain decomposition mode and the size of the applied arrays on the computation time in one compute node

Static arrays 360x360x360				«Dynamic» arrays with regard to decomposition			
p	1d	2d	3d	p	1d	2d	3d
8	9,3	8,1	10,4	8	8,8	7,5	7,3
24	3,1	2,6	3,7	24	3,0	2,5	2,5
64	1,2	1,0	1,6	64	1,2	0,9	1,0

In Table 1 the time is given in seconds required for a processor to perform computations in its grid sub-domain while simulating different ways of the geometric decomposition. The amount of computation work is preserved from computation to computation, but only the character of arrangement of data being processed in the random-access memory is changed.

Having analyzed the obtained results one can come to the conclusion: while developing a parallel computation program it is advantageous to use the dynamic arrays which are adjusted to the chosen number of processors, it is mostly essential for the 3D decomposition.

In order to find the optimized (from the viewpoint of efficiency of parallelizing and the convergence rate) method of solution of the system of linear algebraic equations for finite-difference Poisson equation for pressure the following methods were compared: Seidel's method with the red-black ordering, Buleev's explicit method, the conjugate gradient method and the biconjugate gradient stabilized method [6].

The results of comparison showed that for the problems of large size only Seidel's method with the red-black ordering completely retains, in parallelizing, the sequential algorithm properties, which is expressed in conservation of the number of iterations which are required for the convergence of the method irrespective of the number of processors used and the decomposition method. It is important to point out that such realization of the algorithm is very well scaled to any reasonable number of computational nodes.

The parallel implementation of the methods of conjugate and biconjugate gradients theoretically does not also destroy the properties of the sequential algorithm. But in practice one may find out that the number of iterations is altered at random with different decomposition methods. The numerical experiments showed that this fact is a consequence of non-commutativity of calculation of the sum of a great number of terms in case of machine arithmetic. There arises an insignificant error in the seventh or eighth sign which later leads to the change in the course of the iteration process. It is necessary to point out that the iteration process finally converges, only the sequence of approximations to solution is insignificantly changed.

The parallel algorithm of Buleev's method for the specified problem operates correctly in case if the domain decomposition along the Ox axis is not used and thus it becomes admissible to apply only 1D or 2D decompositions.

Table 2. The run time with different number of processors for the methods under consideration

Method \ Number of processors	1	9	25	64	144	225
Seidel's method	24495	7129	2821	763	313	279
Buleev's method	9823	1929	816	610	256	184
BiCGStab method preconditioned by Buleev's method	476	205	82	41	32	29
BiCGStab method preconditioned by Seidel's method	469	134	47	20	5	4

On the basis of the results obtained one can come to the conclusion that the best way to solve the chosen problem by a multiprocessor computer with the distributed memory is to use the biconjugate gradient stabilized method preconditioned by Seidel's method.

5 The Results of Testing the Numerical Method and the LES Model

We shall consider the turbulent flow over a square cylinder which is located in a channel (Fig. 2). The choice of the flow domain and the value of Reynolds number $Re = U_{in}D/\vartheta$ corresponds to the experiment [7]. Although the geometry of this test case is very simple, an unsteady wake is formed behind the cylinder. The following grid $Nx_1 \times Nx_2 \times Nx_3 = 132 \times 122 \times 22$ was used in the computation.

The experiment of flowing over a square cylinder is well described, there are measurement data presented in the form of the tables. This benchmark test is often used to evaluate different approaches to the turbulence simulation. The oncoming laminar flow impinges on the cylinder front and separates in the form of the leading eddies.

Figure 3 presents an instant picture of the velocity and temperature distributions in the wake of a heated cylinder. It illustrates the way how heat is distributed at the expense of the turbulent motion. One can see in Figure 3 how a large-sized eddy is separated from the surface of the cylinder.

The computations showed a high level of agreement with the experimental data [7]. The profile of the velocity component u_1 averaged over space and time is correlated well with the measurement data. Figure 4 presents the computation results of time-averaged variations of the longitudinal velocity of the flow along the flow-oriented axis passing through the middle of the transverse plane of the computation domain. The Smagorinsky model was applied to simulate the subgrid-scales. In Figure 4 the comparison results of applying different subgrid models are also given in order to describe small-scale turbulences borrowed from [8].

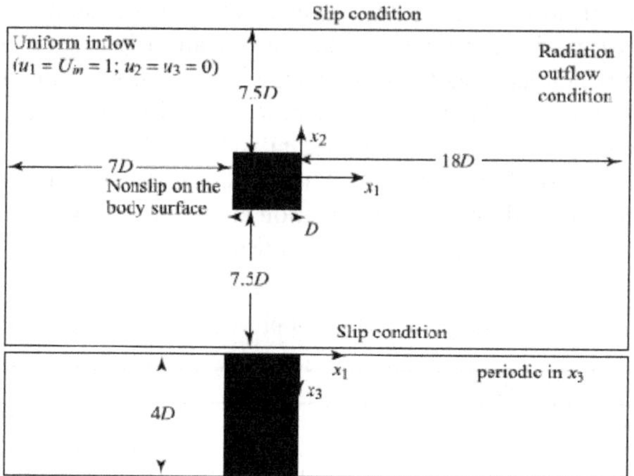

Fig. 2. Computational domain and boundary conditions

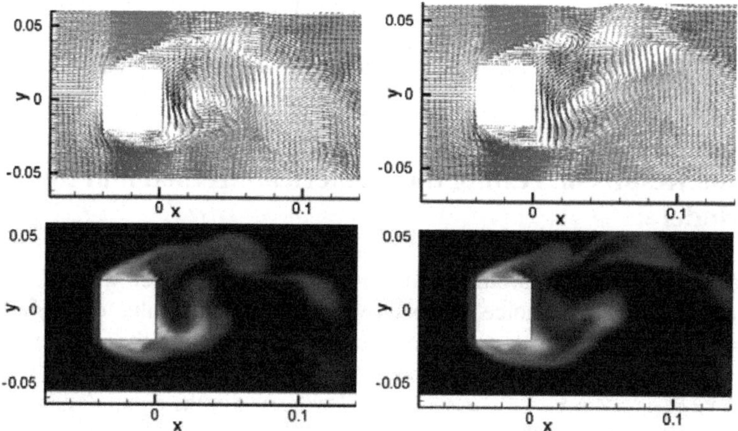

Fig. 3. The flow field and the temperature at the instants t1 < t2, where t2 = t1 +0,2T, T is the period of separating an eddy

The comparison of the results (Fig. 4) shows that Van Leer‘s scheme makes its contribution into the subgrid simulation and allows to obtain a better agreement of the numerical experiment results with the measurements from [7] for the average characteristics of the flow.

Figures 5 and 6 present the time-averaged profiles of the normal turbulent stresses downstream behind the cylinder. One can see from the Figures that rather a good correspondence has been obtained with the data of the experiment [7] for the fluctuations of the velocity components u_1 and u_2. No serious overestimates in $\overline{u_1^2}/U_{in}^2$ intensity and underestimates in $\overline{u_2^2}/U_{in}^2$ intensity along the Ox_1 axis are observed, (which is characteristic of the computation in [8] according to

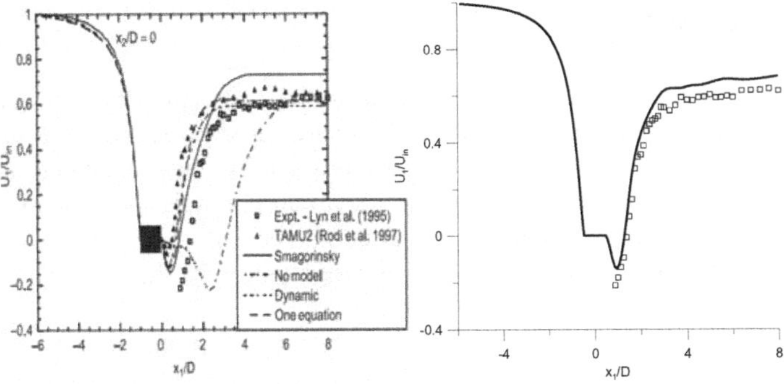

Fig. 4. The profile comparison of the average velocity u_1 while applying different subgrid models [8] (left) and the model used in this study (right)

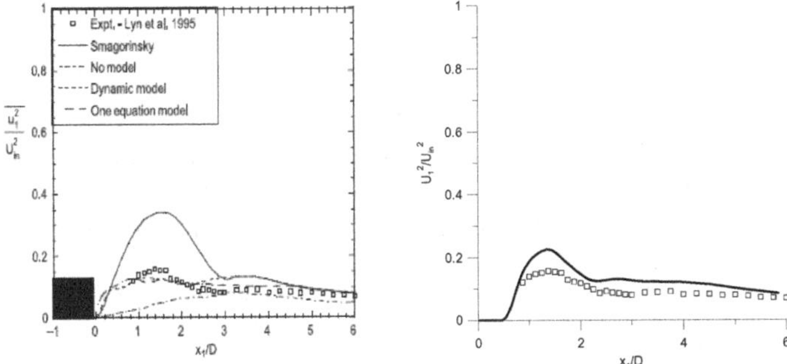

Fig. 5. The turbulent velocity fluctuations u_1 (left), the result obtained in this study (right)

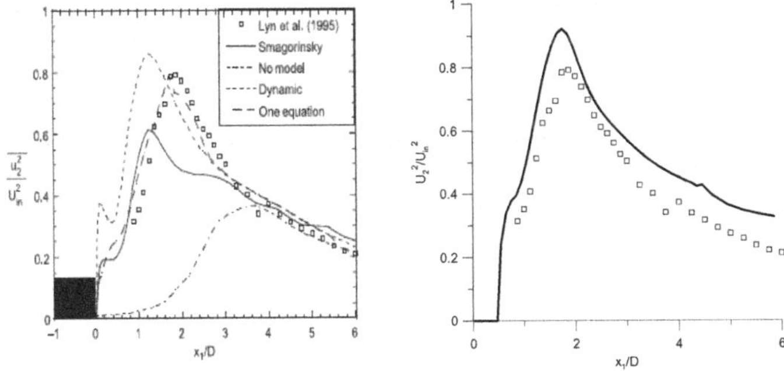

Fig. 6. The turbulent velocity fluctuations u_2 (left), the result obtained in this study (right)

Smagorinsky's model). The turbulent stresses at the cylinder walls are approximately 5 times as high as near the smooth walls of the channel, which points to the cylinder effects on the turbulence generation.

6 Conclusions

The results of the numerical experiments showed that the developed mathematical turbulence model is able to reproduce the characteristic features of a turbulent flow, the unsteady character of the flow under study and the heat exchange in the channel with the obstacles placed inside, it also allows to simulate the process of separating eddies. The use of Smagorinsky's model combined with Van Leer's scheme made it possible to get a good agreement of the average values of the flow with experimental data for the domain under study.

The application of the 2D decomposition in the computations is 76 % effective with the use of 25 computer nodes. With the further increase in the computation nodes number up to 100 with the chosen grid sizes a characteristic value of efficiency about 50 % was obtained for the problems of this class.

The work was performed under the financial support of the "SKIF GRID" program.

References

1. Kolmogorov, A.N.: Local structure of turbulence in the incompressible viscous liquids at very high Reynolds numbers. DAN SSSR 30, 299–303 (1941)
2. Smagorinsky, J.: General circulation experiments with the primitive equations. I: The basic experiment. Month. Weath. Rev. 91(3), 99–165 (1963)
3. Van Leer, B.: Towards the ultimate conservative difference scheme. II. Monotonicity and conservation combined in a second order scheme. J. Comp. Phys. 14, 361–370 (1974)
4. Danilkin, E.A.: On the question of 3D decomposition efficiency for the numerical solution of transport equation with use of multiprocessor computer with distributed memory. Vestnik TGU 3(2), 39–46 (2008)
5. Danilkin, E.A., Starchenko, A.V.: Parallel numerical method of solution of Navier-Stokes equations for large eddy simulation of turbulent flows. Vestnik NGY 7(2), 49–61 (2009)
6. Van der Vorst, H.A.: Bi-CGSTAB: A fast and smoothly converging variant of Bi-CG for the solution of nonsymmetric linear systems. SIAM Journal 13(2), 631–644 (1992)
7. Lyn, D., Einav, S., Rodi, W., Park, J.H.: A laser-Doppler velocimetry study of ensemble averaged characteristics of the turbulent near wake of a square cylinder. J. Fluid Mech. 304, 285–319 (1995)
8. Nakayama, A., Vengadesan, S.N.: On the influence of numerical schemes and subgrid-stress models on large eddy simulation of turbulent flow past a square cylinder. Int. J. Numer. Methods Fluids 38, 227–253 (2002)

ARPP: Ant Colony Algorithm Based Resource Performance Prediction Model

Ce Yu[1], Kelang Xiong[1], Jizhou Sun[1], Yanyan Huang[2], and Jian Xiao[1]

[1] School of Computer Science & Technology, Tianjin University, Tianjin 300072, China
yuce@tju.edu.cn, xiongkelang@hotmail.com

[2] Network Center, Hebei University of Technology, Tianjin, 300130, China

Abstract. Resource performance prediction is the basis of dynamic load balance in distributed computing. A model for resource performance prediction named ARPP is introduced and carried out. ARPP model monitors key parameters of resources and estimates the directions using ant algorithm. The implement and analysis of ARPP is based on GridSim simulator and the process of astronomical image mosaicking application. The experiment result shows the efficiency of the model and the determination of optimized parameters.

Keywords: Ant algorithm, Performance prediction, Dynamic load balance, Distributed computing.

1 Introduction

With the development of distributed computing patterns and technologies such as Grid computing and Cloud computing, dynamic load balance scheduling methods for applications running on multiple computing resources are getting more important, which will improve the process performance and save electricity power. Dynamic load balance require that the tasks be scheduled according to the runtime status of resources, and only with prediction of resource performance, the scheduler may get the most efficient solutions.

Types of traditional performance prediction models have been brought forward. Some of them require the system monitoring the resources to gather history information to do prediction, such as autoregressive model, autoregressive moving average model, autoregressive difference moving average model, regression analysis model and neural network model. While some other model are too simple to meet dynamical characters, such as moving average model, arithmetic moving average model, self-balancing model and trend prediction model.

An ant colony algorithm based resource performance prediction model for dynamic load balance is introduced in this paper, which is named ARPP model. It originated in ant colony algorithm based grid task scheduling method and makes improvements and supplements for performance prediction. ARPP model combined the resource performance prediction and task scheduling and can deal with long term prediction with low system overhead.

C.H. Hsu and V. Malyshkin (Eds.): MTPP 2010, LNCS 6083, pp. 173–181, 2010.

2 Related Works

Ant System is an evolution simulation algorithm which was introduced by M.Dorigo, V.Maniezzo and Colorni[1][2], originally for traveling problems. The algorithm is from the food searching process of ant colony. The behavior of a single ant is simple, but when many ants are grouped, the behavior become more complex. Besides, ant colony can adapt to change of environment rapidly. For example, if a barrier is placed on the road of the ant colony, a new route will soon be established. It is found by biologists and bionics researchers that the ants use pheromone to communicate with each other. Each ant will leave amount of pheromone when it is moving along, while it can apperceive the same pheromone left by other ants and move to the direction which there are more pheromone. The more ants passed by one route, the more ants will select the route.

The application discussed in this paper is based on Montage[6], which is a toolkit for assembling astronomical images into mosaics. It uses algorithms that preserve the calibration and positional (astrometric) fidelity of the input images to deliver mosaics that meet user-specified parameters of projection, coordinates, and spatial scale. It contains independent modules for analyzing the geometry of images on the sky, and for creating and managing mosaics; these modules are powerful tools in their own right and have applicability outside mosaic production, in areas such as data validation.

3 Resource Performance Prediction Model

A resource description model based on ant colony algorithm was established for Grid environment[3][4]. When a new resource is added, the scheduler will initialize its pheromone using the basic parameter(1):

$$\tau_j(0) = m \times p + c/s_j \tag{1}$$

Here m is the processor number of resource j; p is speed of each processor (MIPS); c is the package size of sample data; s_j is the transmission time of sample data. The initial pheromone expresses the inherent capacity for computing and communication.

For each subtask, the scheduler creates a new ant with computing and communication loads. The object of optimization is the minimal execution cost. Initially, the ants are distributed to the resources, and each ant computes the probability of its next subtask allocated to usable resources using current pheromones of resources(2):

$$p_j^k(t) = \begin{cases} \frac{[\tau_j(t)]^\alpha \cdot [\eta_j]^\beta}{\sum_u [\tau_u(t)]^\alpha \cdot [\eta_u]^\beta} & j, u \in UsableResources \\ 0 & others \end{cases} \tag{2}$$

$\tau_j(t)$ is the pheromone density of resource j at time t; η_j is the inherent parameter of the resource and $\eta_j = \tau_j(0)$; α is weight of pheromone and β is the weight of inherent parameters. The scheduler computes the object function for the allocation result of every ant and chose current optimized solution.

When an ant allocates a subtask to some resource, the pheromone of corresponding resource will change (so does the probability that the resource will be selected)(3):

$$\tau_j^{new} = \rho \cdot \tau_j^{old} + \Delta\tau_j \tag{3}$$

$\Delta\tau_j = -K$ and K is the computing and communication loads of the subtask; ρ is the permanence of pheromone.

When a subtask is finished successfully on resource j, the pheromone of j will be increased by $\Delta\tau_j$ where $\Delta\tau_j = Ce \times K$, Ce is encouragement coefficient. But when a subtask is failed on resource j, the pheromone of j will be decreased by $\Delta\tau_j$ where $\Delta\tau_j = Cp \times K$, Cp is amercement coefficient[3][4].

3.1 Prediction Model and Definitions

Three speeds are defined to be predicated: communication speed (between the node which submits the task and each resource in the computing system), computing speed, and data transmission speed (between each two resources).

Let $\tau_{comm}i$ express the communication pheromone of resource i, $\tau_{comp}i$ express the computing pheromone of i, $\tau_{tran}ij$ express the data transmission pheromone between i and j. When a new resource is added, its initial pheromone will be computed(4):

$$\begin{cases} \tau_{comm}i = 1/T_{comm}i \\ \tau_{comp}i = 1/T_{comp}i \\ \tau_{tran}ij = 1/T_{tran}ij \end{cases} \tag{4}$$

$T_{comm}i$ is the time sampled by submitting standard image file to resource i; $T_{comp}i$ is sampled by running standard image composing process on resource i; $T_{tran}ij$ is sampled by transmitting standard image file from resource i to resource j.

When the task is submitted, the scheduler will allocate the subtasks according to the speed of each resource. The speed is predicated by the following formula(5):

$$v = \sqrt[1+\alpha]{\tau \times v_{curr}^{\alpha}} \tag{5}$$

The speed v includes communication speed, computing speed and data transmission speed. τ is the pheromone density (including communication pheromone, computing pheromone and data transmission pheromone). v_{curr} is currently sampled speed of the resource, as mentioned in (4). α is the weight of current sampled speed to the pheromone.

When the subtask is allocated to some resources, the scheduler will reduce its pheromone and update the system globally, to reduce the subtasks allocated to the resources which already have heavy load. Pheromone updating formulas are(6):

$$\begin{cases} \tau_{comm}^{new} = C_d \times \tau_{comm}^{old} \times \frac{L/n}{L_\alpha} \\ \tau_{comp}^{new} = C_d \times \tau_{comp}^{old} \times \frac{L/N_{PE}}{L/n_{PE}} \times \frac{L_\alpha}{C_a} \\ \tau_{tran}^{new} = C_d \times \tau_{tran}^{old} \times \frac{L_\alpha}{C_a} \end{cases} \tag{6}$$

τ^{new}_{comm}, τ^{new}_{comp} and τ^{new}_{tran} are updated pheromones. C_d is a constant and is the decline coefficient of pheromone. τ^{old}_{comm}, τ^{old}_{comp} and τ^{old}_{tran} are original pheromones.

L is the total number of subtasks, n is the resource number occupied by the task, and L_α is the number of subtasks allocated of the resource. N_{PE} is global PE (processor element) number, and n_{PE} is the PE number of current resource. C_a expresses normal number of subtasks on one resource and it is a constant.

When one communication subtask, or computing subtask, or data transmission subtask is finished on some resource, the corresponding pheromone will be increased to allow more subtasks to be allocated to it(7):

$$\begin{cases} \tau^{new}_{comm} = \frac{1}{C_d} \times \tau^{old}_{comm} \times \frac{L_\alpha}{L/n} \\ \tau^{new}_{comp} = \frac{1}{C_d} \times \tau^{old}_{comp} \times \frac{L_\alpha/n_{PE}}{L/N_{PE}} \times \frac{C_a}{L_\alpha} \\ \tau^{new}_{tran} = \frac{1}{C_d} \times \tau^{old}_{tran} \times \frac{C_a}{L_\alpha} \end{cases} \tag{7}$$

After all the subtasks are finished, the scheduler updates the pheromone of each resource according to the performance during the execution(8):

$$\begin{cases} \tau^{new} = \rho \times \tau^{old} + (1-\rho) \times \Delta\tau \\ \Delta\tau_{comm} = C_e \times \frac{L_\alpha}{T_{comm}} \times \frac{\overline{T_{comm}}}{T_{comm}} \times \frac{L_\alpha}{L/n} \\ \Delta\tau_{comp} = C_e \times \frac{L_\alpha}{T_{comp}} \times \frac{\overline{T_{comp}}}{T_{comp}} \times \frac{L_\alpha/n_{PE}}{L/N_{PE}} \\ \Delta\tau_{tran} = C_e \times \frac{L_\alpha}{T_{tran}} \times \frac{\overline{T_{tran}}}{T_{tran}} \times \frac{L_\alpha}{L/n} \end{cases} \tag{8}$$

τ^{new} is the new pheromone, and τ^{old} is the original pheromone. ρ is the permanence of pheromone. $\Delta\tau$ is the change of pheromone. C_e is the encouragement coefficient. T_{comm}, T_{comp}, T_{tran} are the times of communication, computing and data transmission, respectively. And $\overline{T_{comm}}$, $\overline{T_{comp}}$, $\overline{T_{tran}}$ are the average times of communication, computing and data transmission in the system.

L_α/T_{xxxx} indicates the actual speed of the resource. $\overline{T_{xxxx}}/T_{xxxx}$ means that the less time consumed by the resource, the higher performance it has, and more subtasks should be scheduled to this resource. $L_\alpha/n/L$ indicates that if more subtasks were scheduled to some resource, then the resource can finish more subtasks in fixed time and thus has higher performance, and the pheromone should be increased to attract more subtasks. So does $(L_\alpha/n_{PE})/(L/N_{PE})$.

3.2 Prediction Steps

Under the performance predication model defined above, the practical prediction steps are listed as the following:

Step 1: Sample the communication and computing speed of each resource and the data transmission speed between each two resources, initialize corresponding pheromones(4).

Step 2: The scheduler gets the lists of resource and pheromones according to the requirement of submitted task, samples current communication and computing speed of each resource and the data transmission speed between each two resources, and predicts parameters of the resources using resource performance prediction formula (5).

Step 3: The scheduler compute the subtask load for each resource using subtask allocation algorithm according to the predicted parameters of the resources. Update the pheromones of resources involved using resource allocation pheromone updating formula (6).

Step 4: Call the astronomic image process workflow to execute the subtask submitted. Update the communication pheromone using (7) when the input on the resource is completed. Update the computing pheromone using (7) when the computing on the resource is completed. Update the data transmission pheromone using (7) when the data transmission on the resource is completed.

Step 5: After the task submitted is finished, update the pheromones of all resources involved using task return formula (8).

4 Experiment and Results

To evaluate the effect of the performance predication model and to determine the key parameters of ant colony algorithm, a series of simulation experiments are carried out.

The computing model of astronomical image mosaicking process is shown in Fig. 1.

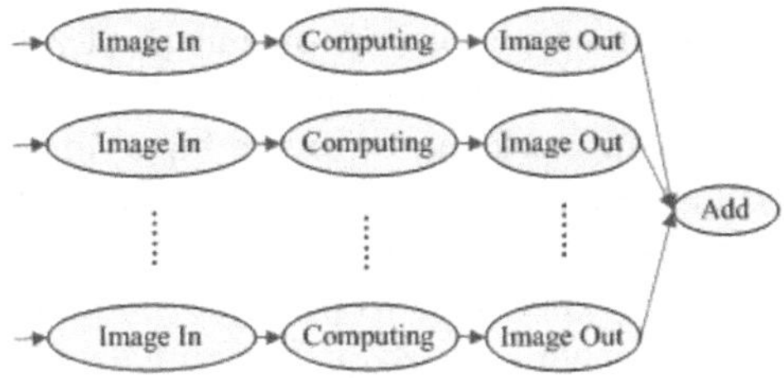

Fig. 1. Astronomical image mosaicking process

Three computing processes are included in the computing phase, namely image reprojection, background modeling and rectification, and co-additions and weighting of output pixel fluxes.

Image reprojection involves the redistribution of information from a set of input pixels to a set of output pixels. For astronomical data, the input pixels represent the total energy received from an area on the sky, and it is critical to preserve this information when redistributed into output pixels.

If several images are to be combined into a mosaic, they must all be projected onto a common coordinate system (see above) and then any discrepancies in brightness or background must be removed. The assumption is that the input images are all calibrated to an absolute energy scale (i.e. brightnesses are absolute

Table 1. Resource Configuration

Resource	Processor Number	Computing ability (MIPS)	Network (Mbps)	Disk capacity (GB)
R0	8	377	50	70
R1	4	515	30	55
R2	2	410	40	45
R3	2	410	100	50
R4	4	380	100	60
R5	8	380	80	75
R6	1	410	50	35
R7	1	410	30	32
R8	4	410	30	50
R9	2	380	50	40

and should not be modified) and that any discrepancies between the images are due to variations in their background levels that are terrestrial or instrumental in origin.

In the reprojection algorithm (described in the pixel overlap discussion above), each input pixel's energy contribution to an output pixel is added to that pixel, weighted by the sky area of the overlap. In addition, a cumulative sum of these sky area contributions is kept for the output pixels (essentially and physically an "area" image). When combining multiple overlapping images, these area images provide a natural weighting function; the output pixel value is simply the area-weighted average of the images being combined.

The nodes which submit tasks and provide input images are out of process system and do no computing. Each computing node executes all the processes of input image files, computing and output result file. Only one result file is needed to be transmitted to the gather node to perform the final co-addition.

The simulation environment is based on GridSim[5] and extended its functions to support paralleled file transmission.

4.1 Simulation Configuration

In the simulation, there are 10 users who will submit tasks to 10 resources, and the parameters of each resource are showed in table 1. The 10 users are connected to the Cloud by Router1. R0 to R4 use Router2, and R5 to R9 use Router3. The three routes are connected by the network with 10Mbps bottleneck.

The speed of resource is computed follow a linear function of subtasks number on it. The communication is calculated from $L_{\alpha}i/v_{comm}i$, where $L_{\alpha}i$ is the number of subtasks on resource i, and $v_{comm}i$ is its communication speed. Computing time is calculated from $L_{\alpha}i/v_{comp}i$, where $v_{comp}i$ is the computing speed of resource i. The output time is calculated from $C/v_{tran}ij$, where $C = 40$ (the size of result file to be transmitted is 40 times of input file size) and $v_{tran}ij$ is the data transmission speed between resource i and j. The time of final adding

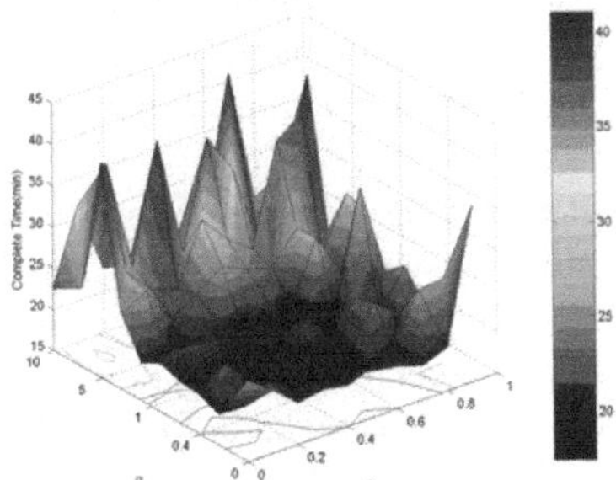

Fig. 2. Light load, Task execution time (ρ,α)

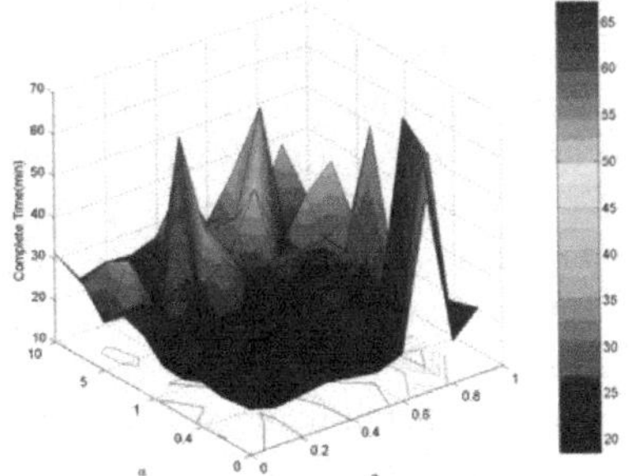

Fig. 3. Medium load, Task execution time (ρ,α)

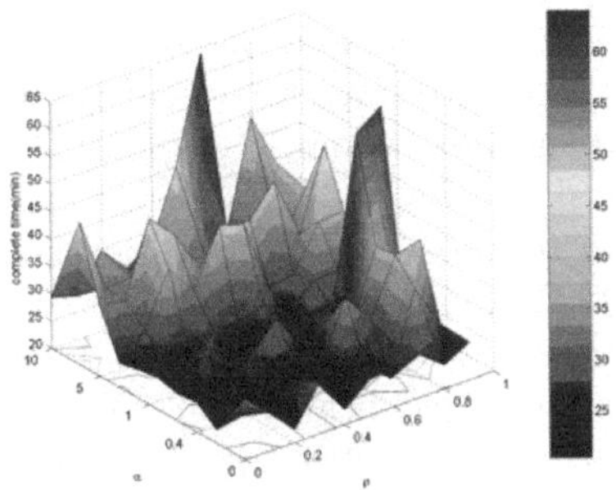

Fig. 4. Heavy load, Task execution time (ρ,α)

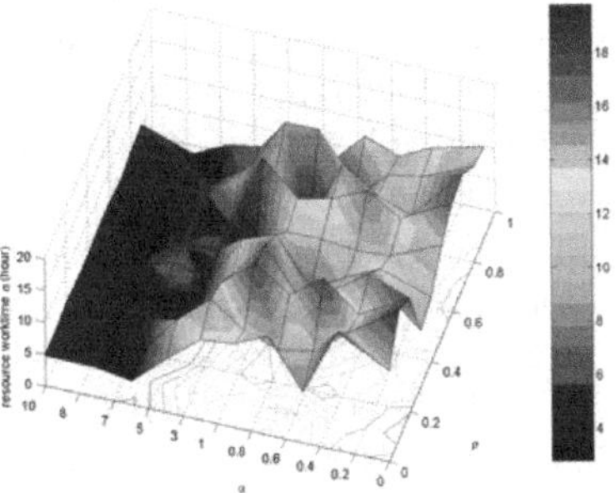

Fig. 5. Light load, Standard deviation of resource working time (ρ,α)

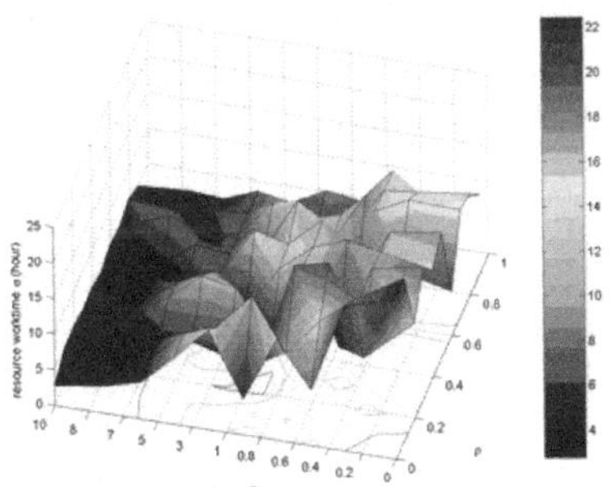

Fig. 6. Medium load, Standard deviation of resource working time (ρ,α)

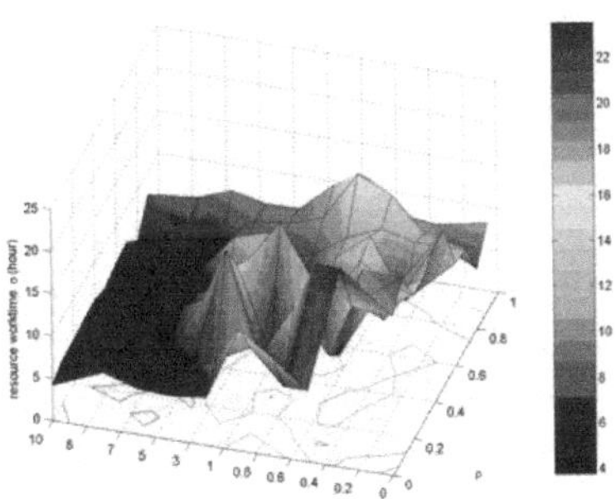

Fig. 7. Heavy load, Standard deviation of resource working time (ρ,α)

operation is $n/v_{comp}i/C2$, where resource number $n = 10$ and $C_2 = 2$ (the computing speed of final adding is twice of $v_{comp}i$).

The bandwidth of 10 users is set to: 100Mbps, 50Mbps, 40Mbps, 30Mbps, 20Mbps, 10Mbps, 10Mbps, 10Mbps, 10Mbps, 10Mbps.

The size of input image file in the simulation is 2MB. Instruction number needed by computing each image file is 8000M. The output file of each resource is 80MB, and the mosaicking of final output file needs 4000M instructions. The storage needed is 42 times of input image file size.

During the simulation, each user first submits 40 random tasks. The file number of each task is following the regular distributing within [20, 400], and the interval of tasks is following the regular distributing within [10, 200] minutes. As for the last 5 random tasks, the file number of each task is following the regular distributing within [280, 320], and the interval of tasks is following the regular distributing within [80, 120] minutes. All 10 users start the submission simultaneously.

4.2 Results Analysis

In following figures, ρ is the permanence of pheromone, and the value series is {0, 0.1, 0.2, 0.3, 0.4, 0.5, 0.6, 0.7, 0.8, 0.9}; α is the weightiness of sampling speed for pheromone when predicating, and the value series is {0, 0.2, 0.4, 0.6, 0.8, 1.0, 3.0, 5.0, 7.0, 10.0}.

Task execution times measured under different load and with different combination of ρ and α are shown in Fig. 2, 3, 4. Task execution time is the average running time of the last 50 tasks submitted by the 10 users. System load is classified by the usage of CPU (light: 10%; medium: 30%; heavy: 60%). When the value of ρ is between 0.1 to 0.8 and the value of α is between 0 to 5, the task execution times are shorter and relatively steady.

Standard deviations of resource working times measured under different load and with different combination of ρ and α are shown in Fig. 5, 6, 7. The average value of standard deviations of resource working times under different are basically equal and are not related to CPU load. When α is greater than 1, standard deviations of resource working times are relatively steady. The effect of value of ρ behaves as wave curve and the preferable values are 0.2 and 0.6.

To summary, task execution time and load are linear relationship, and the degree of load balance has no relationship with load. These results indicate that ARPP is steady and efficiency in the experiment. For optimization of both task execution time and degree of load balance, let $\rho \in [0, 0.8]$ and $\alpha \in [0, 0.8]$; for optimization of task execution time, let $\rho \in [0.5, 0.7]$ and $\alpha = 0.2$; for optimization of degree of load balance, let $\rho = 0.9$ and $\alpha = 0.2$ for light load, let $\rho = 0$ and $\alpha = 10$ for medium load; let $\rho = 0.3$ and $\alpha = 5$ for heavy load.

5 Conclusion and Future Works

ARPP model combines performance prediction and task scheduling bases on ant colony algorithm. Its' steady and efficiency are shown by experiments under various configurations using astronomical image mosaicking application.

Distributed computing, especially Cloud computing, will play more important role in future IT infrastructure. Dynamical load balance is essential for high performance and high efficiency of such environment, which desiderates exact and reliable performance prediction of resources. In future work, ARPP will be improved for service-oriented applications in Clouds or similar environment. The method for determination of optimized parameters and reusable software architecture with support to ARPP will also be investigated.

Acknowledgement. This paper was funded by *National Natural Science Foundation of China* (10978016), *Natural Science Foundation of Tianjin* (08JCZDJC19700), *Key Technologies R & D Program of Tianjin* (09ZCKFGX 00400).

References

1. Dorigo, M.: Optimiztion, Learning and Natural Algorithma (in Italian). Ph.D. thesis, Dipartimento di Elettronica, Politecnico di Milano, IT (1992)
2. Dorigo, M., Maniezzo, V., Colorni, A.: The ant system: Optimization by a colony of cooperating agents. IEEE Transactions on Systems, Man, and Cybernetics Part B 26(1), 29–41 (1996)
3. Xu, Z.-H., Sun, J.-Z.: Ant algorithm based grid computing and task scheduling. Journal of Tianjin University Science and Technology 37(5), 414–418 (2004)
4. Xu, Z., Hou, X., Sun, J.: Ant algorithm based task scheduling in grid computing. CCECE, 1107–1110 (2003)
5. Buyya, R., Murshed, M.: GridSim: a toolkit for the modeling and simulation of distributed resource management and scheduling for Grid computing. Concurrency and Computation: Practice and Experirence 14(3), 13–15 (2002)
6. Montage, `http://montage.ipac.caltech.edu`

Using Molecular Dynamics Simulation and Parallel Computing Technique of the Deposition of Diamond-Like Carbon Thin Films

Jiun-Yu Wu[1], Hui-Ching Wang[2], Jung-Sheng Chen[2], Kuen-Tsan chen[2], and kuen Ting[3]

[1] National Center High-Performace Computing, Taiwan
[2] National Chung Hsing University, Taiwan
[3] Lunghwa University of Science and Technology, Taiwan
adherelinux@hotmail.com,
{hcwang,ktchen}@amath.nchu.edu.tw,
longsnl@yahoo.com.tw,
kuenting@mail.lhu.edu.tw

Abstract. A technique of parallel computing in simulating the deposition of diamond-like carbon thin film by molecular dynamics is proposed. The Tersoff potential which is a multi-body potential is adopted here in determining inter-atomic forces. The deposition of carbon thin film on diamond substrates and silicon substrates under different incident kinetic energies and different substrate temperatures are investigated. The multiprocessor of workstation computer containing 8 cores used for simulating the deposition is based on MPICH2 which is an implementation of message passing interface. The results show that the percentages of deposited sp3 carbon atoms differ from 6.1% to 34.8% depending on the type of substrate, incident kinetic energy and substrate temperature.

Keywords: Molecular Dynamics,Verlet List, Parallel Computing, Leap-Frog Method, Periodic Boundary Condition, Tersoff Potential, MPICH2.

1 Introduction

The semiconductor manufacturing process that produces many electronic products can be more miniaturized, function better and fast, for example: computers, mobile phones, IC and so on. Nanotechnology is based on material in the nano-size under physical and chemical properties of the phenomenon which rearrange atoms or molecules of structure into new materials. In this paper, we adopt molecular dynamics simulation as the deposition of diamond-like carbon thin films. Thin film deposition is usually used for surface treatment method which is applied to tools, dies, semiconductor components, and the surface treatment. The surface of the substrate grows a layer of homogeneous or heterogeneous material during manufacturing process. It got wear-resistant,heat-resistant and heat-resistant [1]. In 1987, Tersoff used an empirical formula to simulate chemical bonds of energy, which interacted forces between atoms [2]. Next year, Tersoff

C.H. Hsu and V. Malyshkin (Eds.): MTPP 2010, LNCS 6083, pp. 182–190, 2010.

[3] proposed for an empirical interatomic potential for multicomponent models. It used geometric mean and arithmetic mean of the concept in the model which is applied to C-Si and Si-Ge models. In 1990, Brenner [4] proposed Brenner potential function to simulate chemical vapor deposition of diamond thin films. In 1991, KauKonen and Nieminen illustrated with dropping atoms of energy influenced deposition of diamond-like carbon thin films. In 2004, Seung-Hyeob Lee Et-Al used Tersoff potential function to deposit thin films and proposed drop atoms of energy were influenced for deposition density. The energy of the drop atoms was 50eV to 75eV which had more sp^3 ratio, and it observed the diamond of stable condition in the distance 2.1 Å. It discussed about the dropping atoms relationship between kinetic energy and residual stress. The molecular dynamics need heavy calculation and amount of time. Hayashia Et Al [7] used 128 multiprocessor cores with molecular dynamics simulation for the deposition thin films.

2 Simulation Method

2.1 Physical Model

The Figure 1 shows the simulation system model. All of the atomic arrangements are on the (111) crystal plane of face-centered-cubic (fcc), and x, y coordinates considered the periodic boundary condition.

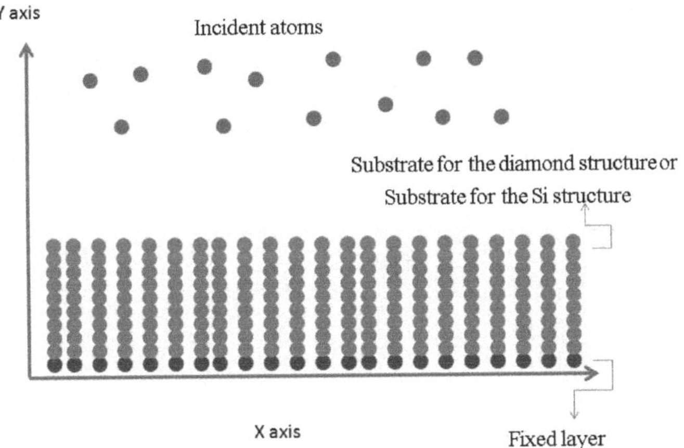

Fig. 1. The Simulation diagram

There are two kinds of atoms in this system: Si,C atoms. The atom positions and velocities were updated by velocity varlet integrator [8]. In the simulation process, the model of lattice is approximate dimensions: 14.268 Å× 14.268 Å× 17.835Å containing 1172 atoms. The atoms at the bottom of the substrate were fixed. The atoms dropped from the center which was under the substrate height of 35.67Å. The initial velocity of atoms are random .

2.2 Molecular Dynamics

Molecular Dynamics (MD) simulation is a good approach for studies of microscopic physical systems. An empirical potential function calculated interactive forces between atoms. The MD considers the simulation of dynamics based on Newton‘s second law:

$$\vec{a} = \frac{\vec{F}}{m}$$

$\vec{a}$:Atomic of acceleration
$\vec{F}$:Atomic of force
m:Atomic of mass

$$\vec{F} = -\vec{\nabla} E$$

E:Potential function
$\vec{\nabla}$:Gradient

The MD is usually adopted an empirical potential function to calculate interactive forces between atoms. In this paper, the Tersoff potential [2] is adopted to model the covalent bonds between atoms in Diamond-like Carbon Thin Films.

2.3 Tersoff Potential

The Tersoff potential is a three-body potential function which includes an angular contribution of the force. It widely used to show in silicon, carbon, germanium, applied to bond tetrahedrally in structure materials and so forth. The model is similar to the Taylor expansion of the energy as a function of atomic positions:

$$\sum_i V_1(r_i) + \sum_{i<j} V_2(r_i, r_j) + \sum_{i<j<k} V_2(r_i, r_j, r_k) + ...$$

r_n:n atomic of position
V_m:m body potential

Tersoff potential formular :

$$\frac{1}{2} \sum_{i \neq j} V_{ij}$$

$$V_{ij} = f_c(r_{ij})\{f_R(r_ij) + b_{ij} f_A(r_{ij})\}$$

$$f_R(r_{ij}) = A_{ij} e(-\lambda_{ij} r_{ij})$$

$$f_A(r_{ij}) = -B_{ij}e(-\mu_{ij}r_{ij})$$

$$f_c(r_{ij}) = \begin{cases} 1 & r_{ij} < R_{ij} \\ \frac{1}{2} + \frac{1}{2}\cos[\frac{\pi(r_{ij}-R_{ij})}{S_{ij}-R_{ij}}] & R_{ij} < r_{ij} < S_{ij} \\ 0 & r_{ij} > S_{ij} \end{cases} \Bigg\}$$

$$b_{ij}(r_{ij}) = (1 + \beta_i^n \zeta_{ij}^n)^{-\frac{1}{2n_i}}$$

$$\zeta_{ij} = \sum_{k \neq i,j} f_c(r_i k) g(\theta_{ijk})$$

$$g_(\theta_{ijk}) = 1 + \frac{c_i^2}{d_i^2} - \frac{c_i^2}{[d_i^2 + (h_i - \cos(\theta_{ijk}))^2]}$$

A, B, λ,μ,β, n, c, d, h, R, S are parameters of the Tersoff potential, r_{ij} is the distance between the atoms i and j , c is the bond-angle.C There are two different materials in this model. λ_{ij} and μ_{ij} are inequalities of arithmetic and A_{ij}, B_{ij}, S_{ij} are geometric means.

$$\mu_{ij} = \frac{\mu_i + \mu_j}{2}$$

$$\lambda_{ij} = \frac{\lambda_i + \lambda_j}{2}$$

$$A_{ij} = \frac{A_i + A_j}{2}$$

$$B_{ij} = \frac{B_i + B_j}{2}$$

$$R_{ij} = \frac{R_i + R_j}{2}$$

$$S_{ij} = \frac{S_i + S_j}{2}$$

Table 1. The parameters for the Tersoff potential, carbon and silicon

	C	Si
A(eV)	1393.6	1830.8
B(eV)	346.7	471.18
$\lambda(\AA^{-1})$	3.4879	2.4799
$\mu(\AA^{-1})$	2.2119	1.7322
β	1.5724×10^{-7}	1.1×10^{-6}
n	0.72751	0.78734
c	3.8049×10^{4}	1.0039×10^{5}
d	4.384	16.217
h	−0.57058	−5.9825
R($\AA$)	1.8	2.7
S($\AA$)	2.1	3.0

3 Parallel Computation

In this paper, the parallel computation in simulating the deposition of diamond-like carbon thin film by molecular dynamics is proposed. We use multiprocessor to calculate a large-scale MD simulation. Our simulation is implemented on a workstation intel(R) xeon(R) XPU E5410 @ 2.33GHz with 8 cores, and contains 6144 KB cache. The workstation supports MPI(Message-Passing Interface), C++ and FORTRAN90. The execution time is measured by the MPI's functions: MPI Wtime. The execution times show in Figure 2:

Total atoms =672			
total steps	8 cores time(s)	1 core time(s)	speed up
100	2.29	8	3.49345
1000	23	80.52	3.50087
10000	230.17	804	3.49307
100000	2307.25	8053	3.490302

Fig. 2. The execution time

4 Results and Discussion

There are three different parameters for temperature adjustment, energy and substrate materials.

4.1 Substrate for the Diamond Structure

The Substrate for the diamond structure of parameter shows in Table 2. The atomic of kinetic energy is 20 eV. The substrates of temperature are 300K, 573K, 673K, 773K, 873K. The result of deposition shows in figures 3(a,b,c,d,e). The figure 2(e) isn't effect. The deposition ratio is decrease in temperature 873K.

Table 2. The diamond substrate

$Timestep$	1030000
$Substrate\ \ for\ \ the\ \ temperature(K)$	$300, 573, 673, 773, 873$
$Electron\ \ Voltag(eV)$	20
$Total\ \ number\ \ of\ \ atoms$	1172
$The\ \ deposition\ \ of\ \ number$	500
$The\ \ deposition\ \ of\ \ rate(atom/ps)$	2000
$\Delta t(fs)$	0.1

4.2 Substrate for the Si Structure

The Substrate for the structure of paremeter is shown in Table 3. The kinetic energy of the atomic are 1eV, 10eV and 20eV,temperature is 573K and 873K. The result of deposition is shown in figure 4(a,d,c,d,e,f).The blue atoms are Si,and red atoms are carbon in this results.

Table 3. The deposition of ratio on diamond substrate

	$300k$	$573K$	$673K$	$773K$	$873K$
$20eV$	34.8%	34.8%	31.4%	34.8%	24.6%

Table 4. The Si substrate

$Timestep$	1030000
$Substrate\ \ for\ \ the\ \ temperature(K)$	$573, 873$
$Electron\ \ Voltag(eV)$	$1, 10, 20$
$Total\ \ number\ \ of\ \ atoms$	1672
$The\ \ deposition\ \ of\ \ number$	1000
$The\ \ deposition\ \ of\ \ rate(atom/ps)$	1000
$\Delta t(fs)$	0.1

Table 5. The deposition of ratio on si substrate

	$1eV$	$10eV$	$20eV$
$573K$	2.7%	18.9%	19.7%
$873K$	6.1%	18.9%	16.7%

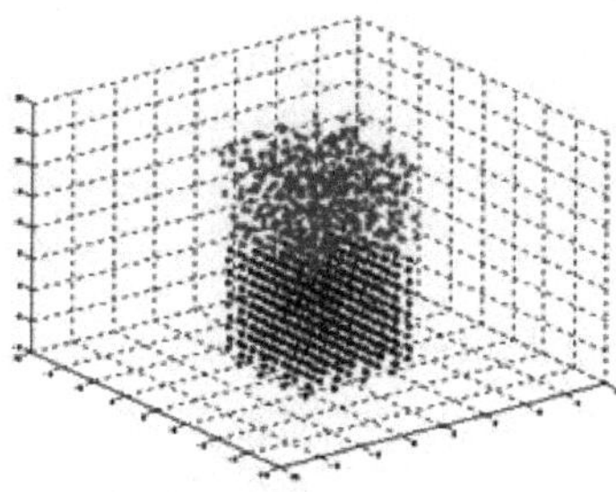

The deposition of proportion:34.8%

(a). The diamond substrate, the temperature 300K and the kenetic energy 20eV.

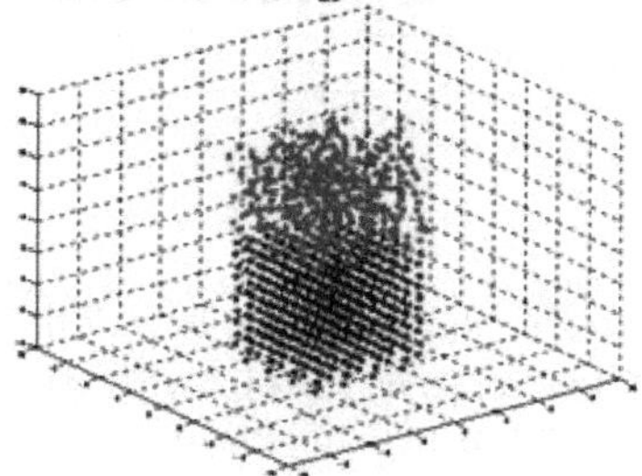

The deposition of proportion:34.8%

(b). The diamond substrate, the temperature 573K and the kenetic energy 20eV.

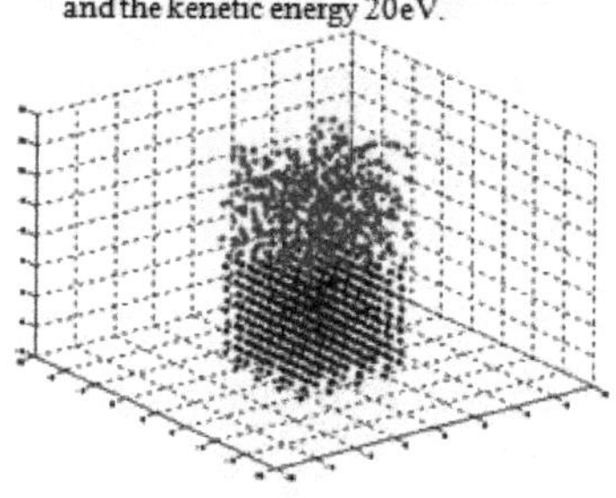

The deposition of proportion:31.4%

(c). The diamond substrate, the temperature 673K and the kenetic energy 20eV.

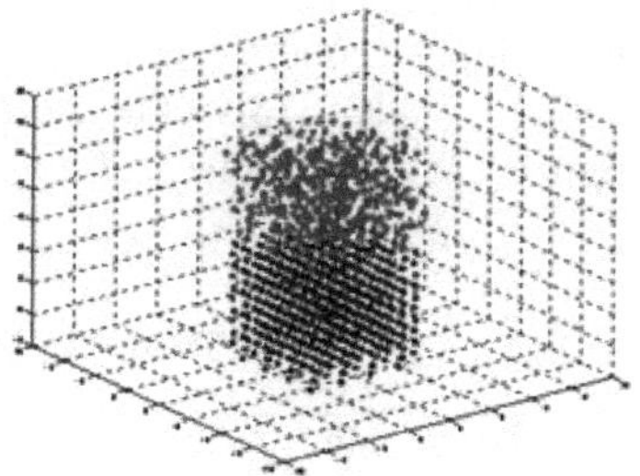

The deposition of proportion:34.8%

(d). The diamond substrate, the temperature 773K and the kenetic energy 20eV.

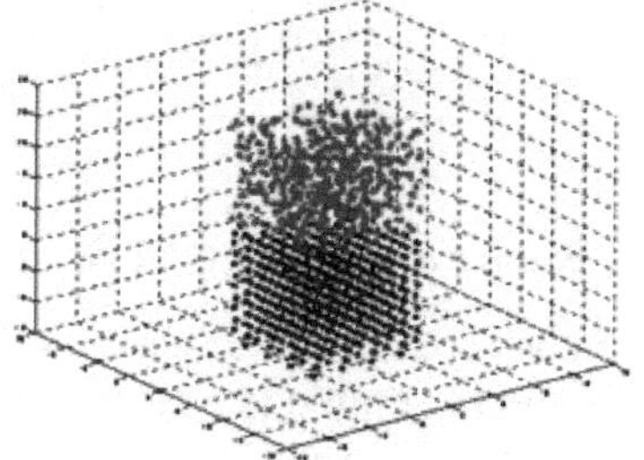

The deposition of proportion:24.6%

(e). The diamond substrate, the temperature 873K and the kenetic energy 20eV.

Fig. 3. The deposition diagram

Figure 4(a,b) shows that the kinetic energy for the 1eV, higher temperature is to increase ratio. Figure 4(c,d) kinetic energy for the 10eV, higher temperature isn't effect. Figure 4(e,f) shows kinetic energy 20eV. The temperature 873K is to decrease ratio.As the kinetic energy increases,the ratio increases in the simulation results.

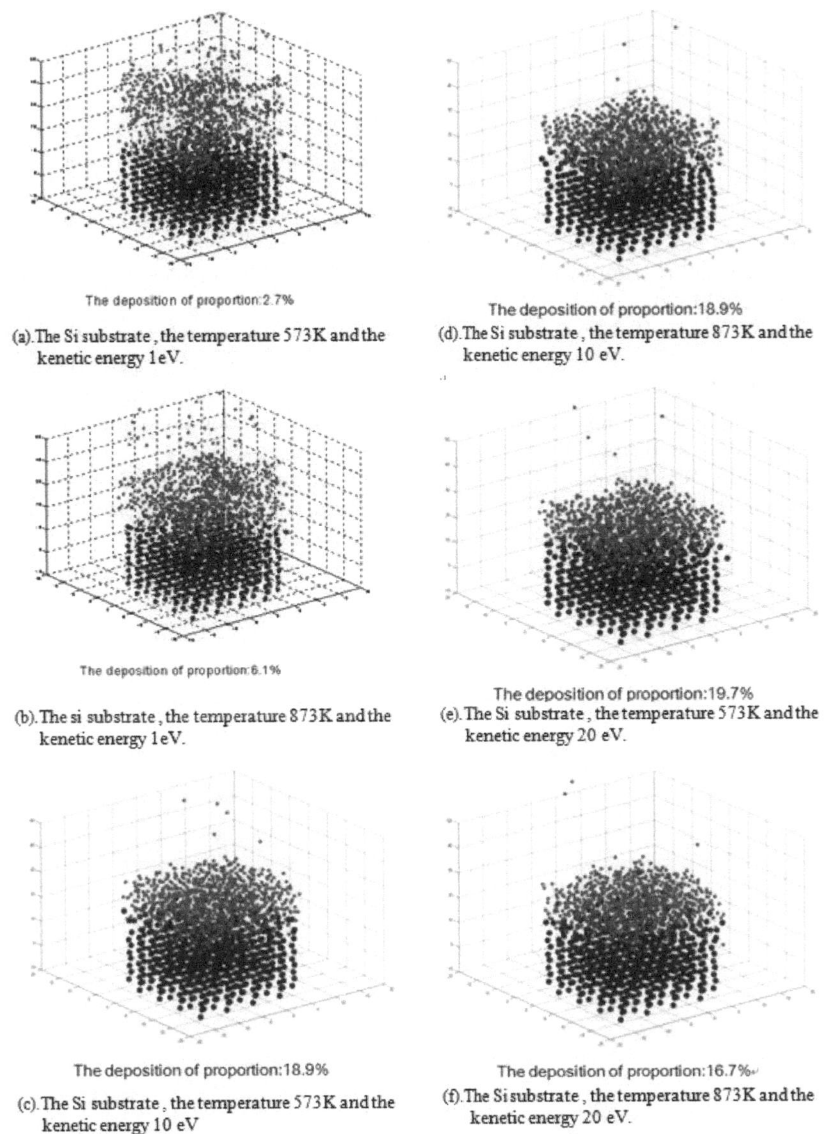

(a).The Si substrate, the temperature 573K and the kenetic energy 1eV.

(d).The Si substrate, the temperature 873K and the kenetic energy 10 eV.

(b).The si substrate, the temperature 873K and the kenetic energy 1eV.

(e).The Si substrate, the temperature 573K and the kenetic energy 20 eV.

(c).The Si substrate, the temperature 573K and the kenetic energy 10 eV

(f).The Si substrate, the temperature 873K and the kenetic energy 20 eV.

Fig. 4. The deposition diagram

5 Conclusion

In this paper, it adopt Tersoff potential function, applied parallel computing, and molecular dynamic simulation of the deposition as diamond-like carbon thin films. It studies under different temperatures and injected energy to influence deposition of thin films.The substrate of the diamond structure increases ratio when temperature is getting higher.

References

1. Juang, D.-R.: Process Technology, GAU Lih Book Co., Ltd. (in Taiwan)
2. Tersoff, J.: New empirical approach for the structure and energy of covalent systems. Phys. Rev. B 37(12) (1987)
3. Tersoff, J.: Modeling solid state chemistry Interatomic potentials for multicomponent systems. Phys. Rev. B 39(8) (1988)
4. Brenner, D.W.: Empirical potential use simulating the chemical vapor. Phys. Rev. B 42(15) (1990)
5. KauKonen, H.-P., Nieminen, R.M.: Molecular-dynamics simulation of the growth of diamondlike films by energetic carbon-atom beams. Phys. Rev. Letts. 68(5) (1991)
6. Lee, S.-H., Lee, C.-S., Lee, S.-C., Lee, K.-H., Lee, K.R.: Structural properties of amorphous carbon films by molecular dynamics. Sand Coatings Technology, 177–178, 812–817 (2004)
7. Hayashia, R., Tanakab, K., Horiguchia, S., Hiwatarib, Y.: A classical molecular dynamics simulation of the carbon cluster formation process on a parallel computer. Diamond and Related Materials 10, 1224–1227 (2001)
8. Rapaport, D.C.: The art of molecular dynamics simulation. Cambridge University Press, Cambridge (2004)

Statistical Modeling of Ocean Surface Noise

Alex Shvyrev and Igor Yaroshchuk

Pacific Oceanological Institute FEB RAS, Baltiyskaya 43,
690041 Vladivostok, Russia
shvyrev@poi.dvo.ru

Abstract. The paper describes investigations of some aspects of creating stochastic wave models of dynamic acoustic noise. The problem about sound radiation by homogeneous and stationary pressure fluctuations on a surface of layered waveguide is considered. The method of statistical modeling based on the randomization of spectral density of surface sources is used. It allows calculating random realizations of sound pressure and particle velocity that are exact solution of the equations of linear acoustics. The software for calculation of statistical characteristics of surface noise in a distributed computing environment is developed. Some examples of its applications are given.

Keywords: ocean acoustics, dynamic noise, statistical modeling, parallel computing.

1 Introduction

The problems of generation and transformation of surface dynamic noise have been studied in ocean acoustics for the past sixty years and still remained very relevant. Successful theoretical and experimental studies of physical fields are mostly attributed to the proper choice of the probabilistic model of signals and noise. One of the possible approaches to the development and study of the stochastic wave models of dynamic noise is the use of various simulation methods, in particular, statistical modeling. The method is based on the using to solve problems (including deterministic) sequences of random or pseudo-random numbers and constructing with their use of random processes with certain parameters equal to the required parameters of these problems.

Method of statistical modeling is ideal for parallel computing. Calculations of separate realizations (estimations) is fairly easily being distributed among the network nodes (processors, workstations), without intensive data exchange. In most cases it is possible almost linear increase in performance depending on the number of computing devices.

For the development of computational schemes for statistical modeling we used the space-time spectral representation of the density function of random surface noise sources. Randomization of the spectral representation [1] allowed to construct statistical schemes [2], which provides the weak convergence of the random wave fields.

C.H. Hsu and V. Malyshkin (Eds.): MTPP 2010, LNCS 6083, pp. 191–199, 2010.

Within the objectives of the study of the dynamic noise transformation deep into the ocean we have taken a number of numerical experiments for typical acoustic-oceanographic situations. All calculation results were obtained using the method of statistical modeling of surface noise [2], which allows calculating random realizations of sound pressure and particle velocity that are exact solution of the equations of linear acoustics. In the numerical experiments was used the software package, developed by the authors in the Scilab [3] environment, that implements the algorithms of statistical modeling in a distributed computing environment with the help of PVM [4].

The purpose of the numerical experiments was to study the influence of the basic acoustic-oceanographic factors on the formation of the noise field structure and acoustic energy transfer processes. Part of the results is given as an example in this paper. A more complete and detailed description is available in [5, 6].

2 Problem Statement

Let $q(t;\vec{r})$ (where $\vec{r}=\{x,y\}$ is the vector on a plane surface) is the homogeneous and stationary Gaussian field of acoustic pressure on ocean surface generated by external forces (wind, sea-way and their interaction). All statistical properties of q are described by the space-time spectral density $\Phi_q(\omega;\vec{\kappa})$:

$$\left\langle q(\omega;\vec{\kappa})q^*(\omega';\vec{\kappa}')\right\rangle = \Phi_q(\omega;\vec{\kappa})\delta(\omega-\omega')\delta(\vec{\kappa}-\vec{\kappa}'), \tag{1}$$

where $q(\omega;\vec{\kappa})$ is the spectral amplitude of the space-time spectrum:

$$\begin{aligned} q(t;\vec{r}) &= \iint q(\omega;\vec{\kappa})\exp\left[i(\vec{\kappa}\vec{r}-\omega t)\right]d\omega d^2\kappa, \\ q(\omega;\vec{\kappa}) &= \frac{1}{(2\pi)^3}\iint q(t;\vec{r})\exp\left[i(\omega t-\vec{\kappa}\vec{r})\right]dt d^2 r. \end{aligned} \tag{2}$$

It is convenient to represent function Φ_q, which we regard as given, examining the problem of noise field transformation, in the following form:

$$\Phi_q(\omega;\vec{\kappa}) = P_q(\omega)S_q(\vec{\kappa};\omega), \tag{3}$$

where P_q – time spectral density, $P_q(\omega)=\int\Phi_q(\omega;\vec{\kappa})d^2\kappa$, S_q – normalized space-time spectral density, $\int S_q(\vec{\kappa};\omega)d^2\kappa = 1$.

The ocean is supposed layered on z and homogeneous on $\vec{r}$. Particle velocity we normalize to the wave admittance of the medium $(\bar{\rho}c)^{-1}$ ($\bar{\rho}$ – water density, c – typical value of water sound speed), as in this case it has the same dimension, as pressure.

Pressure and particle velocity inside the medium can be found under the following formulas

$$\begin{aligned} p(\omega;\vec{r},z) &= \frac{1}{(2\pi)^2}\int q(\omega;\vec{r}')G(\omega;\vec{r}-\vec{r}',z)d^2r', \\ \vec{v}(\omega;\vec{r},z) &= \frac{1}{ik}\nabla p(\omega;\vec{r},z), \end{aligned} \tag{4}$$

where G is the Green function of the corresponding problem, $k=\omega/c$.

3 Statistical Modeling

To solve problems using the method of statistical modeling, we should be able to numerically obtain realizations of the random processes and fields. For this purpose we use the method of randomization of the spectral representation [1].

The wave-vector space $K \equiv R^2$ $(\vec{\kappa} \in K)$ is divided into M nonintersecting domains $K_1, \ldots, K_M$. Then an arbitrary random realization of the surface-source field can be written in the form

$$\begin{aligned} q(\omega;\vec{r}) &= P_q^{1/2}(\omega)\sum_{m=1}^{M} w_m^{1/2}(\omega) a_m \exp\left[i\left(\vec{\kappa}_m \vec{r} + \psi_m\right)\right], \\ w_m(\omega) &= \int_{K_M} S_q(\vec{\kappa};\omega) d^2\kappa, \end{aligned} \tag{5}$$

where the random vectors $\vec{\kappa}_m \in K_m$ are distributed with the probability densities $\left.\frac{S_q(\vec{\kappa};\omega)}{w_m(\omega)}\right|_{\vec{\kappa}\in K_m}$, a_m are independent random quantities with Rayleigh distribution, moreover $\langle a_m^2 \rangle = 1$, ψ_m are independent random quantities, uniformly distributed on $[0, 2\pi]$. The obtained random field (5) is statistically homogeneous for any ω and its space-time spectral density is equal to $\Phi_q(\omega;\vec{\kappa})$.

To calculate the random realizations of the pressure and particle velocity fields inside the medium, we substitute (5) into (4):

$$\begin{aligned} p(\omega;\vec{r},z) &= P_q^{1/2}(\omega)\sum_{m=1}^{M} w_m^{1/2}(\omega) a_m \exp\left[i\left(\vec{\kappa}_m \vec{r} + \psi_m\right)\right] G(\omega;\vec{\kappa}_m, z), \\ \vec{v}_\perp(\omega;\vec{r},z) &= P_q^{1/2}(\omega)\sum_{m=1}^{M} w_m^{1/2}(\omega) a_m \frac{\vec{\kappa}_m}{k} \exp\left[i\left(\vec{\kappa}_m \vec{r} + \psi_m\right)\right] G(\omega;\vec{\kappa}_m, z), \\ v_z(\omega;\vec{r},z) &= P_q^{1/2}(\omega)\sum_{m=1}^{M} w_m^{1/2}(\omega) a_m \exp\left[i\left(\vec{\kappa}_m \vec{r} + \psi_m\right)\right] \frac{1}{ik}\frac{\partial}{\partial z} G(\omega;\vec{\kappa}_m, z), \end{aligned} \tag{6}$$

where $\vec{v}_\perp = \{v_x, v_y\}$.

Equations (5) and (6) represent the statistical-modeling pattern. Note that expressions (6) yield the exact solution of the linear-acoustics equations for the sources of type (5). Now we can obtain an ensemble of realizations (6), and then calculate the necessary statistical estimations.

Below we discuss only the relative amplitudes of the time spectrum of the pressure and the particle velocity. Therefore, without loss of generality, we assume $P_q(\omega) = 1$.

4 Algorithm and Implementation

First of all, it is necessary to randomize the field of surface sources. Bearing in mind that the field inside the medium is formed mainly by homogeneous waves, we divide the space K into two regions – the region of homogeneous wave sources

$K_R = \{\kappa \le k\}$, $\kappa = |\vec{\kappa}|$ (radiating spectrum) and the region of inhomogeneous wave sources $K_E = \{\kappa > k\}$ (evanescent spectrum). Each of these regions we split into nonoverlapping subregions, which are concentric rings:

$$\begin{cases} K_{R,m} = \{k_{m-1} \le \kappa < k_m\}, & m = 1,\dots,M_R, \quad k_0 = 0, k_{M_R} = k, \\ K_{E,m} = \{k_{m-1} \le \kappa < k_m\}, & m = 1,\dots,M_E, \quad k_0 = k, k_{M_E} = \infty. \end{cases} \tag{7}$$

Such partition of the space K seems physically reasonable. First, in each realization of the field (5) will be present both the components that emit homogeneous waves, and the components that emit inhomogeneous waves. If the space K is split evenly, then we need a very large number of regions in order that the contribution of homogeneous wave sources to be significant in the field (6), because the probability measure of the last is very small. Secondly, the choice of decomposition (7) is reasonable from symmetry considerations for the case of layered media, where the Green's function depends only on the wave vector magnitude. And thirdly, for such symmetric domains there are simple and robust algorithms of random vectors simulation for arbitrary probability distribution functions.

The region of homogeneous wave sources (radiating spectrum) also divided into two parts: discrete radiating spectrum and continuous radiating spectrum. A splitting interval in the discrete radiating spectrum region should be sufficiently small because of its resonance structure. Since the angle of total internal reflection does not depend on frequency, then the randomization can be done only once, and then we can use the obtained ensemble for all frequencies.

Distributed computing on formulas like (6) can be organized in several different ways. In this work, we implemented parallelization of the calculation of statistical moments of the field (6) for various frequencies and estimates. This is the most natural way satisfying to the logic of our problem. With this method of parallelization the task is divided into M*N subtasks, where M – the number of the frequencies to be calculated, N – the number of the estimations. The subtasks can be processed in parallel branches of the program without intensive data exchange. The performance of this method linearly depends on the number of used processors (while the latter is not greater than M*N). A monitor program for the parallel execution of the subtasks was written using Scilab [3] and PVM [4]. The program provides dynamic allocation of the subtasks on available computational resources and their execution control. All calculations in our paper were made using this software.

5 Calculation Results: Shallow Water Example

As a model of surface sources was used the isotropic field with the uniform limited spatial spectral density:

$$S_q(\kappa;\omega) = \begin{cases} \dfrac{1}{4\pi k^2}, \kappa \le 2k \\ 0, \kappa > 2k \end{cases}; \quad \kappa = |\vec{\kappa}|. \tag{8}$$

As a typical layered model of shallow water were chosen summer (thermocline) and winter (homogeneous waveguide) sound speed profiles. The summer sound speed profile and medium parameters are given in Fig. 1.

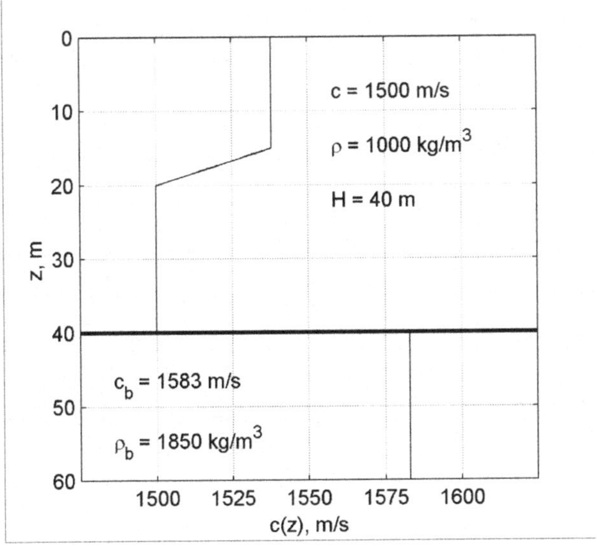

Fig. 1. Parameters of the medium models

The calculations of all statistical characteristics were made for the following four medium models: 1) layered waveguide with homogeneous liquid bottom (water: c(z), $\hat{\rho}$; bottom: c_b, $\hat{\rho}_b$); 2) homogeneous waveguide with homogeneous liquid bottom (water: c, $\hat{\rho}$; bottom: c_b, $\hat{\rho}_b$); 3) layered half-space (water: c(z), $\hat{\rho}$; bottom: c, $\hat{\rho}$); 4) homogeneous half-space (c, $\hat{\rho}$).

Such number of the medium models was used that it was possible to separately study effects of the sound speed profile influence and the bottom influence on the noise field. In all subsequent figures (Fig. 2-4) the numbers of the curves correspond to the numbers of the medium models. The angle of total internal reflection, that divides discrete radiating spectrum (DRS) and continuous radiating spectrum (CRS), for models 1 and 2 is equal approximately 19°. The model 1 corresponds to summer sound propagation conditions in shallow water (thermocline), and model 2 – winter (homogeneous waveguide).

Fig. 2 shows depth dependences of the average square of time spectral amplitude module of sound pressure (further we call this quantity the intensity). On frequency 10 Hz acoustic modes is not generated, therefore difference between the summer (curve 1) and winter (curve 2) field structure almost is not present; the bottom influence also is insignificant. On frequency 40 Hz, when one mode is formed in waveguide, the bottom insertion in the model strongly (at a maximum on 12 dB) changes the level of intensity, and the difference between the winter and summer level of intensity does not exceed 2 dB. On frequency 100 Hz this difference amounts

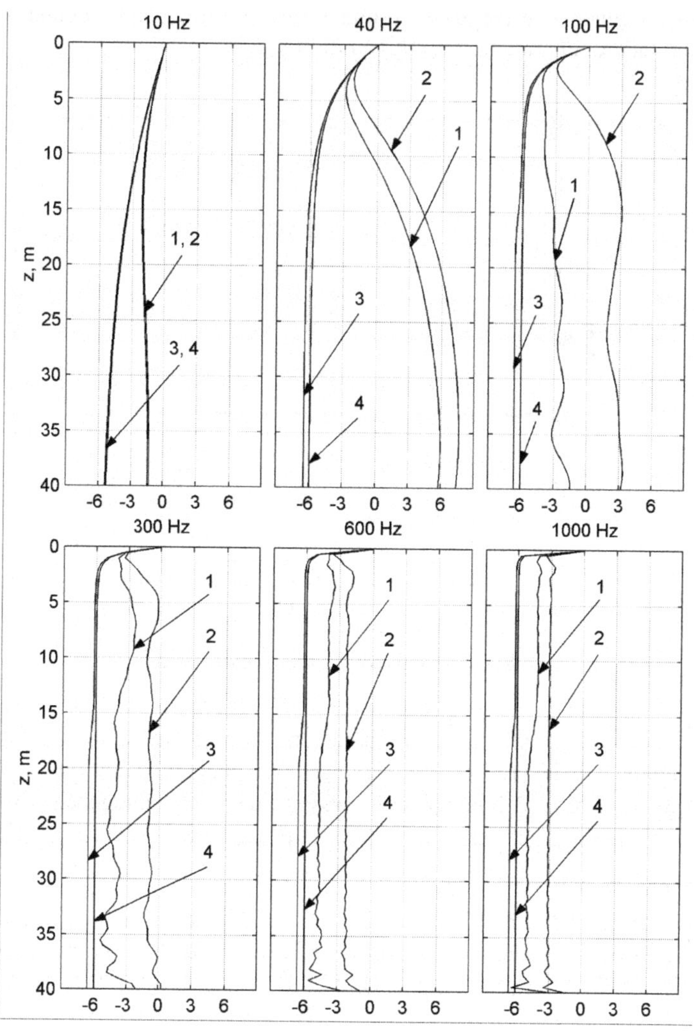

Fig. 2. $\left\langle |p|^2 \right\rangle$, dB

to 6 dB, that is explained by the different behavior of the CRS/DRS ratio (see Fig. 3): in the winter the field is formed substantially by modes, in the summer – by CRS. The same effect is observed on frequency 300 Hz. At the further increase of frequency CRS prevails both summer and winter, that results in reduction of the sound speed profile influence on the intensity level. The bottom influence on intensity structure forming on frequencies more than 100 Hz is comparable to the sound speed profile influence.

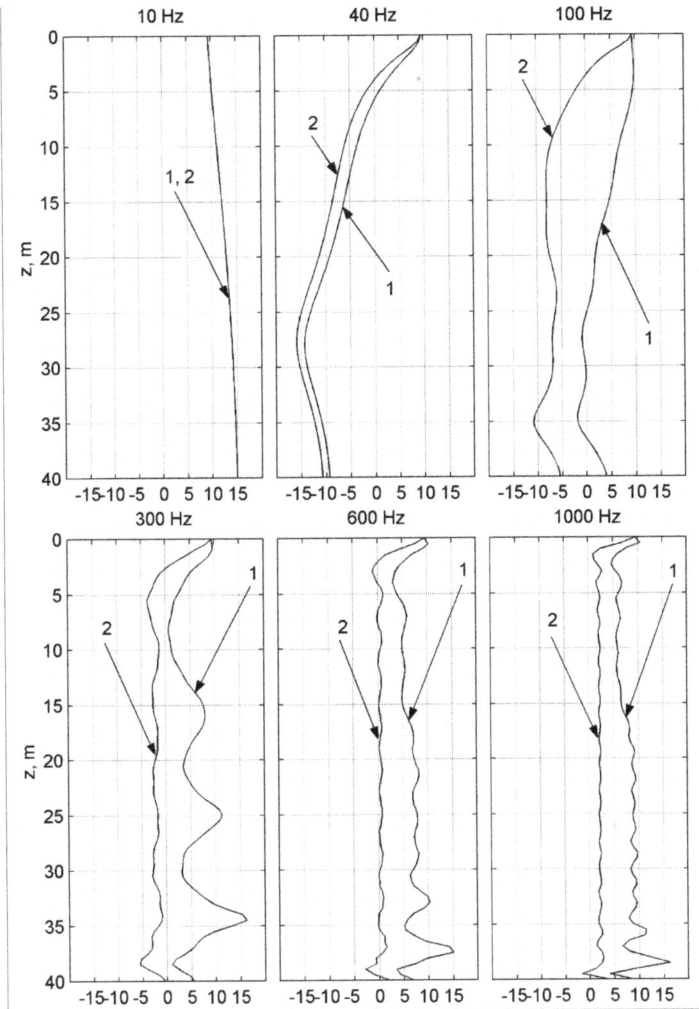

Fig. 3. $\langle |p|^2 \rangle$, CRS/DRS, dB

The real part of time spectrum of vertical component of acoustic energy flux almost does not depend on the bottom parameters, as it follows from Fig. 4, where its frequency dependences for a number of depths are represented. For all medium models

$$\mathrm{Re}\{\langle p v_z^* \rangle\} / \langle |p_0|^2 \rangle \approx 2/3, \tag{9}$$

where p_0 is the sound pressure in homogeneous half-space on the depth, where the influence of the evanescent spectrum is insignificant. The sound speed profile presence weakens the flux level, but it is rather insignificant. Let us note that the value (9) coincides with the value given by the classical model of isotropic surface noise in homogeneous half-space.

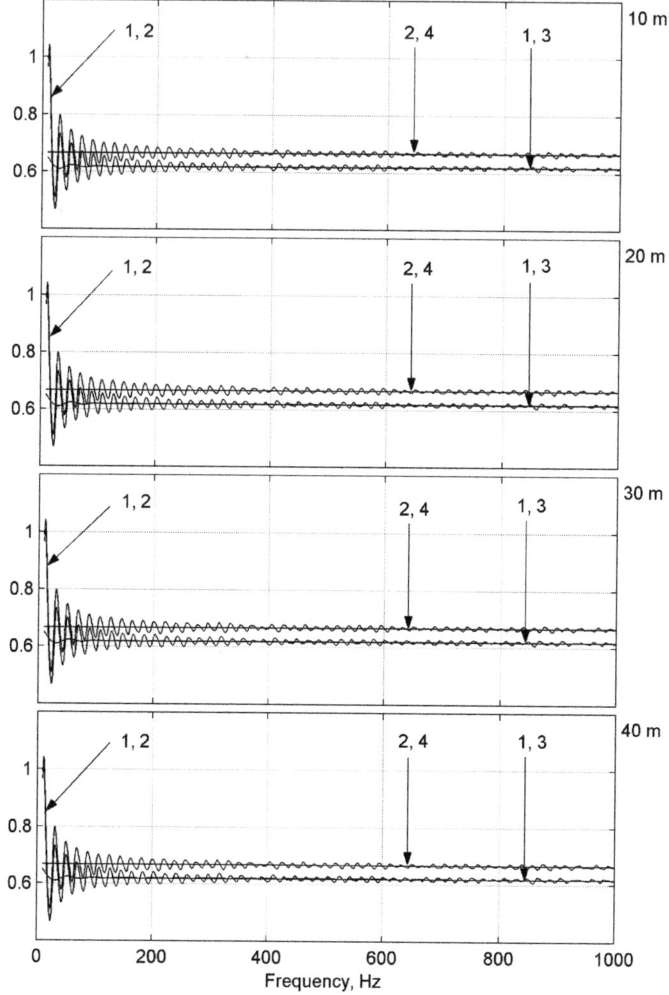

Fig. 4. $\mathrm{Re}\left\{\left\langle p v_z^* \right\rangle\right\} / \left\langle \left| p_0 \right|^2 \right\rangle$

6 Conclusion

In this paper, the example of a fairly simple (and at the same time inaccessible to analytical study and with difficulty accessible to numerical solution) wave problem of transformation of surface noise inside a layered ocean demonstrated the possibility and efficiency of statistical modeling method, based on randomization of the spectral density of pressure fluctuations on ocean surface.

The proposed method of statistical modeling can be extended to more realistic models of the ocean, and also allows obtaining temporal realizations of acoustic fields. There are new opportunities for solving such problems of statistical radiophysics and ocean acoustics, as the study of generation mechanisms of ocean dynamic noise, the parameter estimation of signals from local sources against the noise background, the implementation of marine inhomogeneities monitoring, and much more.

References

1. Mikhailov, G.A.: Numerical Construction of Random Field with the Given Spectral Density. Dokl. Akad. Nauk SSSR 238(4), 793–795 (1978) (in Russian)
2. Shvyrev, A.N., Yaroshchuk, I.O.: Statistical Modeling in the Problem of Excitation of Fields by Random Sources on a Surface. Radiophysics and Quantum Electronics 44(4), 326–331 (2001)
3. Scilab, `http://www.scilab.org`
4. PVM: Parallel Virtual Machine, `http://www.csm.ornl.gov/pvm`
5. Yaroshchuk, I.O., Gulin, O.E.: Statistical Modeling Method for Hydroacoustic Problems, Dalnauka, Vladivostok (2002) (in Russian)
6. Yaroshchuk, I.O., Shvyrev, A.N., Gulin, O.E., Lyashkov, A.S.: Mathematical Modeling for Statistical Hydroacoustic Problems. In: Far Eastern Seas of Russia, book, vol. 4, pp. 396–426. Nauka, Moscow (2007) (in Russian)

On One Parallel Numerical Implementation of a Free-Surface Ocean Dynamics Model

Sergey V. Smirnov and Vladimir A. Levin

Institute of Automation and Control Processes
Far Eastern Branch of Russian Academy of Sciences,
Vladivostok, Russian Federation
smirnoff@iacp.dvo.ru

Abstract. This paper deals with the parallelization of free-surface three-dimensional oceanographic model. The model is based on full nonlinear "primitive" equations of the ocean. Generalized vertical coordinate system is applied for better resolution of main features of the simulated basin. The numerical model is conservative. Numerical integration procedure is based on time-splitting method with Robert-Asselin filtering. Numerical model code is implemented for running on cluster computers. The parallelization is achieved using domain decomposition method and standard MPI to ensure portability of the code.

1 Introduction

Ocean models represent one of the fundamental tools to investigate the physics of the ocean. The numerical methods used by ocean models consist in discretizing the Navier-Stokes equations on a three dimensional grid and computing the time evolution of each variable for each grid point. In order to answer the huge amount of computational demand raised to conduct simulation on high resolution computational grid, a parallelization strategy is needed. In this paper we present parallel version of free-surface three-dimensional oceanographic model. In particular in section 2 we shortly describe the ocean model features and algorithm, in section 3 and 4 we describe the technique used to parallelize and optimize the model code. Conclusions are summarized in section 5.

2 Ocean Model Description

The numerical ocean model is described in [7]. It is an s-coordinate, free surface, primitive equation model, which includes a turbulence sub-model [1], [2]. The equations consist of the Navier-Stokes equations subject to the Boussinesq and hydrostatic approximations. The equation of state relating density to temperature, salinity, and pressure can generally be nonlinear [3], thus representing important aspects of the oceans thermodynamics. Prognostic variables are the two active tracers potential temperature and salinity, the two horizontal velocity components, any number of passive tracer fields, and the height of the free ocean

C.H. Hsu and V. Malyshkin (Eds.): MTPP 2010, LNCS 6083, pp. 200–203, 2010.

surface. The model uses time splitting technique. The external (barotropic) mode portion of the model is two-dimensional and uses a short time step based on the CFL condition. It consists of a set of vertically integrated equations of continuity and momentum to provide free surface variations. The internal (baroclinic) mode is three-dimensional and uses a long time step based on the CFL condition. The horizontal grid uses curvilinear orthogonal coordinates. The external mode grid is staggered as an Arakawa-C scheme [4], the internal mode grid is staggered as an Arakawa B scheme. The generalized vertical coordinate s-system [5] is capable of simultaneously maintaining high resolution in the surface layer as well as dealing with steep and/or tall topography. The horizontal time differencing is explicit leap-frog time stepping with a Robert-Asselin filter [6], whereas the vertical differencing is implicit. The latter eliminates time constraints for the vertical coordinate and permits the use of fine vertical resolution in the surface and bottom boundary layers. The model has a volume conserving mode-splitting [7] due to an accurate coupling between barotropic and baroclinic free-surface elevations.

3 Parallelization

The numerical model code has been parallelized using a Single Program, Multiple Data (SPMD) approach. The domain is decomposed into N partitions (subdomains) where N is the number of processors to be used for the computation. Each processor runs a complete copy of the time integration code in its respective sub-domain. During the calculation, information is exchanged between processors across sub-domain boundaries to ensure correctness of the boundary fluxes. The inter-processor communication is explicitly defined using Message Passing Interface (MPI) calls. The resulting implementation is highly portable and will run efficiently on a variety of parallel computer architectures including both shared and distributed memory systems. Output data is collected from individual processors and reconstructed into a global array before being written to disk.

The geometric decomposition of the ocean computational grid may be a two (or one) dimensional. The grid is horizontally partitioned into rectangular blocks, leaving the vertical dimension unchanged, because the computation in the vertical dimension is hard to parallelize for the presence of an implicit scheme in the code. In order to guarantee the same results of the serial code at the last bit precision of the machine, and to reduce the frequency of communication, all the sub-domains are overlapped at the inner boundaries on a slice of 2 grid point's thickness. These slices, generally known as overlap areas, represent grid points that contain copies of the boundary values stored on neighboring sub-domains in the grid topology. According to the concept of load balancing, each partition will contain roughly the same number of 3D grid nodes. For the case of explicit integration of the primary 3D equations, the work required is roughly proportional to the number of elements (grid nodes) in a domain. Thus to ensure equal workload among the processors, the decomposition must provide the same number of elements to each partition. The total length of the boundary between partitions is to be minimized. It is introduced to reduce communication overhead.

Behavior of warm water flowing into the Japan/East sea model basin filled with cold water was investigated numerically with the numerical model [7]. The 5-min resolution grid for this area is decomposed into 10 partitions (figure 1). On a deep water part of numerical grid, the nodes are allocated mainly on z-levels. So, the size of each sub-domain depends upon coastline and bottom relief features.

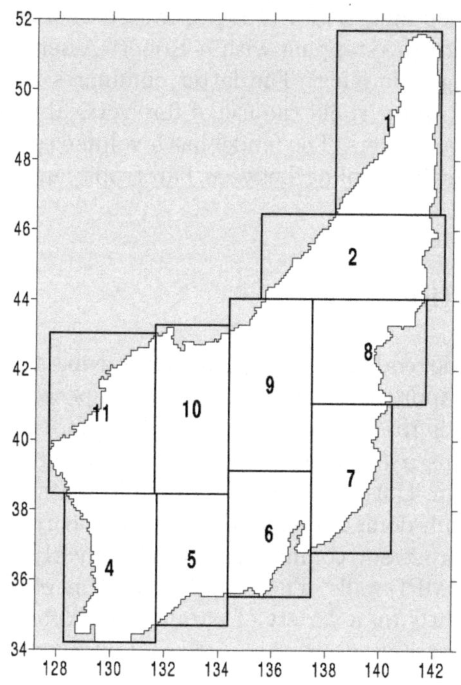

Fig. 1. Sub-domain partitioning

4 Processor-Memory Data Exchange Minimization

In the numerical calculations of large-scale oceanographical problems, memory issues play a very important role, and often impact significantly the numerical models performance, and overall cost of computations. Indeed, as new processor families and processor cores begin to push the limits of high performance, the traditional processor-memory gap widens and often becomes the dominant bottleneck in achieving high performance. So, one need to pay particular attention to issues such as minimizing processor-memory data exchange limited by main memory throughput and using cache memory possibilities. The size of cache has a significant impact on speed. Smaller sized arrays are more likely to fit within available cache than larger sized arrays and this results in speed improvements.

An appropriate technique consists in successive calculations of adjacent vertical slices of 3D cells. Storage for each variable in addition to other quantities such as fluxes through cell faces must be allocated for each cell. Within each slice, data is arranged mainly in neighboring storage locations. The maximum attainable computation speed can be reached if several adjacent slices were loaded in a cash memory. Such an approach is optimal for models with implicit vertical time differencing.

5 Conclusion

The parallel implementation of a free-surface three-dimensional oceanographic model has been successfully constructed. The code is in principle able to efficiently solve large regional oceanographic problems. Portability has been achieved by using MPI standard. The ANSI C code implementation is appropriate for calculation on cluster systems with small cash-size processors.

Acknowledgement. The research was supported by RFBR grant 09-01-12056 and FEB RAS grant 09-I-*П*2-05 (Russia).

References

1. Mellor, G.L., Yamada, T.: Development of a turbulent closure model for geophysical fluid problems. Rev. Geophys. 20, 851–875 (1982)
2. Blumberg, A.F., Mellor, G.L.: A description of a three-dimensional coastal ocean circulation model. In: Heaps, N.S. (ed.) Three-Dimensional Coastal Ocean Models. Coastal Estuarine Science, pp. 1–16. Amer. Geophys. Union (1987)
3. Jackett, D.R., McDougall, T.J.: Minimal adjustment of hydrographic data to achieve static stability. J. Atmos. Ocean. Tech. 12, 381–389 (1995)
4. Mesinger, F., Arakawa, A.: Numerical methods used in atmospheric models. GARP Publications Series, 17, vol. 1, p. 64. World Meteorol. Organization, Geneva (1976)
5. Song, Y.T., Haidvogel, D.B.: A semi-implicit ocean circulation model using a generalized topography-following coordinate system. J. Comput. Phys. 115, 228–244 (1994)
6. Asselin, R.: Frequency filter for time integrations. Mon. Weather Rev. 100, 487–490 (1972)
7. Levin, V.A., Smirnov, S.V.: On one numerical implementation of a free-surface ocean dynamics model. Matem. Mod. 18(4), 19–34 (2006) (in Russian)

Multi-particle Cellular-Automata Models for Diffusion Simulation*

Yu Medvedev

The Institute of Computational Mathematics and Mathematical Geophysics,
Supercomputer Software Dept., pr. Acad. Lavrentiev, 6,
630090, Novosibirsk, Russia
medvedev@ssd.sscc.ru

Abstract. Two new cellular-automata models of the diffusion process are proposed. They are based on integer states of cells instead of Boolean ones in the known models: asynchronous naive diffusion by Toffolli and block-synchronous Margolus diffusion. Computing experiments have been carried out with these models; they demonstrate a good correlation with this physical phenomenon. The main advantages of the proposed models are (i) low automata noise and (ii) variable diffusion rate.

Keywords: Cellular automata, diffusion, simulation.

1 Introduction

Following the common practice, the diffusion process is represented by the differential equation. Usually, computers solve this equation because its analytical solution cannot be obtained. But computers operating with real numbers accumulate round-off errors. In addiction, complicated boundary conditions cause certain difficulties.

Since the end of the last century cellular automata have been used for the diffusion process simulation. One of them is an asynchronous cellular automaton of naive diffusion, proposed by Tomasso Toffolli [1]. Another one is a block-synchronous cellular automaton that was offered by Norman Margolus [1]. Both of them operate over the Boolean alphabet, therefore no round-off errors are available. Cellular-automata simulation of diffusion is actively developing up to our days [2].

But, because of the limitations of the Boolean models, some kind of the result distortion is present [3]. It is aperiodic oscillating of substances concentration, i.e. concentration noise. A noise which is caused by the influence of discrete nature of cellular automata, we call the automaton noise. Such the noise in the integer models is usually less then that in the Boolean ones.

In this paper, an attempt to eliminate the problem of automaton noise is undertaken. For each of the referred Boolean models a new analog, which operates over integers, is proposed. Computer experiments with the proposed models are executed, their behavior is investigated.

---

* Supported by 1) Presidium of RAS, Basic Research Program No 2 (2009), 2) Siberian Branch of RAS, Interdisciplinary Project No 32 (2009).

C.H. Hsu and V. Malyshkin (Eds.): MTPP 2010, LNCS 6083, pp. 204–211, 2010.

2 Main Definitions

2.1 Boolean Models

A 2D cellular automaton of diffusion has the cellular array $W = \{w_{ij}: i \in [1, I], j \in [1, J]\}$; $|W| = I \cdot J$ being the cardinality of the array. The cells $w_{ij} \in W$ are arranged in the Cartesian plane according to Figure 1. Each cell $w_{ij} \in W$ has a state $s(w_{ij}) \in \{0, 1\}$. A value of the state $s(w_{ij})$ shows which of two substances (let us call them "zero" and "one") is present in the cell w. A set of the states $s(w_{ij})$ of all cells $w_{ij} \in W$ at the same instant t is called a global state $\sigma(t) = \{s(w_{ij}): i \in [1, I], j \in [1, J]\}$ of the cellular automaton. In a cellular automaton with a synchronous operation each iteration is a replacement of states $s(t)$ in all cells by the states $s(t + 1)$. The cellular automaton thus changes its global state $\sigma(t)$ by a new one $\sigma(t + 1)$. An iteration of an asynchronous cellular automaton is a sequence of events. Each event is a replacement of a randomly chosen cell state $s(t)$ by the state $s(t + 1)$. The number of the events per iteration is equal to $|W|$.

i \ j	1	2	3	4	5
1	w_{11}	w_{12}	w_{13}	w_{14}	w_{15}
2	w_{21}	w_{22}	w_{23}	w_{24}	w_{25}
3	w_{31}	w_{32}	w_{33}	w_{34}	w_{35}
4	w_{41}	w_{42}	w_{43}	w_{44}	w_{45}
5	w_{51}	w_{52}	w_{53}	w_{54}	w_{55}

Fig. 1. The cellular array W

A model concentration of each substance c_0 and c_1 in any cell $w_{ij} \in W$ is calculated by means of summing up over an averaging vicinity $Av(w_{ij})$ with some radius r. The vicinity $Av(w_{ij})$ consists of the cells which are not further than r from w_{ij}.

$$c_0\left(w_{ij}\right) = \frac{\sum_{w \in Av\left(w_{ij}\right)} \left(1 - s\left(w_{ij}\right)\right)}{\left|Av\left(w_{ij}\right)\right|}, \quad c_1\left(w_{ij}\right) = \frac{\sum_{w \in Av\left(w_{ij}\right)} s\left(w_{ij}\right)}{\left|Av\left(w_{ij}\right)\right|}. \tag{1}$$

2.2 Multi-particle Models

The multi-particle cellular automaton differs from the Boolean one in that it has integer-valued states $s(w_{ij})$ of cells $w_{ij} \in W$; $s(w_{ij})$ represents a number of particles having unit mass. States $s(w_{ij})$ being in the interval $s(w_{ij}) \in [0, S]$, where S is the number of particles, corresponding to a maximum concentration of substance "one" in a cell. The value of $s(w_{ij})$ shows the number of the substance "one" particles in the cell w_{ij}. Concentrations of substances "zero" and "one" in the cell w_{ij} are equal to $S - s(w_{ij})$ and $s(w_{ij})$, respectively.

The model concentrations c_0 and c_1 of each substance, respectively, are calculated as follows:

$$c_0\left(w_{ij}\right) = \frac{\sum_{w \in Av\left(w_{ij}\right)} \left(S - s\left(w_{ij}\right)\right)}{\left|Av\left(w_{ij}\right)\right| \cdot S}, \quad c_1\left(w_{ij}\right) = \frac{\sum_{w \in Av\left(w_{ij}\right)} s\left(w_{ij}\right)}{\left|Av\left(w_{ij}\right)\right| \cdot S}. \tag{2}$$

A radius r of the averaging vicinity Av can be diminished up to zero, and this will not excessively increase the automaton noise, like in the Boolean models. So, it is possible to use an automaton with a smaller in comparison with the Boolean model amount of cells.

3 Naive Diffusion

3.1 Boolean Model

The Boolean asynchronous cellular automaton simulates a diffusion process as follows. Each cell of the automaton $w_{ij} \in W$ is located at some point of a plane and has four neighbors. In Figure 2, a cross template with the cell w_{ij} and its neighboring cells a, b, c, and d, is given. At each iteration, the following steps are done:

1. A random cell $w_{ij} \in W$ is equiprobably selected.
2. One of four neighboring cells $w_k \in \{a, b, c, d\}$ is equiprobably selected.
3. Cells w_{ij} and the selected one w_k exchange their states $s(w_{ij})$ and $s(w_k)$.

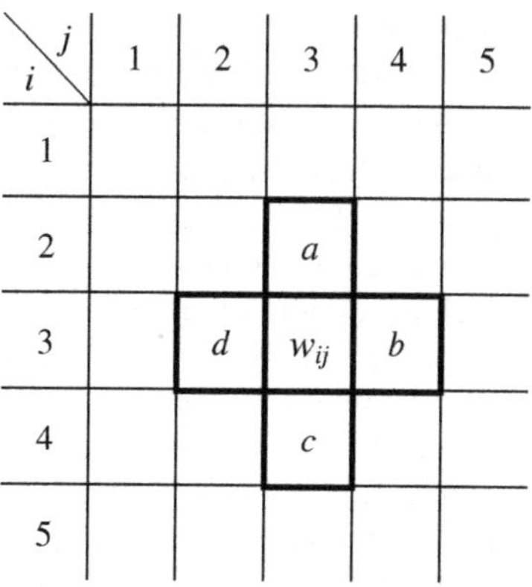

Fig. 2. A template in the naive diffusion cellular automaton

The above iteration is repeated many times. After that, averaging (1) should be executed for obtaining the concentrations c_0 and c_1 for every cell of the array. The effect of operation of such an automaton is infiltration of substance "zero" into substance "one" and vise versa.

3.2 Multi-particle Model

At each iteration, the following steps are done.

1. A random cell $w_{ij} \in W$ is equiprobably selected.
2. One of the four neighboring cells $w_k \in \{a, b, c, d\}$ is equiprobably selected.
3. The quantities of particles s representing the state value in each of the cells w_{ij} and w_k are divided into the two groups s_k and s_{1-k} as follows:

$$\begin{aligned} s_k(w_{ij}) &= \lfloor k \cdot s(w_{ij}) \rfloor + s'(w_{ij}), \\ s_{1-k}(w_{ij}) &= s(w_{ij}) - s_k(w_{ij}), \\ s_k(w_k) &= \lfloor k \cdot s(w_k) \rfloor + s'(w_k), \\ s_{1-k}(w_k) &= s(w_k) - s_k(w_k), \end{aligned} \tag{3}$$

where $k \in (0, 1]$ is a certain parameter determining a model diffusion rate; $s'(w) = 1$ with the probability $k \cdot s(w) - \lfloor k \cdot s(w) \rfloor$ and $s'(w) = 0$ otherwise; $\lfloor x \rfloor$ denotes the floor x.

4. A group of particles $s_k(w_{ij})$ moves from the cell w_{ij} to the cell w_k, while a group of particles $s_k(w_k)$ moves in the contrary direction, where $w_k \in \{a, b, c, d\}$. Thus, their new states are:

$$s(w_{ij}) = s_{1-k}(w_{ij}) + s_k(w_k),$$
$$s(w_k) = s_{1-k}(w_k) + s_k(w_{ij}). \tag{4}$$

After the assigned number of iterations, averaging (2) should be executed for obtaining the concentrations c_0 and c_1. An expected result is the same as in the Boolean model, differing only in that the amount of the automaton noise is less.

4 Margolus Diffusion

4.1 Boolean Model

In the Boolean cellular automaton of the Margolus diffusion, a template is a square of 2×2 cells. In this model, the even and the odd iterations are distinguished. They are executed in a different way. In Figure 3a, the even partitioning (for using at the even iterations) is given; both indices i and j of the upper left cell of the blocks of 2×2 cells are even. In Figure 3b, the odd partitioning (for using at the odd iterations) is given; i and j of the upper left cell are odd. Because of the synchronous mode of the Margolus diffusion, the transition rule in all the blocks is applied simultaneously. At each iteration in each block, the following steps are carried out:

1. The direction of rotation is selected either clockwise or counterclockwise equiprobably, independent of other blocks.
2. The cells a, b, c, and d in each block exchange their values $s(a)$, $s(b)$, $s(c)$, $s(d)$ around a circle, according to a selected direction ($a \to b \to c \to d \to a$ or $a \to d \to c \to b \to a$).

The necessary number of iterations is to be repeated. After that, averaging (1) should be executed for calculating the concentrations c_0 and c_1 in every cell of the array. In [4], it is proved that the Boolean Margolus block-synchronous cellular automaton exactly simulates the diffusion process.

4.2 Multi-particle Model

At each iteration in each block, the following steps are carried out:

1. The direction of rotation is selected either clockwise or counterclockwise equiprobably, independent of other blocks.
2. The number of particles $s(a)$, $s(b)$, $s(c)$, and $s(d)$ in each of the cells a, b, c, and d is divided into the two groups s_k and s_{1-k} as follows:

$$s_k(a) = \lfloor k \cdot s(a) \rfloor + s'(a),$$
$$s_{1-k}(a) = s(a) - s_k(a), \text{ etc.,} \tag{5}$$

where $k \in (0, 1]$ is a parameter of a model diffusion rate; $s' = 1$ with the probability $k \cdot s - \lfloor k \cdot s \rfloor$, and $s' = 0$ otherwise; $\lfloor x \rfloor$ denotes the floor x.

3. The cells a, b, c, and d in each block exchange their values $s_k(a)$, $s_k(b)$, $s_k(c)$, $s_k(d)$ around a circle, according to a selected direction. In the case of the clockwise exchange, their states thus will be equal to:

$$\begin{aligned} s(a) &= s_{1-k}(a) + s_k(d), \\ s(b) &= s_{1-k}(b) + s_k(a), \\ s(c) &= s_{1-k}(c) + s_k(b), \\ s(d) &= s_{1-k}(d) + s_k(c). \end{aligned} \tag{6}$$

In the case of the counterclockwise exchange, their states thus will be equal to:

$$\begin{aligned} s(a) &= s_{1-k}(a) + s_k(b), \\ s(b) &= s_{1-k}(b) + s_k(c), \\ s(c) &= s_{1-k}(c) + s_k(d), \\ s(d) &= s_{1-k}(d) + s_k(a). \end{aligned} \tag{7}$$

i \ j	1	2	3	4	5
1	c	d	c	d	c
2	b	a	b	a	b
3	c	d	c	d	c
4	b	a	b	a	b
5	c	d	c	d	c

a) even partitioning

i \ j	1	2	3	4	5
1	a	b	a	b	a
2	d	c	d	c	d
3	a	b	a	b	a
4	d	c	d	c	d
5	a	b	a	b	a

b) odd partitioning

Fig. 3. Cellular array partitioning for the Margolus Diffusion

After finishing all the iterations, averaging (2) should be executed for obtaining the concentrations c_0 and c_1.

5 Computer Simulation

For validation of the proposed models, their program realization has been constructed. This allows carrying out computing experiments with both models. The code has been written in C. The computing experiments, performed with the multi-particle Margolus diffusion model only, are described below. Results of the experiments with the naive diffusion model are identical to the Margolus one we will not dwell on them here.

5.1 Diffusion of a Square Spot

The first example simulated by the proposed model is rather simple. The size of a cellular automaton is $I = 100$ by $J = 100$ cells. The cells with $i \in [33, 67]$ and $j \in [33, 67]$ are initialized by ones, $s(w_{ij}) = 1$. The rest of the cells are initialized by zeros, $s(w_{ij}) = 0$. So, the substance "one" looks as a square spot on the substance "zero" background. In Figure 4a the array of the averaged concentrations c_1 of the substance "one" in the initial global state of the automaton is shown. Because of the averaging radius $r = 8$ being significant for the selected size of the cellular array, the boundary between the substances is blurred. After the simulation is completed, the spot of the substance "one" diffuse into the substance "zero" as a round-shaped spot (Figure 4b). This effect demonstrates that the model simulates diffusion process correctly.

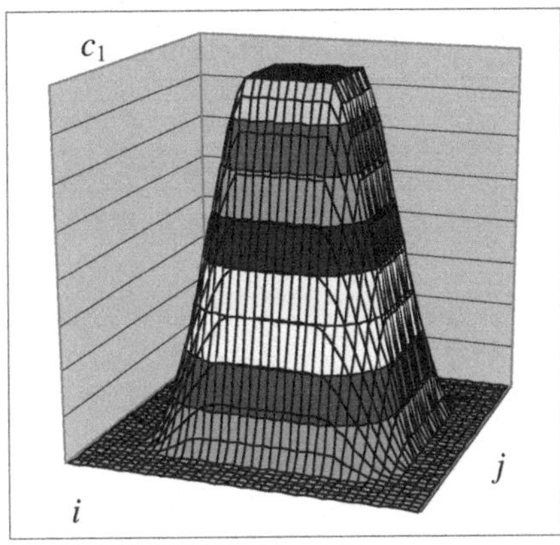

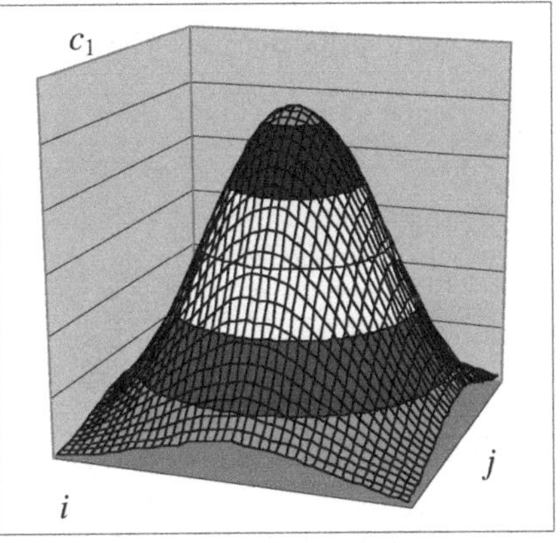

a) initial state b) result of the simulation

Fig. 4. Diffusion of a square spot

5.2 Automaton Noise

For investigation of the automaton noise a following computer experiment has been performed. A size of a cellular automaton is $I = 100$ by $J = 1000$ cells. The cells with $j \in [450, 550]$ are initialized by ones, $s(w_{ij}) = 1$. The rest of the cells are initialized by zeros, $s(w_{ij}) = 0$. So, a stripe of the substance "one" is between of two stripes of the substance "zero". In this case the diffusion process propagates only along j. Therefore it is acceptable to average according to (1) for the Boolean model and to (2) for the multi-particle one with the average vicinity $Av = \left\{ w_{ij} : \forall i \notin (1, I), j = const \right\}$. Thus the blur caused the by average radius vanishes. Averaged concentrations c_1 of the substance "one" from $j = 1$ to $j = 1000$ is given in Figure 5.

In Figure 5a the concentrations with simulation by the Boolean cellular automaton at 0, 1000, 2500, and 5000 iterations are given. The automaton noise is seen clearly. In Figure 5b the results of simulation by the multi-particle cellular automaton with the same initial condition, at 0, 3000, 15000, and 75000 iterations, with $k = 0.5$ are given. The shape of curves and the magnitude of automaton noise turned out to be identical to the Boolean model but diffusion rate was three times less.

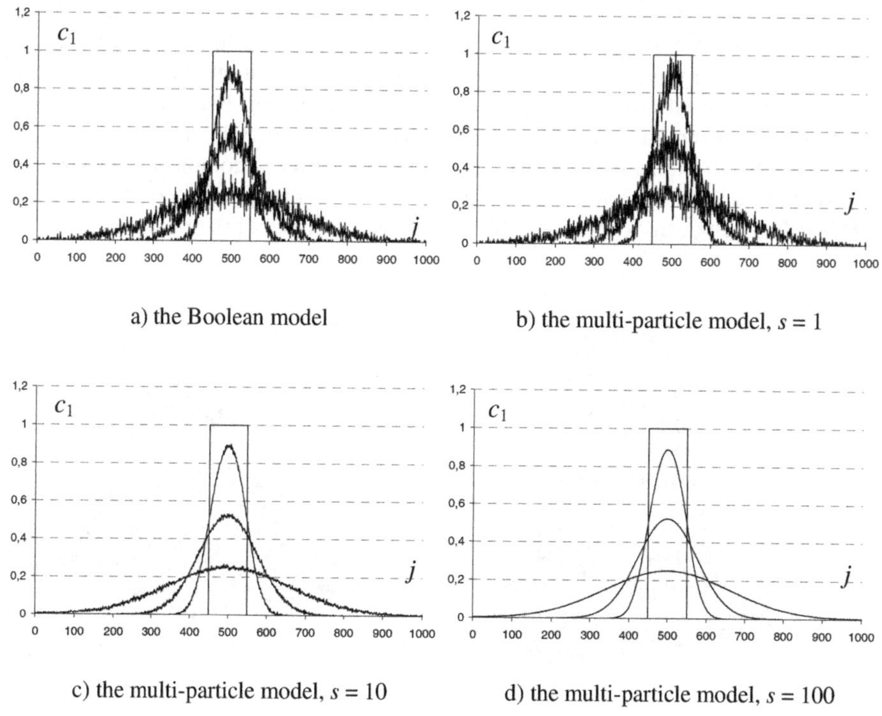

a) the Boolean model

b) the multi-particle model, $s = 1$

c) the multi-particle model, $s = 10$

d) the multi-particle model, $s = 100$

Fig. 5. Automaton noise in different cases

For automaton noise reduction the cells with $j \in [450, 550]$ are initialized by tenths, $s(w_{ij}) = 10$. It is possible for multi-particle model. The results are demonstrated in Figure 5c. They are the same as in the previous case, including rate of process and even a time of calculation. Only the noise level, predictably, turned out lower. In Figure 5d the results with the same initial condition are given, except the cells with $j \in [450, 550]$ which are initialized by hundreds, $s(w_{ij}) = 100$. The results are the same, but the noise is almost invisible. With increasing $s(w_{ij})$ of the cells with $j \in [450, 550]$, the noise diminishes; and the results difference from those of Figure 5d is seen only a microscope is used.

6 Conclusion

The two new cellular-automata models of the diffusion process are proposed. The results of computer simulation allow us to conclude the following:

The proposed models authentically feature the diffusion process because the results of simulation coincide with the results of the known models.

In the models proposed, the automaton noise with respect to the Boolean ones is extremely low that allows one to use a cellular array of a smaller size and to decrease the run time.

The parameter k in the proposed models affects the diffusion rate. Therefore, the models proposed give a possibility to simulate the diffusion process with various diffusivities.

In addition, the proposed models can be used in compositions with other multi-particle cellular-automata models [5].

References

1. Toffolli, T., Margolus, N.: Cellular Automata Machines. MIT Press, USA (1987)
2. Kroc, J., Sloot, P. (eds.): Simulating Complex Systems by Cellular Automata. Springer, Heidelberg (2010) (in print)
3. Bandman, O.L.: Computation Properties of Spatial Dynamics Simulation by Probabilistic Cellular Automata. Future Generation Computer Systems 21, 633–643 (2005)
4. Malinetsky, G.G., Stepantsov, M.E.: The Diffusion Process Simulation by Cellular Automata with the Margolus Vicinity. Journal of computational mathematics and mathematical physics 36(6), 1017–1021 (1998) (in Russian)
5. Bandman, O.L.: Cellular-Automata Models of the Spatial Dynamics. System Informatics 10, 57–113 (2005) (in Russian)

Parallel Implementation of Kinetic Cellular Automata for Modeling CO Oxidation over Pd(110) Surface*

Valentina Markova[1] and Anastasia Sharifulina[2]

[1] Institute of Computational Mathematics and Mathematical Geophysics
Russian Academy of Sciences
Prospekt Akademika Lavrentjeva 6, Novosibirsk, Russia
[2] Novosibirsk State Technical University
Prospekt Karla Marksa 20, Novosibirsk, Russia
{markova,sharifulina}@ssd.sscc.ru

Abstract. For simulating *CO* catalytic oxidation on platinum-group metals asynchronous cellular automata (CA) with probabilistic transition rules (kinetic CA) are used. Based on the properties of surface kinetic CA has to have a huge cellular arrays and long evolution. It is obvious that modeling such processes in real time can only be done with the help of supercomputer. In the paper, parallel implementation of approximation of an kinetic CA with block-synchronous CA is investigated.

Keywords: Catalytic oxidation reaction, kinetic cellular automata, synchronous mode, asynchronous mode, block-synchronous mode.

1 Introduction

CO catalytic oxidation on platinum-group metals [1-2] is a classical model reaction of heterogenic catalysis which, in addition to fundamental interest has an important practical use in connection with an ecological problem of purification of exhaust gases from carbon oxide admixtures. All processes taking place on the catalyst surface consist of a variety of elementary interactions of atoms and molecules. All interactions occur asynchronously and the probability of each interaction is determined by physical parameters. Conventional modeling methods, based on solutions of differential equations, turned out to be inadequate for description and analysis catalytic oxidation reactions [1].

A more effective description of the mechanism of catalytic oxidation is a class of kinetic cellular automata (asynchronous cellular automata with probabilistic transition rules) introduced in [5]. Based on the properties of surface reactions kinetic CA should have huge cellular arrays and a long evolution. It is obvious that simulating such processes in real time can only be done with the help of a

* Supported by 1)Presidium of Russian Academy of Sciences, Basic Research Program N 2 (2009), 2) Siberian Branch of Russian Academy of Sciences, Interdisciplinary Project N 32(2009).

C.H. Hsu and V. Malyshkin (Eds.): MTPP 2010, LNCS 6083, pp. 212–221, 2010.

supercomputer. It has been known that asynchronous CA have no the efficient and simple parallel implementation. To reduce the modeling time, the idea to introduce a partial synchronization of operation in asynchronous CA were conceived [3-6]. Here, approximation of an asynchronous cellular automaton with a block-synchronous CA [5] is used for parallelization of kinetic CA.

This paper is organized as follows. Apart from Introduction and Conclusion the paper contains four sections. CO catalytic oxidation reaction over Pd_{110} surface is given in Section 2. In Section 3, a kinetic cellular automaton is defined and dynamics of oxidation reaction is studied using kinetic cellular automaton. In Section 4, transformation of an asynchronous CA to a block-synchronous CA and comparison of both modes of operation in cellular automata are proposed. Section 5 is dedicated to the parallel implementation of block-synchronous CA.

2 *CO* Oxidation Reaction over Pd_{110} Surface

It is known that CO oxidation reaction over the surface of Pd_{110} is described by the following elementary actions [2].

1. $O_{2(gas)} + 2\star \xrightarrow{k_1} 2O_{ads}$ - oxygen adsorption and dissociation;
2. $CO_{gas} + \star \overset{k_2\ k_{-2}}{\leftrightarrow} CO_{ads}$ - adsorption and desorption of the CO;
3. $CO_{ads} + O_{ads} \rightarrow CO_{2(gas)} + 2\star$ - reaction between the CO_{ads} and the O_{ads};
4. $O_{ads} + \star_v \xrightarrow{k_4} [*O_{sub}]$ - formation of the subsurface oxygen;
5. $CO_{ads} + [*O_{sub}] \xrightarrow{k_5} CO_{2(gas)} + 2\star + \star_v$ - reaction between the CO_{ads} and the O_{sub};
6. $CO_{gas} + [*O_{sub}] \overset{k_6\ k_{-6}}{\leftrightarrow} [CO_{ads} * O_{sub}]$- formation of the complex ;
7. $[CO_{ads} * O_{sub}] \xrightarrow{k_7} CO_{2(gas)} + 2\star + \star_v$ - decomposition of the complex;
8. $CO_{ads} + \star \xrightarrow{k_8} \star + CO_{ads}$; CO_{ads} - diffusion along the surface;
9. $CO_{ads} + [*O_{sub}] \xrightarrow{k_8} \star + [CO_{ads} * O_{sub}]$ - formation of the complex;
10. $[CO_{ads} * O_{sub}] + [*O_{sub}] \xrightarrow{k_8} [*O_{sub}] + [CO_{ads} * O_{sub}]$ - diffusion of the complex.

Here the symbols $\star$ and $\star_v$ are the active sites on the surface and the subsurface layers, respectively, k_i and k_{-i}, $i = 1, 2, -2, 4, 5, 6, -6, 7, 8$, are the rate coefficients for the direct and the reverse i-th action of the reaction. k_8 (the rate of reagents motion along the surface) is defines by $M_{dif} \times \sum_1^7 k_i$, where M_{dif} is a diffusion parameter. All the steps of oxidation reaction occur asynchronously and are carried out with a certain probability $p_i = k_i / \sum_1^8 k_i$.

CO oxidation on platinum group metals is accompanied by oscillations of the reaction rate and adsorbed species concentrations. The driving force of the oscillations is associated with a relatively slow process of formation and consumption of subsurface oxygen, whose periodical transitions of the metal from an inactive to a highly active catalytic state. This transition occurs as follows. At the initial time, the catalyst surface is free, it corresponds to its active state. Carbon monoxide and oxygen are adsorbed on the catalyst surface from the gaseous phase. Then CO_{ads} molecules and atoms of O_{ads} enter into the reaction with CO_2 molecules formation. CO_2 desorbs and two neighboring sites become active. Simultaneously atomic oxygen O_{ads} is transformed into the subsurface oxygen. This is accompanied by a decrease in the value of the sticking coefficient of oxygen. As a result, adsorption of oxygen is blocked, and molecules of CO_{ads} are accumulated on the surface, and the amount of CO_2 formation is decreased. This corresponds to an inactive state of the surface. The obtained molecules of CO_{ads} diffuse over the surface to the sites occupied by O_{sub}. CO_{ads} and O_{sub} enter into the reaction with formation of free sites for the oxygen and carbon monoxide adsorption. Thereafter the surface again becomes active.

3 A Kinetic Cellular Automata for Modeling CO Oxidation

3.1 Definition of a Kinetic Cellular Automaton

A kinetic CA is defined by a multitude $\mathcal{N}_a = \langle A, M, T, \Theta \rangle$, where A is a state alphabet, M is a naming set, T is a set of used templates, Θ is a set of substitutes performing the oxidation reaction stages, a - is an asynchronous function mode. The state alphabet of a cell is the set $A = \{\star, CO_{ads}, O_{ads}, O_{sub}, [CO_{ads}O_{sub}]\}$, whose symbols denote the active surface centers, atoms and molecules taking part in the oxidation reaction. The set $M = Z^2 = \{(i,j) : i = 0, 1, \ldots, I,\ j = 0, 1, \ldots, J\}$ represents the catalyst surface. A pair $(a, (i,j)) \in A \times M$ is called *cell* (a site on the surface). Each cell corresponds to a finite-state automaton with the name (i,j) and the state a. The set of cells $\Omega = \{(a, (i,j))\} \subset A \times Z^2$ forms *a cellular array.* ($\Omega(0)$ is the initial state of the cellular array).

On the set M a naming function $\varphi : M \to M$ is defined. A set of naming functions for a cell with the name $(i,j) \in M$ determines *a template*

$$T(i,j) = \{(i,j), \varphi_1(i,j), \ldots, \varphi_k(i,j)\}. \tag{1}$$

Template (1) enumerates the names of neighbors of the given central cell with the name $(i,j) \in M$ (*neighborhood*). The commonly encountered template is "a cross": $T(i,j) = \{(i,j), (i-1,j), (i,j+1), (i+1,j), (i,j-1)\}$. A set of cells with the names from template (1)

$$S(i,j) = \{(v_0, (i,j)), (v_1, \varphi_1(i,j)), \ldots, (v_r, \varphi_r(i,j))\}$$

is called *a local configuration*, $T(i,j)$ being its *underlying template.* The cell $(v_0, (i,j))$ is further referred to as *a reference cell* for the configuration $S(i,j)$. The two local configurations $S(i,j)$ and $S'(i,j)$ with the same reference cell, whose underlying templates are written in the form of *substitution*

$$\Theta(i,j) : S(i,j) \xrightarrow{p} S'(i,j), \tag{2}$$

where $S'(i,j) = \{(u_0, (i,j)), (u_1, \varphi_1(i,j)), \ldots, (u_q, \varphi_q(i,j))\}$, $l = 0, 1, \ldots, q$, represent an elementary act of the cellular array updating with probability p. In our applications, we shall use substitutions for the left-hand and the right-hand sides for which the condition $T(i,j) = T'(i,j)$ holds.

Substitution (2) is applicable to a cell with the name $(i,j) \in M$ (a cell is randomly chosen) with probability p if $S(i,j) \in \Omega(t)$. If this condition is not true, nothing happens, an attempt to apply this substitution fails. In the opposite case the substitution $\Theta(i,j)$ is applicable with probability p. Application of the substitution $\Theta(i,j)$ to a cell with the name $(i,j) \in M$ is as follows.

- The values $u_k = f_k(v_0, v_1, \ldots, v_q), l = 0, 1, \ldots, r$, are calculated.
- The cell states $(u_k, \varphi_k(i,j))$ are replaced by the obtained values.

The application of substitution $\Theta(i,j)$ to one cell is performed in a certain period of discrete time τ, called *a step*. Application of $\Theta(m)$ to all cells of array transfers its *global state* $\Omega(t)$ to the next *global state* $\Omega(t+1)$ and is called *an iteration*. According to mechanism of CO oxidation reaction, the iteration consists of $p =| M | \times M_{diff}$ steps. The sequence $\Omega(0), \Omega(1), \ldots, \Omega(t), \ldots$, obtained as a result of iterative functioning of kinetic CA is called *an evolution*. The evolution describes the dynamics of catalytic oxidation reaction at a microscopic level.

Let us consider a substitution which models the oxygen adsorption to the catalyst surface. The molecule of oxygen is adsorbed from gas on two neighboring free cells with the probability p. For cell named (i,j), a name of one of its neighbors determined by the following function:

$$\varphi(i,j) = \begin{cases} (i-1,j), & 0 < r \leq 0.25, \\ (i,j-1), & 0.25 < r \leq 0.5, \\ (i+1,j), & 0.5 < r \leq 0.75, \\ (i,j+1), & 0.75 < r \leq 1. \end{cases}$$

As a result, the oxygen adsorption is described by the following substitution

$$\Theta(m) : \{(\star, (i,j))(*, \varphi(i,j))\} \xrightarrow{p} \{(O_{ads}, (i,j))(O_{ads}, \varphi(i,j))\}.$$

3.2 Studding of CO Oxidation Reaction Dynamics

In experiments conducted, the dynamics of an oxidation reaction is represented by the evolution of kinetic CA of size 400×400. The boundary conditions are periodical. In the initial conditions all cells are active. The oscillatory behavior was observed under a certain set of values of the rate coefficients for the reaction actions: $k_1 = 1$, $k_2 = 1$, $k_{-2} = 0.2$, $k_4 = 0.03$, $k_5 = 0.01$, $k_6 = 1$, $k_{-6} = 0.5$, $k_7 = 0.02$ the diffusion parameter $M_{dif} = 50$.

CO Oxidation reaction dynamics is presented Figure 1. This figure shows the oscillations of values of the reagent concentration (O_{ads}, O_{sub}, CO_{ads}) and the reaction rate. The reagent concentration is determined by a number of cells being in a state corresponding to this reagent to the overall amount of cells in the array. The reaction rate is defined by the amount of formed molecules CO_2 in one cell per iteration. It has experimentally been found that an amplitude and oscillations of a concentration period for each reagent and the reaction rate remain constant. Let us consider, the reaction dynamics in within the period

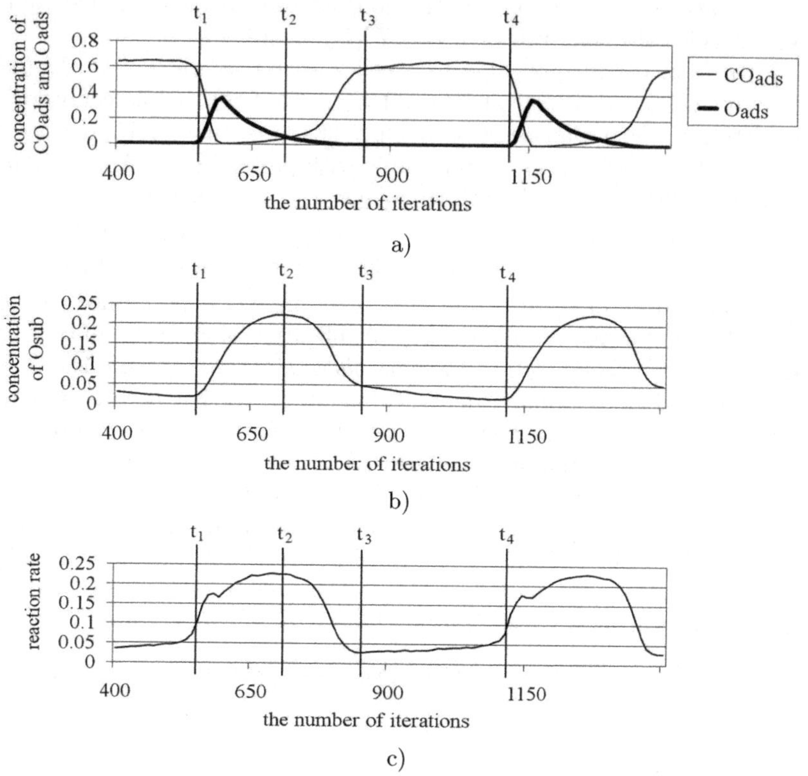

Fig. 1. The oscillations of concentrations of reagents in the oxidation reaction: O_{ads}, CO_{ads} (a), O_{sub} (b) and the reaction rate (c)

(time $t_1 - t_4$). Beginning with the moment of the time t_1, the concentration of CO_{ads} starts to decreasing and the concentration of O_{ads} sharply rises. Changes in the coverages $CO_{ads} \rightarrow O_{ads}$ occur via formation of a wave, whose front is characterized by the high concentration of catalytically active sites responsible for a maximal rate of CO_2 (Fig.1.a). The obtained O_{ads} and CO_{ads} react to form CO_2 (Fig.1.b). At the same time the concentration of adsorbed oxygen redistributes $O_{ads} \rightarrow O_{sub}$. The position of a maximum on the curve of the concentration O_{sub}, which is achieved at the time t_2 (Fig.1.c), determines the moment of a decrease in the reaction rate. Simultaneously, the catalyst surface accumulates CO_{ads}, this process being accompanied by the complete removal of O_{ads}. A minimal value of the reaction rate is stipulated by the interaction of CO_{ads} only with subsurface oxygen. At the moment of the time (t_3), CO_{ads} concentration reaches its maximum value. It is conserved for a time ($t_3 - t_4$). A decrease in the O_{sub} concentration towards a critical value (the time t_4) again creates favorable conditions for the reaction, and this completes the oscillation cycle. The results of the numerical calculation experiments correspond to the results obtained in work [2].

4 Block-Synchronous Cellular Automata for Modeling CO Oxidation Reaction

4.1 Algorithm for Transformation of Asynchronous CA into Block-Synchronous CA

The algorithm for transformation of an asynchronous cellular automaton into a block-synchronous one is as follows.

1. On the set M, a *block template* $T_{B(i,j)} = \{(i,j), \varphi_1(i,j), \ldots, \varphi_r(i,j)\}$ is defined. Here the naming functions $\varphi_1(i,j), \ldots, \varphi_r(i,j)$ enumerate r neighbors of a reference cell with the name (i,j) involved in substitutions from Θ. Denote the block by $B_{\varphi_k(i,j)} = B_k$, where $\varphi_k(i,j)$ is the reference cell name. The block B_k is characterized by the following features.

- $T(i,j) \subseteq T_{B_k}$, where $T(i,j)$ is an underlying template of substitution from the set of substitutions.
- On the naming set M, the block template B_k defines a set of partitions $\mathcal{M} = \{\mathcal{M}_1, \mathcal{M}_2, \ldots, \mathcal{M}_r\}$, where $\mathcal{M}_k = \{B_k^1, B_k^2, \ldots, B_k^G\}$. Each partition consists of $\frac{|M|}{r}$ blocks. For all partitions from $\mathcal{M}$, the following relations are carried out

$$\bigcup_g B_k^g = M, \qquad \bigcap_g B_k^g = \emptyset, \qquad g = 1, 2, \ldots, G.$$

Here the second condition is the correctness one. It requires that any two substitutions do not change the same cell state at the same time.

2. Each iteration is divided into r synchronous stages. On the k-th stage, $k = 1, 2, \ldots, r$, substitutions are applied synchronously to the k-th reference cells of all blocks belonging to the k-th partition.

Figure 2 demonstrates constructing a block template for our case. From the reaction mechanism it follows that the underlying template is "a cross" and there are two substitutions which are applied one after the other. Indeed, a substitution corresponding to oxygen adsorption may change (if such the substitution is successful) the state of a randomly chosen cell with the name (i,j) and one from four its neighboring cells $\{(i-1,j), (i+1,j), (i,j-1), (i,j+1)\}$ from the state $\star$ to the state O_{ad}. As soon as the given substitution is applied, a substitution corresponding to stage 3 is always applied to two cells updated their previous states, which in turn, select their neighbors according to "cross" template. As a result, we obtain a block template consisting of thirteen cells.

The partition $\mathcal{M}_6$ is shown in Figure 3. It includes blocks for which the cell with the number 6 is the reference one.

4.2 A Comparison of Block-Synchronous and Asynchronous Modes of Operation in CA

In the experiments carried out, a catalytic reaction is represented by the evolutions of asynchronous (as) and block-synchronous (bs) CA of of size 400×400.

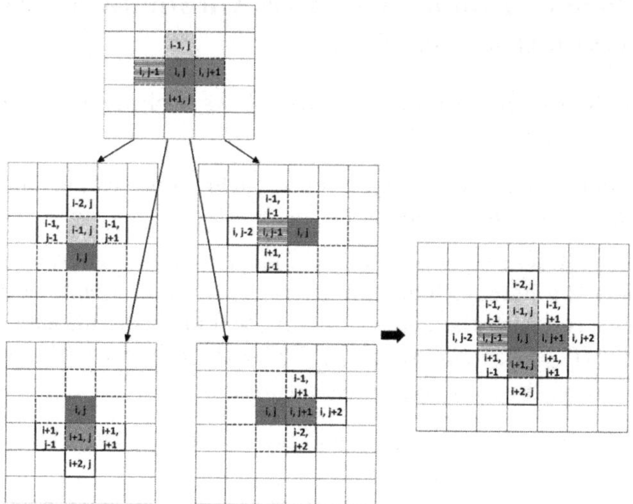

Fig. 2. The block template

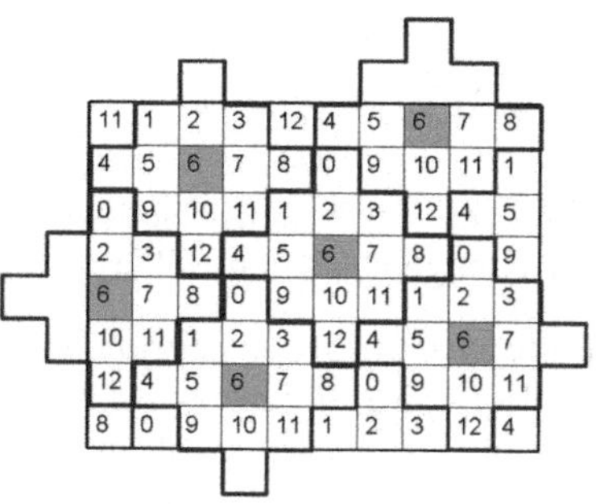

Fig. 3. The partition M_6

Reaction steps are carried out with the following values of rate: $k_1 = 1$, $k_2 = 1$, $k_{-2} = 0.2$, $k_4 = 0.03$, $k_5 = 0.01$, $k_6 = 1$, $k_{-6} = 0.5$, $k_7 = 0.02$, the diffusion parameter $M_{dif} = 50$. The same random-number generator is used in cellular automata with both operation modes. Because of this the substitution application sequences coincide. The experiments demonstrate that asynchronous CA and block-synchronous CA display a similar behavior of oxidation reaction. Figure 4 exhibits the oscillations in concentration of O_{sub} for both operation modes. A mean-square error of approximation for concentrations of the reagents (CO_2, CO_{ads}, O_{ads}, O_{sub}, $[CO_{ads}O_{sub}]$) amount to the following values: $d_{CO_2} = 0.54\%$, $d_{CO_{ads}} = 2.00\%$, $d_{O_{ads}} = 1.15\%$, $d_{O_{sub}} = 0.52\%$, $d_{[CO_{ads}O_{sub}]} = 0.2\%$. This proves that the approximation of the kinetic CA by a block-synchronous CA is correct. The time complexity of block-synchronous CA evolution is less than the time complexity of asynchronous CA evolution. This fact is associated with a

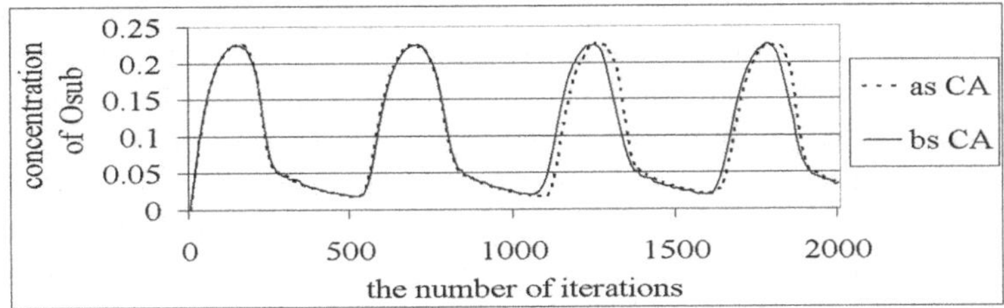

Fig. 4. The oscillations of concentrations of O_{sub} for asynchronous CA simulating oxidation reaction and its approximation by block-synchronous CA

different amount of generated random numbers. Indeed, in the asynchronous mode, all array cells are randomly chosen at each iteration. In the block-synchronous mode, only thirteen cells are randomly chosen at each iteration.

5 Parallel Implementation of Block-Synchronous Cellular Automata

Parallel implementation of the block-synchronous CA is performed by the domain decomposition method and consists in the following.

1. A cellular array of size $\mid M \mid$ is divided into non-intersecting equal parts, called *domains* (*Dom*), $\mid Dom \mid = K \times K = |M|/n$, where n is the number of processors used. The domains obtained are allocated to supercomputers processors. The memory of each processor holds its own domain of cells and copies of boundary cells of the neighboring processors. In the initial conditions all domain cells are active.

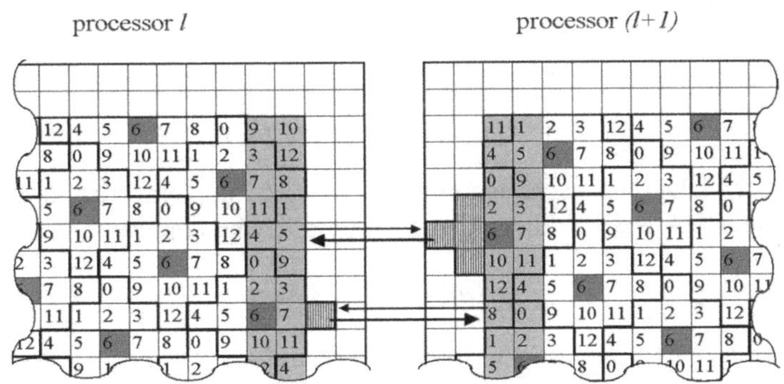

Fig. 5. Structure of the domains in the neighboring processors

Figure 5 shows the domains in the l-th and the $(l+1)$-th processors (here, on the naming set, the partition $\mathcal{M}_6$ is defined). It is evident, that in the course of the decomposition of cellular array by rectangle domains, certain blocks are shared among the neighboring processors. As a result, a part of substitutions cannot be applied to the reference cell with number 6 if it is a boundary one (Fig.5, the l-th processor) or if it is a near-boundary one (Fig.5, the $(l+1)$-th processor). Hence, to apply substations to the cell with number 6, the states of the cells with numbers 1, 4, 5, 9 must be sent from the l-th processor to the $(l+1)$-th processor and should be written into the corresponding cells to the left of the domain in the first case. And in the second case, the state of the cell with number 8 must be sent from $(l+1)$-th processor to l-th processor and should be written into the corresponding cells to the right of the domain.

2. The evolution iteration of the block-synchronous cellular automaton consists of 13 stages. At each stage, the partition $\mathcal{M}_l, l = 1, \ldots, r$, is randomly chosen. The substitutions of the set Θ are applied synchronously to the cells with a chosen number l in all blocks of the partition $\mathcal{M}_k$ in all the processors. (The quantity of such cells will be $\mid Dom \mid /n$.) After application of substitutions, each processor gives back to the neighboring processors the updated values of their cells. In addition, each processor sends updated values of cells its own boundary and near-boundary rows (columns) into the neighboring processors. The values of those cells are arranged around the perimeter of the domain. Those four rows (columns) are intended for storage of copies of updated values of the neighboring processors. The total volume V of the data exchange, which is sent by each processor to the four neighboring processors, is equal to $8K$ cells. As a result of the data exchange, each domain has values of all cells which are necessary for execution of the next stage.

The cluster "$MVS{-}100K$" of the Joint SuperComputer Center of the Russian Academy of Sciences is used for parallel implementation of the block-synchronous CA. The basic technical characteristics "$MVS - 100K$" are the following: 1460 nodes consisting of two quad-core microprocessors Intel Xeon, the time of transfer of one byte $t_b = 0.7ms$, the latency time $t_{lat} = 3.2ms$. The experiments have shown that the time of application of a substitution to a cell for the supercomputer "$MVS - 100K$" is $\tau = 0.58ms$.

According to characteristics "$MVS - 100K$" and a well-known condition of efficiency

$$\tau \cdot |Dom| > h \cdot (t_{lat} + V \cdot t_b),$$

where h=13 (the data exchange number), parallel implementation of the block-synchronous CA will be efficient under the following condition

$$|Dom| > 128 \times 128. \tag{3}$$

Table 1 contains the experimental results of parallel implementation of the block-synchronous CA of sizes 9000×9000 and 12000×12000 cells. The quality of parallel implementation is estimated by efficiency of parallelization $Q(n) = \dfrac{T_1}{T_n \cdot n}$, where T_1 is computation time on one processor, T_n is computation time on n processors. The experiments have demonstrated that the parallel implementation of the block-synchronous cellular automaton efficiency is greater than 90% if condition (3) is met.

Table 1. Efficiency $Q(n)$ of the parallel implementation

$I \times J$	Parameters	n					
		1	4	16	32	64	128
9000×9000	T_n, s $Q(n)$	568.87 1	146.37 0.97	37.48 0.95	18.71 0.95	9.56 0.93	4.96 0.89
12000×12000	T_n, s $Q(n)$	1005.54 1	256.35 0.98	66.97 0.94	33.89 0.93	16.98 0.93	8.49 0.92

6 Conclusion

In this paper, the results of investigation of kinetic cellular automaton simulating oxidation CO on surface Pd_{110} are presented. Experiments showed the following.

- Asynchronous and block-synchronous modes of CA demonstrate identical character of behavior. Approximation error is admissible for probabilistic algorithm.
- Time complexity of block-synchronous CA evolution less than time complexity of asynchronous CA evolution on a single processor. This fact is associated with different amount generated random numbers.
- Parallel implementation of block-synchronous cellular automaton has the efficiency above of 90%.

References

1. Latkin, E.L., Elokhin, V.I., Matveev, A.V., Gorodetskii, V.V.: Manifestation of the adsorbed co diffusion anisotropy caused by the structure properties of the Pd(110) - (1x2) surface on the oscillatory behavior during co oxidation reaction - Monte-Carlo model. Chemistry for Sustainable Development 11, 173–180 (2003)
2. Elokhin, V.I., Latkin, E.I., Matveev, A.V., Gorodetskii, V.V.: Application of statistical lattice models to the analysis of oscillatory and autowave processes in the reaction of carbon monoxide oxidation over platinum and palladium surfaces. Kinetics and Catalysis 44(5), 692–700 (2003)
3. Toffolli, Ò., Margolus, N.: Cellular Automata Machines. MIT Press, USA (1987)
4. Nedea, S.V., Lukkien, J.J., Hilbers, P.A.J., Jansen, A.P.J.: Methods for parallel simulations of surface reactions. Advances in computation: Theory and practice, Parallel and distributed scientific and engineering computing 15, 85–97 (2004)
5. Bandman, O.L.: Synchronous versus asynchronous cellular automata for simulating nano-systems kinetics. Bulletin November Computer Center, Computer Science, vol. 25, pp. 1–12. NCC Publisher, Novosibirsk (2006)
6. Bandman, O.L.: Parallel simulation of asynchronous cellular automata evolution. In: El Yacoubi, S., Chopard, B., Bandini, S. (eds.) ACRI 2006. LNCS, vol. 4173, pp. 41–48. Springer, Heidelberg (2006)

Parallel System for Abnormal Cell Growth Prediction Based on Fast Numerical Simulation

Norma Alias[1], Md. Rajibul Islam[2], Rosdiana Shahir[3], Hafizah Hamzah[3], Noriza Satam[3], Zarith Safiza[3], Roziha Darwis[3], Eliana Ludin[1], and Masrin Azami[1]

[1] Ibnu Sina Institute, Faculty of Science, University Technology, Malaysia
[2] Faculty of Information Science and Technology, Multimedia University, Malaysia
[3] Department of Mathematics, Faculty of Science, University Technology, Malaysia
norma@ibnusina.utm.my, md.rajibul.islam05@mmu.edu.my, {zarithsafiza.ag,norizasatam, roziha.darwis,norhafizahamzah}@gmail.com, supida_machine@yahoo.com

Abstract. The paper focuses on a numerical method for detecting, visualizing and monitoring abnormal cell growth using large-scale mathematical simulations. The discretization of multi-dimensional partial differential equation (PDE) is based on finite difference method. The predictor system depending on users input data via a user interface, generating the initial and boundary condition generated from parabolic or elliptic type of PDE. The processing large sparse matrixes are based on multiprocessor computer systems for abnormal growth visualization. The multi-dimensional abnormal cell has produced the numerical analysis and understanding results at the target area for the potential improvement of detection and monitoring the growth. The development of the prediction system is the combinations of the parallel algorithms, open source software on Linux environment and distributed multiprocessor system. The paper ends with a concluding remark on the parallel performance evaluations and numerical analysis in reducing the execution time, communication cost and computational complexity.

Keywords: parallel system, abnormal cell growth simulation, IADE method, AGE method, distributed memory systems.

1 Introduction

This paper focuses on the development of mathematical modelling and simulation for the human tumour growth involving for simulation up to 10^{-9} computational skill. The implementation of parallel algorithm of the large sparse matrix is visualised and monitor the tumour growth characteristics. The development of the parallel system is suitable as the prediction of the abnormal cell growth such as breast, brain and thyroid tumor. Based on the initial and boundary condition on the properties of tumor cells growth, some mathematical models have been developed to quantify the proliferation and invasion dynamics of tumor within anatomically accurate heterogeneous human cells [7][4]. Andrews et al. [1] and Angelis et al. [2] was presented mathematical

C.H. Hsu and V. Malyshkin (Eds.): MTPP 2010, LNCS 6083, pp. 222–231, 2010.

model of glioblastoma tumor spheroid invasion and an advective-diffusion model for solid tumor evolution in vivo respectively. The contribution of this paper is a fast numerical method to visualize abnormal cell growth in multidimensional space. This research relates to visualizing abnormal cells growth depending on users input data by the implementation of fast numerical methods in providing the high speed and accurate prediction.

1.1 Problem Statement

The mathematical modeling and simulation on one, two and three dimensional of PDE for abnormal tumor growth involving parameter estimation, growth rate, diffusion coefficient, initial and boundary condition. The model of Swanson, Alvord and Murray [10], Bellomo and Preziosi [11], 3D Parabolic Equation (3D) are discretized using finite difference method. Next, numerical schemes called New Iterative Alternating Decomposition Explicit (IADE), [5], New Alternating Group Explicit Methods (AGE) [7], BRIAN [7][4] and Red Black Gauss Seidel methods (RBGS) are used for solving linear system of equation generated by the finite difference method. The flowchart (see Figure 1) below illustrates the machine interface of multi-dimensional mathematical simulation for the abnormal cell growth.

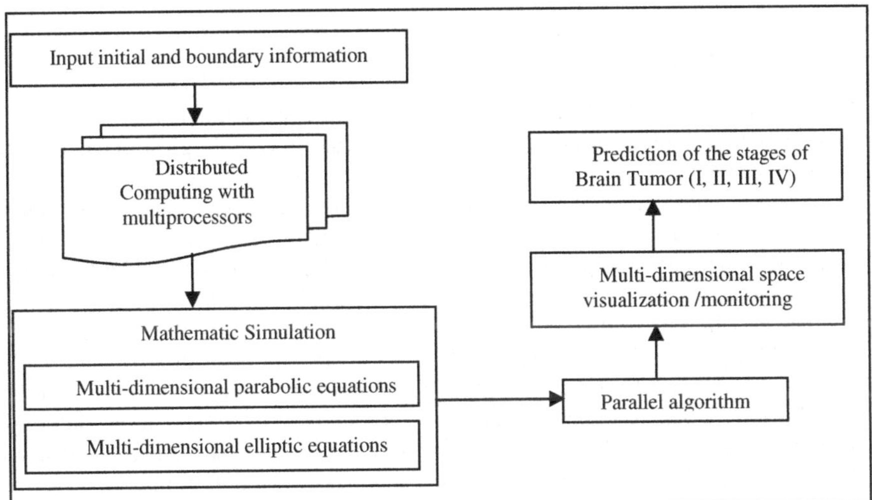

Fig. 1. Flowchart illustrating machine interface of the abnormal cell growth visualization software of the present invention

2 Solving the Cell Growth Model

The model of Swanson, Alvord and Murray [10] is as below:

$$\frac{\partial c}{\partial t} = \nabla \cdot (D(x)\nabla c) + \rho c \tag{1}$$

where is the concentration of cells at any position x and time t, ρ is the units of per day and represents the net rate of growth of abnormal cells, including proliferation, loss and death, D denotes the units of cm2 per day and represents the diffusion coefficient of cells in brain tissue, = D_g (constant for x in grey matter) and = D_w (constant for x in white matter).

∇ represents the spatial gradient. The diffusion term describes the active migration of the glioma cells using a simple Fickian diffusion [12] where cells move from regions of higher to lower densities. Abnormal cells are assumed to grow exponentially.

The 2D PDE with parabolic type is referred to as the model implemented by Bellomo and Preziosi [11]:

$$\frac{\partial u}{\partial t} = -\nabla \cdot (Wu) + \nabla \cdot (Q\nabla u) + \tau - Lu \tag{2}$$

Where, $\tau = \tau(u)$ is the proliferation coefficient,

$L = L(u)$ is the death coefficient,

W is the diffusion coefficient,

Q is the drift velocity field and in two dimensions, $W = (P, R)$

The equation (equation 3) below is a three Dimensional Parabolic Equation (3D) that is implemented in this study,

$$\frac{\partial u}{\partial t} = -\frac{\partial (Pu)}{\partial x} - \frac{\partial (Ru)}{\partial y} - \frac{\partial (Su)}{\partial z} + Q\frac{\partial^2 u}{\partial x} + \frac{\partial Q \partial u}{\partial x \partial x} + Q\frac{\partial^2 u}{\partial y^2} + \frac{\partial Q \partial u}{\partial y \partial y} + Q\frac{\partial^2 u}{\partial z^2} + \frac{\partial Q \partial u}{\partial z \partial z} + \tau - Lu \tag{3}$$

2.1 Parallel Systems

The abnormal cell mathematical model involves some parameters need to be counted. Obviously the problems need huge repetitive calculations on large amounts of data to produce the valid results.

To achieve implementation in parallel, suitable software and hardware must be provided as a platform that supports the simultaneous execution of heterogeneous distributed memory computer systems. The software services are built on Linux platform which is based on open source technology. The algorithm is programmed in C language while the software tools involve a web browser, Apache web server, Perl-CGI, HTML pages, PHP, XML, UML, PVM, MPI, C programming and MySQL database. A related factor is that multiple computers often have more total main memory than a single computer, enabling problems that require larger amounts of main memory to be tackled. The software is designed and analysed for its efficiency, distributional, robustness, adaptiveness and stability of the algorithms.

3 Fast Numerical Simulation

Like many other numerical methods, the approach begins with the domain discretization of multivariable the parabolic equation. The implement of fast numerical simulation under consideration are some explicit methods called IADE, AGE and Brian [9]. RBGS is chosen as the comparison method among the IADE, AGE and Brian methods. The originality of this research shows that these methods have found a generally applicable high-level expression of parallelism in solving a large system of heat transfer of multidimensional PDEs. Those numerical methods are straightforward and well suit implemented into the distributed memory parallel computer platform [8]. The numerical analysis and the parallel performance evaluations are analyzed from the aspect accuracy, convergence efficiency, effectiveness and temporal performance.

3.1 AGE Fractional Scheme on Based on the Douglas-Rachford Formula

$$(G_1 + rI)u_{(r)}\left(k + \tfrac{1}{4}\right) = (rI - G_1 - 2G_2 - 2G_3 - 2G_4)u_{(r)}^{(k)} + 2f \tag{4}$$

$$(G_2 + rI)u_{(r)}^{\left(k+\frac{1}{2}\right)} = G_2 u_{(r)}^{(k)} + ru_{(r)}^{\left(k+\frac{1}{4}\right)} \tag{5}$$

$$(G_3 + rI)u_{(r)}^{\left(k+\frac{3}{4}\right)} = G_3 u_{(r)}^{(k)} + ru_{(r)}^{\left(k+\frac{1}{2}\right)} \tag{6}$$

$$(G_4 + rI)u_{(r)}^{(k+1)} = G_4 u_{(r)}^{(k)} + ru_{(r)}^{\left(k+\frac{3}{4}\right)} \tag{7}$$

It is notable that the Brain Method is dependent on the modification values of the acceleration parameter *r with* the time step up to *(k+1.5).*

3.2 Interactive Alternating Decomposition Explicit Method [10]

At time level$\left(k + \frac{1}{2}\right)$,

$$(I + \alpha G_1)u^{k+\frac{1}{2}} = (I + (\alpha + 2r)G_1)(I + 2rG_2) \tag{8}$$

At time level $(k + 1)$,

$$(I + \alpha G_2)u^{k+1} = u^{k+\frac{1}{2}} + \alpha G_2 u^{(k)} \tag{9}$$

With $\alpha = \frac{1}{12} - \frac{2}{3}r$ and $\beta = \frac{2r(3v-2r)}{3}$

4 Parallel Algorithms

The IADE, AGE, Brian and RBGS iterative using parallel computing converging very fast compares to the its sequential algorithm. The fast numerical methods are allowed to divide arrays among local processors and to minimize the communication as shown in Figure 2. The implementation of domain decomposition technique requires the non-overlapping sub-domain. The parallel algorithm of fully explicit of IADE, AGE, BRIAN and RBGS can therefore be used to maximum advantage on distributed memory system. The flowchart in Figure 2 illustrates the parallel algorithm and the communication activities among the server and client processors. The data structure has to be decomposed where given set of ranges assigned to particular processors must be physically sent to those processors in order for the processing to be done. The result must be sent back to whichever processors responsible for coordinating the global result. As the parallel computing executes the same task on multiple processors simultaneously, it can reduce the time execution of computational complexity.

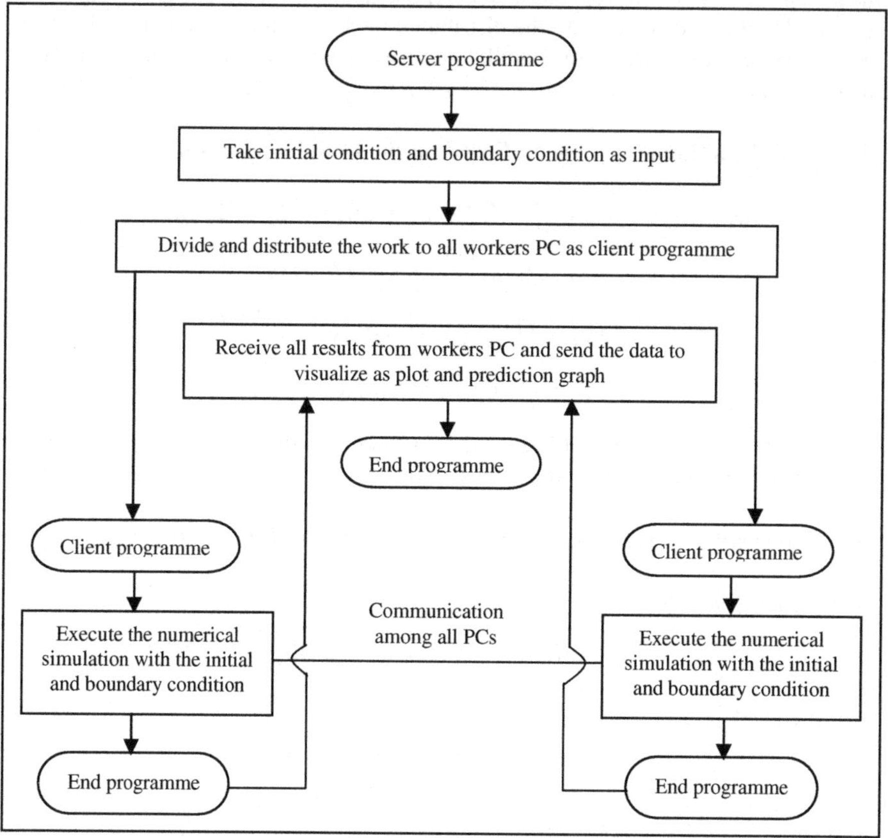

Fig. 2. Parallel algorithms of server and client procedures for the mathematical simulation appeared on the workflow of the high performance computing system

5 Results and Discussion

Based on the initial and boundary condition on the properties of abnormal cells growth, some mathematical models have been modified to quantify the proliferation and invasion dynamics of tumor within anatomically accurate heterogeneous human cells [7][4]. The results obtained reflect the numerical analysis and the parallel performance evaluations are analyzed of accuracy, convergence, stability, errors, residual, speedup, efficiency, effectiveness and temporal performance. The parallel system development are presented the abnormal cell growth graphically and in highly accurate prediction based on the IADE, AGE, Brian and RBGS methods.

The growth rates increase in the first 24 days consistently. After 24 days, the abnormal cells become highly active in evolution. The abnormal cell will grows more than 1000 cells after 30 days.

By data experiment, the values used in solving the mathematical problem are:

- Drift coefficients of P, R and S are 10-5, 10-7 and 10-9 respectively, while the diffusion coefficient, Q is 10-3.
- For each of the proliferation coefficient, and death coefficient, L the values which have been taken are 10^{-2} and 10^{-8}.

A heat capacity for 3000 J/Kg°C and a density of 920 Kg/m3 were used for both normal and cancerous tissue. The metabolic heat generation was 450 W/m3 and blood perfusion rate for normal tissue was considered to be 0.00018 ml/s/ml, and for cancerous tissue values of 29,000 W/m3 and 0.009 ml/s/ml were used to account for the higher blood perfusion rates and metabolic heat generation respectively. An effective thermal conductivity of 0.42 W/m, as estimated by Chato J. C., [3], was used for both normal and cancerous tissue. 3D visualization in Figure 3 shows that heat transfer coefficient of 5 W/m2 K was used as a convective boundary condition at the skin surface to account for natural convection.

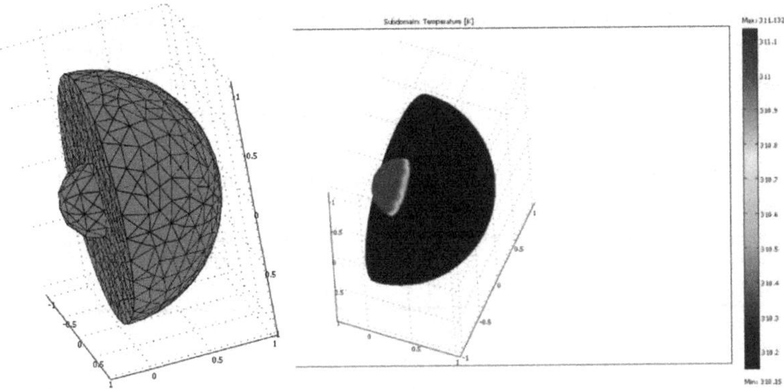

Fig. 3. 3D models of the Mesh of abnormal cell (left) and 3D sub-domain model of abnormal cell visualization by heat detection (right)

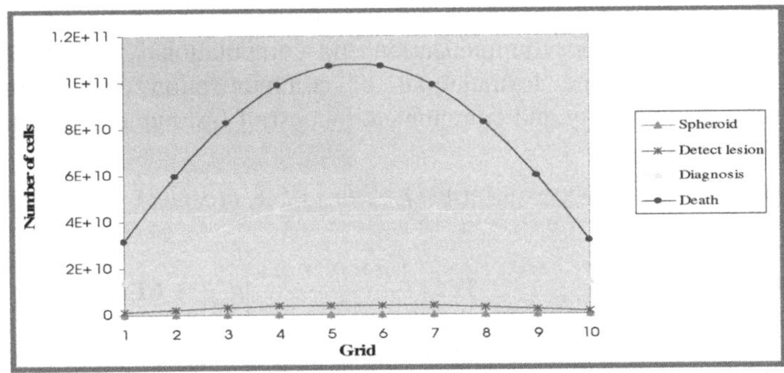

Fig. 4. The number of cell abnormal cell growth in 100 days and visualization for cell growth of untreated patient

In Figure 4, the concentration gradient of abnormal cell growth cells is represented by the curves, the spheroid and macroscopic levels involving the detected lesion, diagnosis and the death. The curve for spheroid level shows that the abnormal cells form only a small dense lesion. However, the cells are highly diffuse after have been detected and when the patients are untreated, the diffusion of these abnormal cells will caused death in only a short term.

5.1 Parallel Performance Evaluation

The speedup, efficiency and effectiveness for the parallel algorithm of fast numerical method are obtained in Table 1. The parallel performance measurements of the parallel algorithm are improved by the increasing of the number of processors. The speedup is increasing as the number of processors, p increasing. Comparable speedups are obtained for all applications with 20 processors. This phenomenon as stated in Amdahl's law shows that the communication cost will eventually become dominant over local computation cost after a certain stage. The efficiency of the parallel algorithm decreases versus number of processors. This could be explained by the fact that several factors lead to the increasing of the communication cost, delay, idle time and load balancing of computational complexity. The effectiveness is increasing up to the optimum number of processors. Then it decreases as number of processors goes beyond 12 processors due to message latency within the processors.

Table 1. Time execution, speedup, efficiency, and effectiveness in respect of the quantity of processors

Num of proc	execution	Speedup	Efficiency	Effective
1	0.0119	1	37.45315	83.4376
4	0.0025	4.739	20.771075	468.4684
8	0.0012	50.649	11.83382	1008.369
12	0.00017	70.005	9.206275	3076.425
16	0.00015	76.337	8.184757	2369.456
20	0.00001	83.062	7.025689	2298.36

In order to assess the communication and computational ratio of the proposed parallel algorithm, some formulations of communication cost is utilized. The computational complexity and communication cost of fast numerical method is given as the following,

Computation complexity: $(15(m-1)^3+14m^3)\, T + (13(m-1)^3+12m^3)\, D$ (10)

Communication cost: $$12L\left(\frac{(m\times m\times m)}{2}\right)t_{data} + 6L\left(t_{starup} + t_{idle}\right) \quad (11)$$

Where m is the size of matrix, T and D are arithmetic operations, L is a number of iterations.

Table 2 shows the communications cost depends on number of processors. When the numbers of processors increase, consequently the communication time will also increase and the computational time will decrease. The communication cost is decrease when the numbers of data-sending m and number of iterations to the convergence criteria are increases. The ratio of computational and communication times show that the computational time is much greater that communication time. The domain decomposition strategy is efficiently utilized, straightforward run on parallel computer systems. As the results, sending a larger value of messages and the behavior of communication cost are reflected the communication cost. Figure 5 shows the parallel performance evaluations for IADE, AGE and Brian methods in terms of time execution, speedup, efficiency, effectiveness and temporal performance.

Table 2. Parallel execution, communication, idle times, the ratio of computational and communication time for (100 X 100 X 100) and (140 X 140 X 140) size of matrices

M	(100 X 100 X 100)					
p	**parallel**	**comp**	**ratio**	**comm**	**comm1**	**idle**
5	99.9	58.59	1.42	41.31	29.43	11.87
%		58.7		41.4	29.5	11.9
10	73.01	29.29	0.67	43.71	29.43	14.27
%		40.1		59.9	40.3	19.6
15	69.11	19.73	0.4	49.27	29.43	19.84
%		28.6		71.4	42.7	28.8
20	63.99	15.15	0.31	48.85	29.43	19.42
%		23.7		76.3	46.0	30.3

M	(140 X 140 X 140)					
p	**parallel**	**comp**	**ratio**	**comm**	**comm1**	**idle**
5	193	111.8	1.38	81.25	49.14	32.11
%		57.9		42.1	25.5	16.6
10	136.2	55.89	0.7	80.34	49.14	31.21
%		41.0		59.0	36.1	22.9
15	121	36.66	0.43	84.29	49.14	35.15
%		30.3		69.7	40.6	29.1
20	110.7	27.15	0.32	83.55	49.14	34.41
%		24.5		75.5	44.4	31.1

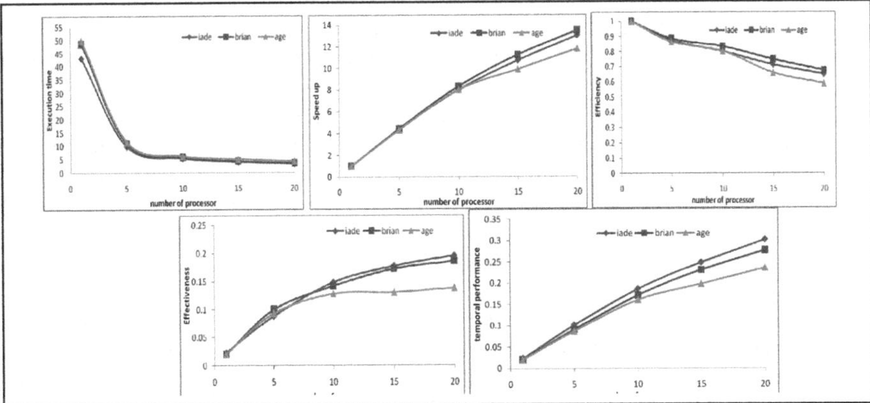

Fig. 5. Parallel performance evaluations for IADE, AGE and Brian in terms of time execution, speedup, efficiency, effectiveness and temporal performance

6 Conclusion

In this paper, the parallel system of this study is being able to visualize the abnormal cell growth in multi-dimensional space using distributed computer systems. . The finite difference method and a weighted approximation of parabolic equation have been formed accurately to obtain the volume of the abnormal cell and rate of cell growth in time and space. The alternative parallel algorithms; IADE, AGE and Brian methods are a well suite and relatively efficient for parallel system development on the distributed parallel computer systems. These schemes are extremely effective in reducing data storage accesses, computational and communication cost on a distributed computer systems. As a conclusion, the parallel system of fast numerical methods can be concluded as a well suite performance tools in solving the grand challenge application of multi-dimensional PDE problems for abnormal cell growth prediction. The future research will be the implementation the higher order of parabolic-elliptic type for the parallel system optimization in terms of numerical analysis and parallel performance evaluations.

References

1. Andrew, M.S., Demuth, T., Mobley, D., Berens, M., Leonard, M.S.: A Mathematical Model of Glioblastoma Tumor Spheroid Invasion in a Three-Dimensional In Vitro Experiment. Biophysical Journal 92, 356–365 (2007)
2. Angelis, E.D., Preziosi, L.: Advection-Diffusion Models for Solid Tumour Evolution in Vivo and Related Free Boundary Problem. Mathematical Models and Methods in Applied Sciences 10(3), 379–407 (2000)
3. Chato, J.C.: Fundamentals of bioheat transfer. In: Gautherie, M. (ed.) Thermal dosimetry and treatment planning, pp. 1–56. Springer, Berlin (1990)
4. Norma, A., bin Masseri, M.I.S., Rajibul Islam, M., Khalid, S.N.: The Visualization of Three Dimensional Brain Tumors' Growth on Distributed Parallel Computer Systems. Journal of Applied Sciences. Asian Network for Scientific Information (ANSINET) 9(3), 505–512 (2009)
5. Norma, A., Norfarizan, M.S., Pheng, H.S.: High performance simulation for brain tumors growth using parabolic equation on heterogeneous parallel computer systems. Journal of Information Technology and Multimedia 4, 39–52 (2007)
6. Sahimi, M.S., Mansor, N.A., Nor, N.M., Nusi, N.M., Norma, A.: A High accuracy variant of the Iterative Alternating Decomposition Explicit Method for Solving the Heat Equation. International Journal of Simulation and processing Modelling 2(½), 45–49 (2006)
7. Norma, A., Norfarizan, Hidayah, S.N.K., Dolly, S., Phang, T.I.: The parallelism algorithm of human tumor growth using high performance computing. In: The 2nd International Conference on Pervasive Computing Technologies for Healthcare, Tampere, Finland (2008)
8. Norma, A., Norfarizan, M.S., Hidayah, S.N.K.: High Performance Visualization of Human Tumor Growth Software. In: Palma, J.M.L.M., Amestoy, P.R., Daydé, M., Mattoso, M., Lopes, J.C. (eds.) VECPAR 2008. LNCS, vol. 5336, pp. 591–599. Springer, Heidelberg (2008)

9. Evans, D.J., Sahimi, M.S.: The Alternating Group Explicit Iterative method for Solving Parabolic Equations. Intern. J. Computer math. 24, 127–145 (1988)
10. Swanson, K., Alvord, E., Murray, J.D.: A Quantitative Model for Differential Motility of Gliomas in Grey and White Matter. Cell Prolif. 33, 317–329 (2000)
11. Bellomo, N., Preziosi, L.: Modeling and mathematical problems related to tumor evolution and its interaction with the immune system. Math. Comput. Modelling 32, 413–452 (2000)
12. Murray, J.D.: Mathematical Biology, 2nd edn. Biomathematics, vol. 19. Springer, Berlin (1993)

Parallel Scalable Algorithms with Mixed Local-Global Strategy for Global Optimization Problems*

Konstantin Barkalov, Vasily Ryabov, and Sergey Sidorov

Nizhni Novgorod State University, pr. Gagarina. 23,
603950 Nizhni Novgorod, Russia
barkalov@fup.unn.ru, {vasily.v.ryabov,sidorov.sergey}@gmail.com

Abstract. This paper continues development of information-statistical approach to minimization of multiextremal functions in the case of non-convex constraints. Proposed approach is called index method. Solving multidimensional problem is reduced to solving equivalent single dimensional one. Dimension reduction is based on Peano curves that allow mapping multidimensional hyper cube onto the segment on real axis. We also use rotating Peano curves that allowed effectively parallelize algorithm to use hundreds of processors. Special attention was paid to mixed local-global strategy for algorithm convergence acceleration.

Keywords: global optimization, parallel computing, index method, local-global strategy.

1 Introduction

Following research work continues development of well-known approach for minimization of multiextremal functions in the case of non-convex constraints, described in [1–6] and called index method of global optimization. This approach takes into account every constraint separately and it does not use penalty functions method. Solving multidimensional problem is reduced to solving equivalent single dimensional one. Reduction is based on Peano curves (called Peano evolvents). It allows one-to-one mapping multidimensional hyper cube onto the segment on real axis. We used set of evolvents scheme ("rotated evolvents") that allows effectively parallelize algorithm to use hundreds of processors. Global search algorithm was extended by mixed local-global strategy for algorithm convergence acceleration. Global and local-global strategies are compared in experiments section of the paper. We also provide experiments results of DIRECT[9] global search method on the same problems. Experiments prove value of parallel index method with mixed local-global strategy. Experiments were performed on cluster of Nizhni Novgorod State University (NNSU).

* Supported by grants counsel of President of Russian Federation (grant № MK-1536.2009.9), Supported by federal Agency of Science and Innovations and state contract № 02.740.11.5018.

C.H. Hsu and V. Malyshkin (Eds.): MTPP 2010, LNCS 6083, pp. 232–240, 2010.

2 Problem Definition

Lets us consider the following global optimization problem

$$\varphi^* = \varphi(y^*) = \min\{\varphi(y): y \in D,\ g_j(y) \le 0,\ 1 \le j \le m\},$$
$$D = \{y \in R^N: a_i \le y_i \le b_i,\ 1 \le i \le N\}, \qquad (1)$$

where target multiextremal function $\varphi(y)$ (further called $g_{m+1}(y)$) and left parts of constraints $g_j(y), 1 \le j \le m$, that satisfy Lipschitz condition with constants $L_j, 1 \le j \le m+1$, i.e.

$$|g_j(y_1) - g_j(y_2)| \le L_j |y_1 - y_2|,\ 1 \le j \le m+1,\ y_1, y_2 \in D.$$

Using Peano curve $y(x)$ for one-to-one mapping $[0,1]$ onto the N-dimensional hypercube D

$$D = \{y \in R^N: -2^{-1} \le y_i \le 2^{-1},\ 1 \le i \le N\} = \{y(x): 0 \le x \le 1\},$$

we can reduce current problem to the single dimensional problem:

$$\varphi(y(x^*)) = \min\{\varphi(y(x)): x \in [0,1],\ g_j(y(x)) \le 0,\ 1 \le j \le m\}.$$

Described scheme reduce multidimensional problem with Lipschitz minimized function and Lipschitz constraints to single dimensional problem where functions satisfy Holder conditions (look. [1]), i.e.

$$|g_j(y(x')) - g_j(y(x''))| \le K_j |x' - x''|^{1/N}, x', x'' \in [0,1], 1 \le j \le m+1,$$

where N is dimension of initial multidimensional problem and coefficients K_j are connected with Lipschitz constants L_j of initial problem by inequalities $K_j \le 4 L_j \sqrt{N}$.

Both different index method modifications and convergence theory can be found in [3], [5].

3 Multiple Mapping

Dimension reduction of multidimensional problem to single dimensional one based on evolvents has the following major properties: continuity and keeping argument variation uniformly bounded.

However in this case some information about points proximity in multidimensional space looses because point $x \in [0,1]$ has only left and right neighbors but corresponding point $y(x) \in R^N$ has 2^N directions for neighborhood points. If 2 neighbor points y', y'' in N-dimensional space mapping by Peano evolvent onto the segment then mapped points x', x'' can be far from each other on [0,1]. As a result several local externals (less than 2^N) in single dimensional problem correspond to one global minimum in multi dimensional problem. This fact makes worse properties of single dimensional problem. Multiple mapping

$$Y_L(x)=\{y^1(x),\dots,\ y^L(x)\} \tag{2}$$

allows storing partial neighborhood information instead of usage single Peano curve $y(x)$ ([2], [4]).

Every Peano curve $y^i(x)$ from $Y_L(x)$ can be obtained as a result of shifting along main diagonal of hyper interval D. A set of Peano curves created in that way allows to get close x', x'' in one of $y^i(x)$ for every images y', y'' differ in one dimension.

3.1 Rotated Evolvents

One of disadvantages of classical shifted evolvents (S-evolvents) approach is appearance of additional constraint that creates difficult search area in single dimensional segment. At the second, the number of evolvents L (as a result the number of solved problems in parallel) depends on required accuracy ε of solved problem for S-evolvents. Usage more than $\lceil \log_2(\varepsilon^{-1}) \rceil$ evolvents is not effective from the resource economy point of view. For example, it is well-handled to choice less than 10 evolvents to solve a problem with accuracy 10^{-3} for every coordinate.

Scheme of evolvent creation proposed in [7] allows to overcome these disadvantages. The differentiation of this scheme is evolvents development of set of Peano curves by rotation across point of origin. A set of Peano curves created in that way allows to get quite close x', x'' in one of $y^i(x)$ for every images y', y'' differ in one dimension.

Built this way evolvent will be called rotated evolvents or R-evolvents.

The maximum number of different evolvent rotations that map N-dimensional hyper cube onto single dimensional segment is equal to 2^N. It is superfluity to use all of them. It is necessary to choice a part them. We propose to rotate evolvents on $\pm\pi/2$ angle for every coordinate plane. The number of such rotation pairs is defined by the number of space planes and is equal to $C_N^2 = N(N-1)/2$. Current approach allows to build up to $N(N-1)+1$ mappings of N-dimensional area taking initial mapping into account. Additional constraint existing in S-evolvents [2] does not appear in R-evolvents. Current approach is scalable to 2^N if it is necessary.

3.2 Parallel Index Method

A set of single dimension multiextremal problems forms by a set of mappings $Y_L(x)=\{y^1(x),\dots,\ y^L(x)\}$

$$\min\{\varphi(y^l(x)){:}x\in[0,1],\ g_j(y^l(x))\le 0,\ 1\le j\le m\},\ 1\le l\le L. \tag{3}$$

Every problem from the set can be solved independently, and every calculated value $z=g_\nu(y')$, $y'=y^i(x')$ of function $g_\nu(y)$ in i-th problem can be interpreted as calculation of $z=g_\nu(y')$, $y'=y^s(x'')$ values for any s-th problem without repeated time-consuming computations of function $g_\nu(y)$. This informational unity allows to solve current problem (1) by parallel solving of L problems (3) on the set of segments [0,1]. Every single dimension problem can be solved on separate processor.

Every processor uses L queues for inter-process communication to store here information about its own iterations for other processors. The scheme does not contain any master processor that increases computations reliability.

More detailed description of solving rules of parallel index algorithm of global optimization is presented in [8].

3.3 Local Adaptive and Mixed Algorithms

Local adaptive algorithm is a modification of index method of global search. The main assumption is usage of current estimation of probability density of current optimum point distribution for choosing the next point beginning from some iteration. Density estimations are determined by values of problem functionals calculated in iteration points. Consequently density is re-estimated every iteration. Density maximums correspond to current optimal values neighborhoods.

Decision rule of local adaptive algorithm is presented in [6]. Significant method parameter is an integer $0 \le \alpha \le 30$ that determines behavior of convergence. With $\alpha = 0$ search procedure has a global behavior, with $\alpha = 30$ it has a local one.

Mixed algorithm is a modification of index method of global search where iteration defined by index method alternates with iterations defined by local adaptive algorithm beginning from the certain step. Alternation frequency is a method parameter.

Presented scheme is adapted to the parallel index method with rotated evolvents.

4 Experiments Results

Global Expert software system for parallel solving multi dimensional global optimization problems is under development at the department of Computer Science of Nizhni Novgorod State University. Effective processing large volume of search data is one of the Global Expert features. Search data processing subsystem stores all calculated trial points and has its own swapping mechanism that takes into account global search algorithm features. Support of parallel mixed local-global method is implemented in this subsystem. Also mixed scheme is supported for resuming computations after the stopped previous search run (for example, after exhaustion of computational resources).

4.1 Test Functions Classes

The problem of test cases choice for global search methods comparison oriented on definite functions class is very important. There are a lot of different classes of Lipchitz multiextremal functions for global search methods comparison in the literature. Amount of them are generated automatically. One of such known generators is GKLS, described, for example, in [12] and free for download. The kind of GKLS generator is a clear way to set generated problems difficulty: the number of local minima, size of attraction areas and so on. That was a reason to use GKLS for the experiments.

Paper [13] describes 6 classes of 100 functions for global search methods comparison. Table 1 has parameters of GKLS-generator for every class. Where N – problem dimension, M – number of local minima, f^* - global optimum value, d – distance from global minimum point to basic paraboloid vertex, r_g – radius of global optimum attraction area. Search area is $[-1.0, 1.0]^N$.

Table 1. GKLS-generator parameters for test functions classes

Class	N	M	f^*	d	r_g
1-simple	2	10	-1.0	0.66	0.33
2-hard	2	10	-1.0	0.90	0.20
3-simple	3	10	-1.0	0.66	0.33
4-hard	3	10	-1.0	0.90	0.20
5-simple	4	10	-1.0	0.66	0.33
6-hard	4	10	-1.0	0.90	0.20

4.2 Comparison with DIRECT and LBDIRECT Algorithms

Well-known global search method DIRECT and its modification LBDIRECT (locally-biased DIRECT) are described in [9–11]. Experiments results for DIRECT and LBDIRECT were used from [13] for the same test functions classes. Moreover results for well-known base of index method (Strongin Algorithm(AG)) were collected in the same paper. Let Index method with rotated evolvents be *AG-R*, index method with rotated evolvents and mixed strategy be *AG-R-mixed*, index method with single evolvent and mixed strategy be *AG-mixed.*

We use the same as [13] stop condition rule to $\|y - y^*\| \le \rho$. Let's $\rho = 0.01 \cdot \sqrt{N}$ for N=2,3 problems and $\rho = 0.02 \cdot \sqrt{N}$ for N=4 problems.

Table 2. AG-R-mixed and other methods comparison in the worst case

Class	N	Maximum number of iterations							
		Direct	LB Direct	AG	AG-mixed r=4.3	AG-R L=2 r=3.8	AG-R L=6 r=3.2	AG-R-mixed L=2 r=3.8	AG-R-mixed L=6 r=3.2
1-simple	2	127	165	239	207	587	-	221	-
2-hard	2	1159	2665	938	90000 (2)	90000 (2)	-	90000 (1)	-
3-simple	3	1179	1717	3945	1287	3505	4276	1301	1013
4-hard	3	77951	85931	26964	90000 (2)	11672	18535	7513	90000 (1)
5-simple	4	90000 (1)	90000 (15)	27682	9684	31611	17399	7057	9697
6-hard	4	90000 (43)	90000 (65)	90000 (1)	82923	90000 (2)	90000 (3)	90000 (2)	81081

Index method and its modifications parameters: density of evolvent m=12, reliability parameter for AG-mixed – r=4.3, for AG-R-mixed – r=3.8 and r=3.2 (depending on number of evolvents L).

Maximum and average numbers of iterations that methods used for global optimum achievement with established accuracy are presented in tables 2 and 3. The number of non-solved problems is presented in the brackets in case of solving not all 100 problems. Dash means non-applicability of the method with certain number of evolvents because maximum number of them should be no more than $L=N(N-1)$.

Table 3. AG-R-mixed and other methods comparison upon the average

Class	N	Average number of iterations							
		Direct	LB Direct	AG	AG-mixed r=4.3	AG-R L=2 r=3.8	AG-R L=6 r=3.2	AG-R-mixed L=2 r=3.8	AG-R-mixed L=6 r=3.2
1-simple	2	68.1	70.7	90.1	82.3	198.1	-	94.4	-
2-hard	2	208.6	304.3	333.1	1968.1	2168.7	-	1080.0	-
3-simple	3	238.1	355.3	817.7	380.2	1359.1	1239.7	346.7	336.8
4-hard	3	5857.2	9990.6	3541.8	3809.6	4483.3	4125.5	1798.9	2543.1
5-simple	4	>12206	>23452	3950.4	1644.1	4179.2	3470.0	1387.5	1315.6
6-hard	4	>57333	>65326	>22315	19788	>24038	>28791	>18642	16934.9

Achieved results show advantage of serial mixed index method with rotated evolvents especially in case of dimension increasing.

4.3 Cluster Experiments

Appreciably multi dimensional problems solutions is required to research the effectives and scalability of parallel mixed index method with rotated evolvents (AG-R-mixed) because current parallelization scheme allows computational resources distribution maximum on $N(N-1)$ computational cores. Two problems have been chosen to research algorithm scalability.

The first one is conditional multiextremal optimization problem based on Rastrigin function (N=8), $y \in [-1.0,\ \ 1.0]^N$:

$$f(y)=\sum_{i=1}^{N}\left(y_i^2-10\cdot\cos(2\cdot\pi\cdot y_i)\right)\rightarrow \min$$

$$g_1(y)=\sum_{i=1}^{N} y_i-0.5\le 0$$

$$g_2(y)=\sum_{i=1}^{N} y_i^2-1.0\le 0$$

$$g_3(y)=\sum_{i=1}^{N}\sin(i\cdot\pi\cdot y_i)-0.3\le 0.$$

The second one is unconstrained optimization problem on Rastrigin's function (N=11):

$$f(y)=\sum_{i=1}^{N}\left(y_i^2-\cos\left(18\cdot y_i^2\right)\right)\rightarrow \min . \quad (7)$$

Global minima is in point (0,0,…,0) for both problems. Minimal function value is f^*=0.0 for the first problem, and $f^*= -N= -11.0$ for the second one.

Table 4 presents parallel and serial index numerical experiments results. Algorithm stopping rule was achieved accuracy $\varepsilon = 0.0035$ in both cases. If we increase accuracy then serial index method will require to use swapping to store all search data. As a result it will lead to increasing computational costs. It was a motivation to choose given accuracy.

Table 4. Serial and parallel AG-R-mixed method comparison

	serial, 56 evolvents	parallel, 56 processors	
		Total (maximum per process)	Speed up (taking into account different number of iterations) *
Number of iterations	309 640	244 356 (4891)	
Number of criteria calculations	33 874	22 722 (789)	
Speed up for iteration number		63.3	49.95
Speed up for criteria calculations		42.93	28.8
Overall Calculation time	40006.6 sec	626.8 sec.	
Speed up for calculation time		63.83	50.37
Found global optimum value	0.016412	0.022182	

It becomes quite difficult to compare parallel and serial AG-R-mixed because all search data is stored on the one computer for serial index method. Search data is distributed via processes in parallel index method that allows not using swapping mechanism.

We propose a workaround: compare serial AG-mixed (with single evolvent and parallel AG-R-mixed (with p evolvents, where p – is number of processors). Different robust parameters (r=3.0 for AG-mixed, r=2.5 – for AG-R-mixed) and different relative accuracy in stopping rules were chosen. Consequently correct comparison of speed up by trial number is difficult. However we can compare calculation time, speed up and effectiveness of global optimum search in the same stopping conditions for number of iterations. Tables 5 and 6 present unconstrained global optimization results for Rastrigin's function

* Repairing in this case is multiplication on p/s, where p – number of iterations executed by parallel method, and s – number of iterations executed by serial method.

$$f(y)=\sum_{i=1}^{N}\left(y_i^2-\cos\left(18\cdot y_i^2\right)\right)\rightarrow \min$$

with N=11, 15 и 20. Global minimum is achieved in $(0,\ldots,0)^N$ point with a value $f^* = -N$.

Table 5. Function (7) minimization results using AG-mixed method

	Iterations number	Time, sec.	Achieved estimation of f^*
N=11	5 000 000	544 068.8	-10.997543
N=15	3 000 000	605 788.4	-14.835308
N=20	3 000 000	605 956.7	-19.989635

Table 6. Function (7) minimization results using AG-R-mixed method

	Iteration number	time, sec. (num processors)	Achieved estimation of f^*	Calculation time speed up
N=11	4 993 599	9 321.5 (110)	-10.995598	112.01
N=15	3 000 016	7 414.6 (100)	-14.925409	81.7
N=20	3 000 056	6 654.7 (100)	-19.048833	91.06

Non round number of iterations in parallel method is a result of its asynchronies and, hence, non-determined check of stop condition for all processors.

5 Conclusion

Local-global strategy was implemented for parallel and serial index method with rotated evolvents. Research of effectiveness of given strategy in serial and parallel case was performed. Achieved results prove convergence speed up and effective parallelism allows solving more difficult problems. Moreover there are parallel versions of DIRECT and its modifications. Future work may be focusing on comparison of parallel index method and parallel modifications of DIRECT.

References

1. Strongin, R.G.: Global Optimum Search. M.: Znanie (1990)
2. Strongin, R.G.: Parallel multiextremal optimization using multiple evolvents. J. Computational mathematics and mathematical physics 31(8), 1173–1185 (1991)
3. Strongin, R.G., Barkalov, K.A.: About convergence of index method in problems of conditional optimizations with ε-reserving solutions Mathematical issues of cybernetics. M.: Nauka, pp. 273–288 (1999)
4. Strongin, R.G., Sergeyev, Y.D.: Global optimization with non-convex constraints. Sequential and parallel algorithms. Kluwer Academic Publishers, Dordrecht (2000)

5. Barkalov, K.A., Strongin, R.G.: Global optimization method with adaptive order of checking constraints. J. Computational mathematics and mathematical physics 42(9), 1338–1350 (2002)
6. Barkalov, K.A.: Convergence acceleration for constrained global optimization problems. Printed Nizhni Novgorod State University, Nizhni Novgorod (2005)
7. Barkalov, K.A., Ryabov, V.V., Sidorov, S.V.: Using Peano curves in parallel global optimization. In: Materials of 9th International Conference-Seminal "High-Performance Computing on Cluster Systems", Vladimir, pp. 44–47 (2009)
8. Strongin, R.G., Gergel, V.P., Barkalov, K.A.: Parallel methods of global optimization problems solving. Priborostroenie 52(10), 25–32 (2009)
9. Jones, D.R., Perttunen, C.D., Stuckman, B.E.: Lipschitzian optimization without the Lipschitz constant. Journal of Optimization Theory and Applications (79), 157–181 (1993)
10. Gablonsky, M.J.: Modifications of the DIRECT Algorithm. Ph.D. thesis, North Carolina State University, Raleigh, NC (2001)
11. Gablonsky, M.J., Kelley, C.T.: A locally-biased form of the DIRECT Algorithm. Journal of Global Optimization 21, 27–37 (2001)
12. Gaviano, M., Kvasov, D.E., Lera, D., Sergeyev, Y.D.: Software for generation of classes of test functions with known local and global minima for global optimization. ACM TOMS 29(4), 469–480 (2003), http://si.deis.unical.it/~yaro/GKLS.html
13. Lera, D., Sergeyev, Y.D., Lipschitz, Hölder: Global optimization using space-filling curves. Applied Numerical Mathematics 60(1-2), 115–129 (2010)

A Fast Parallel Genetic Algorithm for Traveling Salesman Problem

Chun-Wei Tsai[1,2], Shih-Pang Tseng[1,3],
Ming-Chao Chiang[1], and Chu-Sing Yang[2]

[1] Computer Science and Engineering, National Sun Yat-sen University
Kaohsiung 80424, Taiwan
cwtsai87@gmail.com, tsp@mail.tajen.edu.tw, mcchiang@cse.nsysu.edu.tw
[2] Electrical Engineering, National Cheng Kung University
Tainan 70101, Taiwan
csyang@ee.ncku.edu.tw
[3] Computer Science and Information Engineering, Tajen University
Pingtung 90741, Taiwan

Abstract. In this paper, we present a fast scalable method to reduce the computation time of genetic algorithms for traveling salesman problem, called the Parallel Pattern Reduction Enhanced Genetic Algorithm (PPREGA). The general idea behind the proposed algorithm is twofold: (1) Eliminate the *redundant computations* of GA on its convergence process by pattern reduction and (2) Minimize the *completion time* of GA by parallel computing. Our simulation result shows that the proposed algorithm can significantly reduce not only the computation time but also the maximum completion time of GA. Moreover, our simulation result shows further that the loss of the quality of the end result is small.

1 Introduction

The metaheuristic algorithms [1] have attracted particular attentions of researchers from different backgrounds in recent years. Their works have shown fruitful results in solving several complex problems [1,2]. Of them, GA is no doubt one of the most important and well known metaheuristic algorithms because it is very easy to implement. Another reason that makes GA so successful and powerful is its global search ability. The complex problems that GA can solve include combinatorial optimization problems, multi-objective problems, nonlinear optimization problems, and so on. Simply speaking, for any optimization problem the solution of which can be represented by chromosomes and the fitness function of which can be defined, GA can solve it.

Although GA and its variants have been successfully applied to several problem domains [3] such as NP-complete problems and large scale problems, especially in solving the combinatorial optimization problems, there all face the following three major problems:

1. **Encoding the solutions:** First, we have to take into account how the solution of a problem is encoded. The encoding is not necessarily unique. Moreover, all the encodings have their pros and cons. For example, for the

C.H. Hsu and V. Malyshkin (Eds.): MTPP 2010, LNCS 6083, pp. 241–250, 2010.

data clustering problem, the chromosome can be used to encode the centroids or the cluster numbers of all the patterns.[1]

2. **Selecting the operators:** Selection, crossover, and mutation are the major operators to *transit* or *move* the solution structure from one state to the next. The selection operator is used to decide the search directions. The crossover and mutation operators are used to recombine or rearrange the solutions to create new search directions that are more or less close to the current search directions. When GA is applied to a particular problem, the operators of GA will directly or indirectly influence its performance.
3. **Local search:** The local search ability of simple GA (sGA) is always a problem. Although GA can provide an approximate solution, the quality of the end result is generally poor for some complex problems. The local search (LS) methods [4] have been widely used in enhancing the quality of the end result of GA. However, it is always a trade-off. In general, the longer the LS takes, the better the quality of the end result. However, there is no guarantee that this is always the case.

The above three problems are very important when we apply GA to a new problem. Our observation [5] shows that another problem also exists in the convergence process of GA and the other metaheuristic algorithms. The problem is that many computations of GA are redundant. In other words, some of the genes are common to all the chromosomes and may end up being part of the final solution. For this reason, we present an efficient algorithm to detect these common genes (or partial solutions), and then by compressing and removing these common genes—which we refer to as pattern reduction (PR) [5]—many of the redundant computations on the convergence process of the metaheuristic algorithms can be eliminated. In this paper, we present an extended version of the PR, by combining the island model of GA [6,7] with PR, called the Parallel Pattern Reduction Enhanced Genetic Algorithm (PPREGA).

The remainder of the paper is organized as follows: Section 2 briefly discusses the genetic algorithm, parallel genetic algorithm, pattern reduction algorithm, and how to reduce the computation time of GA. Section 3 presents the proposed algorithm and explains the concept and how the proposed algorithm is designed. The data sets we evaluated and the parameter settings are given in Section 4. The experimental results are also discussed in this section. Section 5 concludes the work.

2 Related Work

2.1 GA and Parallel Computing

Parallel genetic algorithm [6,8,7,9] is one of the most important techniques for reducing the computation time of large problems, such as TSP [10]. The three

[1] In other words, the former encodes the coordinates of the centroids as a vector of floating-point or integer values while the latter encodes the cluster numbers as integer values. In this case, encoding the solution as the cluster numbers seems to be more space efficient than encoding the solution as the centroids, especially for large scale data.

major distribution models used by the parallel GA are: master-slave model, fine-grained model (cellular model) and coarse-grained model (island model). We will discuss them one at a time below.

1. **Master-slave model:** In the master-slave model [6], there exists a single population as in the case of traditional GA, divided into s set of sub-populations. It is the set of sub-populations and the computations of all the operators that are distributed among several processors.
2. **Fine-grained model:** Fine-grained model [6] has been designed for a certain class of machines. In the fine-grained model, each machine is assigned a small number of chromosomes each of which can only compete and mate with its neighbors.[2]
3. **Coarse-grained model:** Coarse-grained model [9] is employed to avoid the propagation of local minimum into the whole population. Some of the individuals in each island will be migrated into another island for the exchange of information after a predefined number of generations.

In summary, the major difference among these three models is how the chromosomes communicate and exchange information. The coarse-grained model (island model) has become the most popular method as far as the research on GA is concerned because not only can it save the computation time, but it can also improve the quality of the end result by using some sub-populations.[3] In other words, this model can be employed to avoid falling into local minima too early. However, many researchers [11,12] indicate that the migration rate and strategy of the island model may affect its performance.

2.2 GA and Pattern Reduction

In general, Pattern Reduction Enhanced Genetic Algorithm (PREGA) [5] adds two operators—CGD and CGC—to the TGA. If we disregard these two operators, it will fall back into TGA. The underlying idea of PREGA is to *detect* (CGD) and *remove* (CGC) genes common to all the chromosomes. More precisely, PREGA uses a method such as voting to determine which genes are common to all the chromosomes and thus can be considered as partial solution and removed. Because the removed genes are common to all the chromosome, most of the computations are essentially duplicated. For instance, if there are five pairs of genes all of which represent node a to node b, then not only the distance between a and b but also the fitness value will be computed five times, four of which are obviously redundant. For this reason, PREGA attempts to speed up the performance of GA by eliminating most, if not all, of the redundant computations as early as possible on the fly so that they do not participate in any computations at later generations on the convergence process. Our simulation results show that the pattern reduction algorithm not only can cut down the

[2] In other words, the machines have to be connected if their chromosomes are to be mated.

[3] Each island of coarse-grained model owns a population itself.

computation time of genetic algorithm, but it can also reduce the computation time of other metaheuristic algorithms such as k-means, tabu search, ant colony system, and particle swarm optimization.

3 The Proposed Algorithm

The parallel genetic algorithm (PGA) is one of the most important methods for reducing the computation time of GA and its variants. The general idea of the proposed algorithm describes herein is to use PGA [10] to enhance the performance of PREGA [13], which by itself can be used to reduce a very high percentage of the computation time of GA, as our simulation results show. This is somehow complicated by the fact that PREGA uses a voting method to detect the genes common to all the chromosomes. As such, PREGA has to centralize all the information of chromosomes in order to determine the redundant computations to be removed. In other words, the problem we attempt to solve is how to parallelize PREGA as far as this paper is concerned.

Fig. 1 shows two different methods for combining PREGA with PGA (island model). To simplify the discussion that follows, we will assume that there are $p = 4$ islands and use the population shown in Fig. 1(a), which contains n chromosomes each of which has ℓ genes, as an example, assuming that n and ℓ can

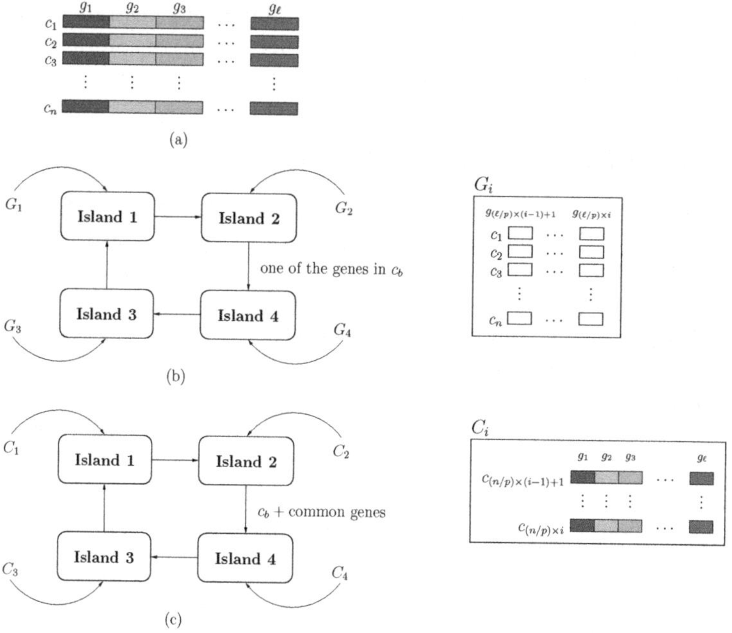

Fig. 1. Two of the many methods for combining PREGA with PGA (island model). (a) the population; (b) the population partitioned by genes, as G_i shows; and (c) the population partitioned by chromosomes, as C_i shows. Note that the subscript i is the island number. In addition, c_b indicates the current best chromosome on a particular island.

be evenly divisible by p. Otherwise, some islands may be assigned more genes or chromosomes than the others. Fig. 1(b) describes the first method that partitions the population by genes. In other words, it distributes genes at loci in the range $G_i = [(\ell/p) \times (i-1)+1, (\ell/p) \times i]$ of all the chromosomes to island i, for $i = 1, 2, \ldots, p$. That is, each island performs GA on the genes at the loci in the range G_i. The downside is that this kind of method has several disadvantages. It may lose some information because some of the relationships between genes may disappear after partitioning, or it may increase the communication cost if the genes (sub-solutions) assigned to each island affect one another. In addition, some problems can not be parallelized this way. Fig. 1(c) describes the second method that partitions the population by chromosomes. In other words, it distributes chromosomes numbered in the range $C_i = [(n/p) \times (i-1)+1, (n/p) \times i]$ to island i, for $i = 1, 2, \ldots, p$. That is, the population is divided into p sub-populations. Then, sub-population i which is composed of chromosomes in the range C_i is mapped to island i. Each island can now evolve independently. The only requirement is that each island sends a chromosome (such as the current best chromosome on this island) to other islands every M iterations.

Now the question is how to exchange the information about the common genes to other islands. The original idea of PR is to check genes that are common to all the chromosomes, and for the common genes, it will assume that all but one of the computations are redundant. In addition, PR will be performed at each iteration of GA. For island model, PR checks whether some of the genes are common to all the chromosomes or not every iteration, but it does not compress and remove the common genes found before it receives the information about the common genes from the other islands. Only when all the common genes (genes at the same loci having the same values) in all the islands are the same, will the PR compress and remove these genes. In other words, a subtle difference between PREGA and PPREGA is that PREGA detects, compresses, and removes the common genes every iteration whereas PPREGA detects, compresses, and removes the common genes once every M iteration. More precisely, when chromosomes begin to migrate from island to island, the information about the common genes will be migrated too, as shown in Fig. 1(c).

3.1 The Algorithm

Fig. 2 gives an outline of the PPREGA. The main difference between PREGA and PPREGA is in that the proposed algorithm adds two conditions to determine when to broadcast the information to other islands, as shown in Fig. 2. First, PPREGA uses sampling to build a better initial population to start with on line 3. Then, PPREGA will partition the initial population into N sub-populations, where N is the number of islands. Then, the sub-population will be sent to islands. The process of parallel GA, on lines 7 through 17, is basically composed of two processes. The first is the standard GA process, which contains the fitness function, selection, crossover, and mutation operators, as shown on lines 9 and 10. The second is the pattern reduction process, which consists of the CGD and CGC operators, as shown on lines 13 through 16. In general, CGD is responsible

```
1  Algorithm PPREGA
2  {
3     Generate an initial population of chromosomes by sampling
4     Partition the population generated above into N sub−populations
5     Send the sub−populations to islands
6     t = 1
7     For each island do
8        /* The process of GA */
9        Use the fitness function to select the fitter chromosomes.
10       Apply the crossover and mutation operators in order.
11       /* The process of PR */
12       If t % M = 0
13          Apply common genes detection (CGD) algorithm to find the common genes
14          Send the information about the common genes and a chromosome to other islands
15       If some of the genes are common to all the chromosome
16          Apply common genes compression (CGC) algorithm to reserve the common genes
17    End
18    t = t + 1
19    If a stopping criterion is satisfied, then stop and output the best chromosome
20    Else go to line 7
21 }
```

Fig. 2. The proposed algorithm—Parallel Pattern Reduction Enhanced Genetic Algorithm (PPREGA)

for detecting the common genes of each sub-population every M iteration on line 13. Accordingly, CGC is carried out once every M iterations on line 16 if common genes are detected on line 13. In other words, once every M iterations, the proposed algorithm will send the information about the common genes to all the other islands. At the same time, it also selects one of the chromosomes to migrate to the other islands, as given on lines 12 and 14. Only when all the islands received the information about the common genes, will PPREGA be invoked to compress and remove the redundant computations of GA. In other words, the CGC operator will be carried out once every M iterations, to reduce the communication cost.

4 Empirical Analysis

In this section, we evaluate the performance of the proposed algorithm by using it to solve the traveling salesman problem. The empirical analysis was conducted on an IBM X3400 machine with 2.0GHz Xeon CPU and 8GB of memory using CentOS 5.0 running Linux 2.6.18. Moreover, all the programs are written in C++ and compiled using g++ (GNU C++ compiler). The benchmarks used in this paper are all from TSPLIB [14]. Unless stated otherwise, all the simulations are carried out for 30 runs, with the population size fixed at 80, the crossover probability at 0.5, the mutation probability at 0.01, the number of generations at 100, and the tournament size at 3 (i.e., 1 out of 3). In addition, for all the simulations, PPREGA is started at the second generation. The nearest-neighbor method is used for creating the initial solution for all the algorithms involved in the simulation. 2-opt is the local search method that is used to fine tune the quality of the end results. Unless stated otherwise, all the simulations use heuristic crossover (HX) as the default crossover operator. For all the simulations, the

Table 1. Simulation results of Parallel GA and Parallel PPREGA

Data set	PGA(1)		PGA(2)		PGA(4)		PGA(6)		PGA(8)	
	D	T	Δ_D	Δ_T	Δ_D	Δ_T	Δ_D	Δ_T	Δ_D	Δ_T
ch130	6,232.42	0.13	-0.66	-46.15	0.59	-76.92	0.17	-84.62	0.55	-84.62
ch150	6,576.88	0.17	-0.24	-52.94	-0.24	-76.47	-0.24	-82.35	-0.24	-88.24
d198	15,978.60	0.26	-0.01	-50.00	0.03	-73.08	0.18	-84.62	0.23	-88.46
a280	2,663.13	0.46	-0.03	-50.00	0.14	-76.09	0.18	-82.61	0.02	-86.96
pcb442	54,964.50	1.16	-3.88	-51.72	-5.11	-76.72	-4.46	-84.48	-5.17	-88.79
d493	37,046.20	1.41	-1.63	-51.77	-2.21	-77.30	-2.52	-85.11	-2.58	-88.65
u574	41,685.30	1.87	-4.05	-51.34	-5.06	-76.47	-6.73	-84.49	-7.42	-88.24
u724	47,229.80	3.02	-4.42	-49.34	-5.89	-77.15	-5.83	-85.10	-6.17	-89.07
pr1002	295,672.00	7.39	-3.66	-52.10	-5.69	-76.73	-6.26	-84.57	-6.62	-88.50
u1060	259,169.00	8.69	-3.05	-51.67	-5.48	-76.18	-6.16	-84.35	-6.72	-88.15
d1291	55,711.30	13.52	-3.45	-50.07	-4.03	-74.63	-4.38	-82.54	-3.80	-86.46
u1432	175,539.00	18.26	-3.53	-50.93	-5.74	-75.08	-6.78	-82.80	-6.70	-86.64
d1655	71,188.50	24.81	-3.65	-49.66	-5.16	-74.16	-5.26	-81.30	-5.97	-85.77
u2152	72,758.20	43.38	-2.06	-49.61	-3.38	-73.63	-3.59	-81.03	-3.69	-84.99
Average			-2.45	-50.52	-3.37	-75.76	-3.69	-83.57	-3.88	-87.40

Data set	PPREGA(1)		PPREGA(2)		PPREGA(4)		PPREGA(6)		PPREGA(8)	
	D	T	Δ_D	Δ_T	Δ_D	Δ_T	Δ_D	Δ_T	Δ_D	Δ_T
ch130	6,313.89	0.03	-0.50	-66.67	-0.47	-66.67	-0.13	-66.67	0.04	-66.67
ch150	6,596.95	0.03	-0.01	-66.67	-0.01	-66.67	-0.01	-66.67	-0.48	-66.67
d198	16,087.70	0.06	0.31	-50.00	-0.12	-83.33	-0.60	-83.33	-0.17	-83.33
a280	2,708.71	0.09	-0.98	-55.56	0.00	-77.78	-1.61	-88.89	-0.33	-77.78
pcb442	53,048.50	0.19	-0.34	-57.89	0.37	-78.95	0.05	-84.21	0.18	-84.21
d493	36,185.50	0.27	0.43	-55.56	0.60	-81.48	-0.02	-88.89	0.16	-88.89
u574	38,950.00	0.28	0.90	-50.00	1.13	-78.57	0.63	-82.14	0.65	-85.71
u724	44,227.90	0.54	0.36	-55.56	0.60	-79.63	0.85	-87.04	0.59	-87.04
pr1002	273,844.00	1.10	0.00	-54.55	-0.09	-80.00	-0.30	-86.36	-0.03	-87.27
u1060	241,742.00	1.31	0.01	-51.91	-0.48	-78.63	-0.57	-84.73	-0.91	-87.02
d1291	53,604.90	1.39	0.05	-49.64	1.31	-74.10	0.27	-79.86	-0.04	-81.29
u1432	162,238.00	3.39	0.65	-57.52	0.67	-82.01	0.97	-87.91	1.05	-89.68
d1655	67,409.00	2.89	-0.55	-49.13	-0.53	-75.43	-2.12	-82.70	-1.10	-83.74
u2152	69,415.70	4.99	0.12	-49.30	0.22	-75.95	0.36	-81.56	-0.45	-84.17
Average			0.03	-55.00	0.23	-77.09	-0.16	-82.21	-0.06	-82.39

T: Time in seconds.

best chromosome of each sub-population on all the islands will be migrated to the other islands once every 5 generations. Also, to save the synchronization time, the CGD and CGC operators are invoked once every 5 generations rather than once every generation.

To simplify the discussion of the simulation results, we will use the following conventions. Let $\beta \in \{D, T\}$ denote either the traveling distance ($\beta = D$), and the computation time ($\beta = T$). Let Δ_β denote the enhancement of β_ϕ with respect to β_ψ in percentage. In other words, Δ_β is defined as follows:

$$\Delta_\beta = \frac{\beta_\phi - \beta_\psi}{\beta_\psi} \times 100\% \tag{1}$$

where β is either D or T for the TSP.

Tables 1 and 2 compare the performance of PGA with different number of islands, PPREGA with different number of islands, and PGA versus PPREGA for which the same number of islands is used. On the top of Table 1 is the experimental results of PGA with 1, 2, 4, 6, and 8 islands each of which is implemented as a thread. On the bottom of Table 1 is the experimental results of PPREGA with exactly the same number of islands as PGA. In Table 1, Δ_D

Table 2. Simulation results of Parallel PPREGA versus Parallel GA

Data set	$i = 1$		$i = 2$		$i = 4$		$i = 6$		$i = 8$	
	Δ_D	Δ_T	Δ_D	Δ_T	Δ_D	Δ_T	Δ_D	Δ_T	Δ_D	Δ_T
ch130	1.31	-76.92	0.8	-92.31	0.84	-92.31	1.17	-92.31	1.35	-92.31
ch150	0.31	-82.35	0.29	-94.12	0.29	-94.12	0.29	-94.12	-0.18	-94.12
d198	0.68	-76.92	1.00	-88.46	0.57	-96.15	0.08	-96.15	0.51	-96.15
a280	1.71	-80.43	0.72	-91.30	1.71	-95.65	0.08	-97.83	1.37	-95.65
pcb442	-3.49	-83.62	-3.82	-93.10	-3.13	-96.55	-3.44	-97.41	-3.31	-97.41
d493	-2.32	-80.85	-1.90	-91.49	-1.73	-96.45	-2.34	-97.87	-2.16	-97.87
u574	-6.56	-85.03	-5.72	-92.51	-5.50	-96.79	-5.97	-97.33	-5.95	-97.86
u724	-6.36	-82.12	-6.02	-92.05	-5.79	-96.36	-5.56	-97.68	-5.81	-97.68
pr1002	-7.38	-85.12	-7.39	-93.23	-7.46	-97.02	-7.66	-97.97	-7.41	-98.11
u1060	-6.72	-84.93	-6.71	-92.75	-7.17	-96.78	-7.25	-97.70	-7.58	-98.04
d1291	-3.78	-89.72	-3.73	-94.82	-2.52	-97.34	-3.52	-97.93	-3.82	-98.08
u1432	-7.58	-81.43	-6.97	-92.11	-6.96	-96.66	-6.69	-97.75	-6.61	-98.08
d1655	-5.31	-88.35	-5.83	-94.07	-5.81	-97.14	-7.32	-97.98	-6.35	-98.11
u2152	-4.59	-88.50	-4.48	-94.17	-4.38	-97.23	-4.25	-97.88	-5.02	-98.18
Average	-3.58	-83.31	-3.55	-92.61	-3.36	-96.18	-3.74	-96.99	-3.64	-96.98

and Δ_T, as defined in Eq. (1), represents, respectively, the enhancement of the traveling distance and the computation time compared to PGA and PPREGA with a single island. As Table 1 shows, PGA can reduce the computation time by 50.52%, 75.76%, 83.57%, and 87.40% using, respectively, two, four, six, and eight islands (processors) in the convergence process of GA. These results are easily justified because in the ideal case, the computation time required T' is inversely proportional to the number of processors p. That is, $T' = T/p$ where T is the computation time of a single processor. Thus, the computation time reduced is $R = T - T/p = (1 - 1/p)T$. For instance, given two processors, the computation time reduced will be 50% (i.e., $R = (1 - 1/2)T = T/2$), given four processors, the computation time reduced will be 75% (i.e., $R = (1 - 1/4)T = 3T/4$), given six processors, the computation time saved is about 85% (i.e., $R = (1 - 1/6)T = 5T/6$), and given eight processors, the computation time saved is about 87.5% (i.e., $R = (1 - 1/8)T = 7T/8$). Table 1 also shows that PPREGA can further save the computation time by about 55%, 77.09%, 82.21%, and 82.39% using, respectively, two, four, six, and eight islands.

In Table 2, Δ_D and Δ_T, again as defined in Eq. (1), indicates, respectively, the enhancement of the traveling distance and the computation time by PPREGA[4] with respect to PGA[5] with the same number of islands. As Table 2 shows, compared with parallel GA, PPREGA can reduce the computation time by 83.31%, 92.61%, 96.18%, 96.99%, and 96.98% using, respectively, single, two, four, six, and eight islands (processors) in the convergence process of GA. In other words, Table 1 shows that PPREGA can even be used in a distributed computing environment to enhance the performance of GA or GA-based algorithms. Fig. 3 compares the performance of Parallel GA and PPREGA. Fig. 3(a) shows that for Parallel GA, as the number of islands (threads) increases, the quality of the end results is improved by about 2.45% to 3.88%. On the other hand, although the quality of the end result of PPREGA does not change as the number of

[4] As shown in the bottom half of Table 1.

[5] As shown in the top half of Table 1.

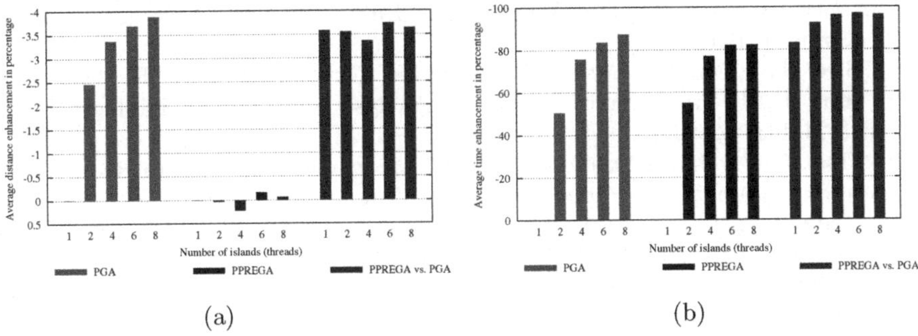

(a) (b)

Fig. 3. A summary of the enhancement of PPREGA with respect to TGA, for both the sequential and parallel version of the algorithms. (a) by distance and (b) by time. Note that the more negative the percentage, the larger the enhancement.

islands increases, PPREGA can improve the quality of the end result of Parallel GA by about 3.36% to 3.74% for all the configurations; that is, from one up to eight islands. Fig. 3(b) shows that the amounts of computation time saved for both Parallel GA and PPREGA are very close to each other. However, even though PGA using two islands can save about 50% of the computation time (compared to PGA using single island), PPREGA using two islands can further save the computation time by about 55% (compared to PPREGA with single island). Thus, PPREGA can save about 92.61% of the computation time in total compared to PGA with single island.

5 Conclusion

In this paper, we proposed a fast scalable algorithm, called PPREGA, that uses the parallel GA (island model) and pattern reduction methods to reduce the computation time of GA. PPREGA was designed to solve the problem of how to efficiently integrate these two methods to speed up the performance of PREGA. We presented in detail how the proposed algorithm is designed and implemented. Our simulation results showed that the proposed algorithm can efficiently reduce not only the computation time but also the completion time of GA. In the future, our goal is to further enhance the performance of PPREGA in terms of not only the computation time but also the quality of the end result.

Acknowledgment

This work was supported in part by National Science Council, Taiwan, ROC, under Contract No. 98-2811-E-006-078.

References

1. Blum, C., Roli, A.: Metaheuristics in combinatorial optimization: Overview and conceptual comparison. ACM Computing Surveys 35(3), 268–308 (2003)
2. Martí, R.: Multi-start methods. In: Glover, F.W., Kochenberger, G.A. (eds.) Handbook of Metaheuristics, pp. 355–368. Kluwer Academic Publishers, Boston (1993)

3. Michalewicz, Z.: Genetic Algorithms + Data Structures = Evolution Programs. Springer, Berlin (1996)
4. Lee, J.S., Oh, I.S., Moon, B.R.: Hybrid genetic algorithms for feature selection. IEEE Transactions on Pattern Analysis and Machine Intelligence 26(11), 1424–1437 (2004)
5. Tsai, C.W.: On the Study of Efficient Metaheuristics via Pattern Reduction. PhD thesis, National Sun Yat-sen University, Taiwan, R.O.C (2009)
6. Cantú-Paz, E.: A survey of parallel genetic algorithms. Calculateurs Paralleles, Reseaux et Systems Repartis 10(2), 141–171 (1998)
7. Mühlenbein, H.: Parallel genetic algorithms, population genetics and combinatorial optimization. In: Proceedings of the Third International Conference on Genetic Algorithms, San Francisco, CA, USA, pp. 416–421 (1989)
8. Kohlmorgen, U., Schmeck, H., Haase, K.: Experiences with fine–grained parallel genetic algorithms. In: Annals of Operations Research (1996)
9. Maeda, Y., Ishita, M., Li, Q.: Fuzzy adaptive search method for parallel genetic algorithm with island combination process. International of Journal of Approximate Reasoning 41, 59–73 (2005)
10. Wang, L., Maciejewski, A.A., Siegel, H.J., Roychowdhury, V.P., Eldridge, B.D.: A study of five parallel approaches to a genetic algorithm for the traveling salesman problem. Intelligent Automation and Soft Computing 11(4), 217–234 (2005)
11. Cantú-Paz, E., Goldberg, D.E.: Efficient parallel genetic algorithms: theory and practice. Computer Methods in Applied Mechanics and Engineering 186(2-4), 221–238 (2000)
12. Cantú-Paz, E.: Efficient and Accurate Parallel Genetic Algorithms. Kluwer Academic Publishers, Norwell (2000)
13. Tseng, S.P., Tsai, C.W., Chiang, M.C., Yang, C.S.: Fast genetic algorithm based on pattern reduction. In: Proceedings of the IEEE International Conference on Systems, Man and Cybernetics, pp. 214–219 (2008)
14. TSPLIB: (2009), `http://www.iwr.uni-heidelberg.de/groups/comopt/software/TSPLIB95/tsp/`.

Parallel Algorithms for Solution of Air Pollution Inverse Problems

Alexander Starchenko and Elena Panasenko

Tomsk State University, 36 Lenin Prospekt, 634050, Tomsk, Russia
starch@math.tsu.ru, ea@math.tsu.ru

Abstract. Parallelization of Marchuk's method for solution of inverse problems based on adjoint equations and dual representation of contaminant concentration functional is considered here. There are N individual adjoint equations independently solved at each time step. Such conditions of numerical investigation allow application of high performance computations. For this purpose the following ways of parallelization are used: geometrical decomposition, functional decomposition and combination of geometrical and functional decompositions.

Keywords: air pollution, inverse problems, parallel algorithms, functional decomposition, domain decomposition.

1 Introduction

At present along with the tool control of the air quality the methods of mathematical modeling are widely used in the analysis of air impurity. The methods of mathematical modeling allow to carry out the forecast of a detailed picture of distributing impurities in the surface layer of the atmosphere quickly and at a low cost. But the complexity of the processes of distribution and transferring the polluting substances make the models of estimating the air quality very cumbersome and specify stringent requirements to the computational resources. A promising way to solve these problems is to apply modern high-performance multiprocessor computational systems which ensure a considerable speed of obtaining the computation results and improve the quality of the numerical weather forecast.

2 Physical Statement of a Direct Problem of Impurity Transport

The authors discuss the transport and scattering of the inert isothermal impurity coming from the source disposed at the altitude $z=h$ in the surface-layer of the atmosphere. The values of wind speed components, turbulent diffusion coefficients and the intensity of the Q source are known. There occurs a dry polluted fallout near the surface. It is need to determine a contaminant concentration at N points.

2.1 Mathematical Statement of the Direct Problem of Impurity Transport

The governing equation for modeling of gaseous contaminant dispersion with proposed physical assumptions is given as follows:

C.H. Hsu and V. Malyshkin (Eds.): MTPP 2010, LNCS 6083, pp. 251–259, 2010.

$$\frac{\partial C}{\partial t}+U\frac{\partial C}{\partial x}+V\frac{\partial C}{\partial y}+W\frac{\partial C}{\partial z}=\frac{\partial C}{\partial x}\left[D\frac{\partial C}{\partial x}\right]+\frac{\partial C}{\partial y}\left[D\frac{\partial C}{\partial y}\right]+ \\ +\frac{\partial C}{\partial z}\left[K_z\frac{\partial C}{\partial z}\right]+Q(t,x,y,z), \tag{1}$$

where C is an impurity concentration; U,V,W are components of the velocity vector of the atmosphere air; D,K_z are the coefficients of the turbulent diffusion; Q is the intensity of the incoming impurity from sources.

The initial and boundary conditions are:

$$\begin{aligned} &t=0:\ C=0;\\ &x=0:\frac{\partial C}{\partial x}=0;\ x=L_x:\frac{\partial C}{\partial x}=0;\\ &y=0:\frac{\partial C}{\partial y}=0;\ y=L_y:\frac{\partial C}{\partial y}=0;\\ &z=0:K_z\frac{\partial C}{\partial z}=v_{dep}C;\ z=L_z:\frac{\partial C}{\partial z}=0. \end{aligned} \tag{2}$$

3 Physical Statement of the Inverse Problem of Impurity Transport

Basing on atmospheric meteorological fields and results of field measurements of air pollution at N points during certain period of time T, it is needed to define coordinates and response time of instantaneous pollution point source (fig.1)

Fig. 1. Determination of the source emission location

3.1 Mathematical Statement of the Inverse Problem

The solution of inverse problem is based on Marchuk method [1]. A mathematical statement of the inverse impurity-transport problem includes several adjoint equations of the same type and a search for a minimum of a certain functional in the process of solving the adjoint equations.

The adjoints to (1) and (2) is obtained in the following way: the equation (1) is multiplied by some adjoint function $C^* = C^*(t,x,y,z)$ and integrated in time and space. After integrating in parts we arrive at the expression:

$$-\frac{\partial C_k^*}{\partial t}-\frac{\partial UC_k^*}{\partial x}-\frac{\partial VC_k^*}{\partial y}-\frac{\partial WC_k^*}{\partial z}-\frac{\partial}{\partial x}\left(D\frac{\partial C_k^*}{\partial x}\right)-\frac{\partial}{\partial y}\left(D\frac{\partial C_k^*}{\partial y}\right)- \frac{\partial}{\partial z}\left(K_z\frac{\partial C_k^*}{\partial z}\right)=P_k; k=1,...,N; \tag{3}$$

having the corresponding initial and boundary conditions:

$$\begin{aligned} &C_k^*(T,x,y,z)=0;\\ &x=0: UC_k^*+D\frac{\partial C_k^*}{\partial x}=0; x=L_x: UC_k^*+D\frac{\partial C_k^*}{\partial x}=0;\\ &y=0: VC_k^*+D\frac{\partial C_k^*}{\partial y}=0; y=L_y: VC_k^*+D\frac{\partial C_k^*}{\partial y}=0;\\ &z=0: K_z\frac{\partial C_k^*}{\partial z}=\alpha C_k^*;\ z=L_z: \frac{\partial C_k^*}{\partial z}=0, \end{aligned} \tag{4}$$

where $P_k = \delta(x-x_k)\delta(y-y_k)\delta(z-z_k)\delta(t-t_k)$, N is the number of the measurement points having the coordinates (x_k, y_k, z_k), t_k is the time moment of the concentration measurement.

The following expression is considered for dual representation of functional [2]:

$$Q_0 C_k^*(t_{00}, x_{00}, y_{00}, z_{00}) = C_k;\ C_k = C(t_k, x_k, y_k, z_k);\ k=1,...,N. \tag{5}$$

In this way the finding of the solution for equation system (5) enables one to establish parameters of the impurity source (the coordinates (x_{00}, y_{00}, z_{00}), the operation time of the source t_k and the intensity Q_0).

4 Numerical Solution

For the numerical realization of (3)-(4) problem the finite-volume method and the explicit difference schemes were used. Van Leer's scheme MLU [3] was applied to approximate of the advection terms.

5 Parallel Realization of the Adjoint Problem

While solving the impurity transport problems the two principal methods of the parallel realization are distinguished: the parallelization according to the physical processes (the functional decomposition) and the geometrical decomposition of the computational region [4].

5.1 Functional Decomposition

The parallelization of the method of the numerical solution of the problem to determine the instantaneous point sources was performed by using "master-slave" principle. In applying this approach the master-process sends the values of the meteorological parameters necessary for solution of the adjoint problems (3)-(4) to each slave processor. In their turn, the subordinate slave-processors, after receiving the information, begin to perform the computations independently of each other and then return their approximate solutions of the dual problem (3)-(4) found at each time step to the master-processor. Which seeks the global minimum of the functional:

$$\sum_{k=1}^{N}\left(Q_0 C_k^*(t,x,y,z)-C_k\right)^2, Q_0>0. \tag{6}$$

The above parallel algorithm for solving the inverse problem using the principle of the functional decomposition can be estimated by the following time expenditures:

$$\begin{aligned} &T_p^{F1} \approx m\cdot t_a \cdot N_x^2 \cdot N_z \cdot M + 5\cdot t_{comm}\cdot N_x^2\cdot N_z\cdot M + t_{comm}\cdot M + t_a\cdot(M+3N),\\ &N = p-1 \end{aligned} \tag{7}$$

Here t_a is the time of one arithmetic operation, t_{comm} is the time to forward one number en route «master-slave» or back, N is the number of the adjoint problems being solved. The first term in (7) is the time expenditure on the simultaneous solution of the adjoint problems, the second term is the time spent on the transfer of the meteorological fields to the slave-processors at each time step, the third term is the time spent to transfer the computed values $Q_0 C_k^*(t,x,y)$ to the master-processor, the forth term is the computation performed by the master-process. It should be noted that the main contribution to (7) is provided by the first and second terms, since the size of the computational grid $N_x^2 \cdot N_z$ include several hundred thousand nodes.

Thus, for the typical values of the characteristic parameters $(m \approx 10 \div 100, N_x^2\cdot N_z \approx 0.5\div 1\cdot 10^6, M \approx 100 \div 3000, N\approx 5\div 20)$:

$$\begin{aligned} &T_p^{F1} \approx m\cdot t_a\cdot N_x^2\cdot N_z\cdot M + 5\cdot t_{comm}\cdot N_x^2\cdot N_z\cdot M/k,\\ &(M/k\approx 5\div 50), N = p-1 \end{aligned} \tag{8}$$

where k is the frequency of the inter-processor transfer of the meteorological parameters.

For the case of sequential computations while solving the inverse problem of the pollutant transport the time cost, in view of the accepted model of estimation, are presented as:

$$T_1 \approx N\cdot m\cdot t_a\cdot N_x^2\cdot N_z\cdot M + t_a\cdot(M+3N) \tag{9}$$

Since the second term in (9) is essentially smaller than the first one then me obtain the following estimates of the acceleration and efficiency for the considered algorithm in solving the inverse problem of the impurity transport for p processor system the estimations of the acceleration and the efficiency can be written as

$$S_p^{F1} = \frac{T_1}{T_p^{F1}} \approx \frac{N \cdot m \cdot t_a \cdot N_x^2 \cdot N_z \cdot M}{m \cdot t_a \cdot N_x^2 \cdot N_z \cdot M + 5t_{comm} \cdot N_x^2 \cdot N_z \cdot M / k} = \frac{p-1}{1+\frac{5\chi}{m \cdot k}},$$
$$E_p^{F1} = \frac{S_p^{F1}}{p} \approx \frac{1-1/p}{1+\frac{5\chi}{m \cdot k}}, \ \chi = t_{comm} / t_a > 1. \tag{10}$$

One can see from (10) that it is impossible to achieve a perfect parallelism by applying this method of implementing the parallel computations, because there is no good load balancing of the processors applied in the computations [4,6]. This is due to the fact that the master-processor does not participate in the general computations (does not compute a single adjoint problem).

To improve the efficiency of the algorithm for the parallel solution of the inverse problem concerning the impurity transport one can consider an approach when the master-processor one of N adjoint problems (3)-(4). In this case one can use estimate (8), but for the case $N = p$. Then

$$S_p^{F2} \approx \frac{p}{1+\frac{5\chi}{m \cdot k}}, E_p^{F2} \approx \frac{1}{1+\frac{5\chi}{m \cdot k}} \tag{11}$$

Therefore one hope that the developed parallel program for solving the inverse problem of the pollutant transport will have higher rates of the execution time.

Time consumption of parallel inverse runs for different variants of functional decomposition with different quantities of available measurement points of pollution ($N = 5,10,20$) are presented in table 1.

The first variant of functional decomposition of solution algorithm has minor advantage in run time due to appearance of additional possibilities to merge of work of active processes, while the quantity of the ones for the same conditions is more in one when compared with the second variant. However, the ration of effectiveness of parallel programs

$$e^F = \frac{E_p^{F1}}{E_p^{F2}} = \frac{T_1 / \left(T_p \cdot p\right)^{(F1)}}{T_1 / \left(T_p \cdot p\right)^{(F2)}} = \frac{\left(T_p \cdot p\right)^{(F2)}}{\left(T_p \cdot p\right)^{(F1)}} < 1$$

displays more scalability of the second variant, which is in a good agreement with theoretical estimations (10), (11), i.e.

$$\left(e^F\right)_{theoretic} \approx \frac{p}{p+1} \text{ (see Table 2).}$$

Table 1. The run of the parallel program (sec) for the functional decomposition

The number of processes used (F1)	Time, sec	The number of processors used (F2)	Time, sec
$p = 6$ ($N = 5$)	312	$p = 5$ ($N = 5$)	323
$p = 11$ ($N = 10$)	316	$p = 10$ ($N = 10$)	337
$p = 21$ ($N = 20$)	322	$p = 20$ ($N = 20$)	354

Table 2. The efficiency correlation of the parallel programs e^F for two variants of the functional decomposition

The number of concentration measurements N	$\left(e^F\right)_{theoretic} \approx \dfrac{p}{p+1}$	$\left(e^F\right)_{experimental}$
5	0,83	0,86
10	0,91	0,97
20	0,95	1,01

5.2 Geometric Decomposition

Parallel solution of each adjoint problem (3)-(4) is based on one dimensional decomposition (along Oy axis, fig.2) of computational domain. In this case the computations will be made in the following way: N adjoint problems (3)-(4) are

Numerically solved successively one after another in a sub-domains of the computer network allocated to every processor. The computations during the time interval $[0,T]$ are performed by every simultaneously. But for a correct computation of the network values of the function C^* in the vicinity of the sub-domains boundaries (fig.2) some network values are required from the near-by sub-domains. This can be provided only by means of an inter-processor data transfer using the function of a blocking exchange MPI_SENDRECV of MPI standard [5].

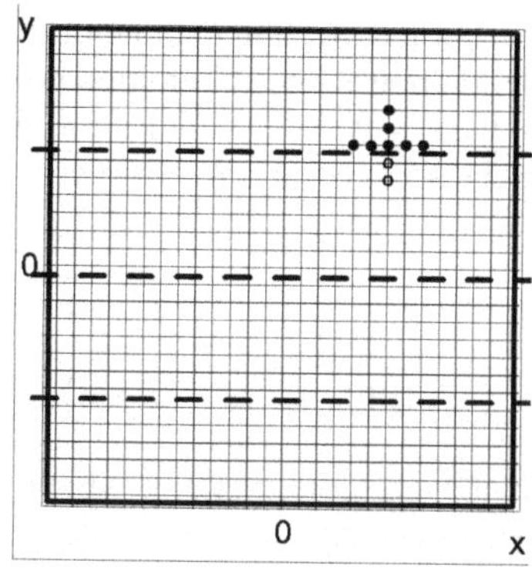

Fig. 2. Computational grid in the plane xOy with indication of sub-domains for four processors

To estimate time consumption of the proposed geometrical decomposition like it was done for functional decomposition, let is write the following expression:

$$T_p^G \approx \frac{N \cdot m \cdot t_a \cdot N_x^2 \cdot N_z^2 \cdot M}{p} + \frac{5 \cdot t_{comm} \cdot N_x^2 \cdot N_z \cdot M / k}{p} + \\ +4 \cdot N_x \cdot N_z \cdot N \cdot M \cdot t_{comm} + N \cdot M \cdot t_{comm} + t_a (M + 3N). \tag{12}$$

The first term here is time consumption for solution of N adjoint equations in sub-domains of entire computational domain, the second one is time consumption for transfer of meteorological fields from the master-process (which also solves adjoint problems) to all slave-processes with frequency of k steps, the third term is time consumption for data transfer between processes with neighbouring sub-domains (fig.2), the fourth and fifth terms are time consumption for obtaining and estimation of functional (6) by the master-process. Let is write the following expression considering specific values N_x, N_z, M, k, m, N for the problem:

$$T_p^G \approx \frac{N \cdot m \cdot t_a \cdot N_x^2 \cdot N_z \cdot M}{p} + \frac{5 \cdot t_{comm} \cdot N_x^2 \cdot N_z \cdot M / k}{p} + \\ +4 \cdot N_x \cdot N_z \cdot N \cdot M \cdot t_{comm}. \tag{13}$$

and taking into account time consumption of sequential program (9) we have

$$S_p^G = \frac{T_1}{T_p^G} \approx \frac{N \cdot m \cdot t_a \cdot N_x^2 \cdot N_z \cdot M}{\frac{N \cdot m \cdot t_a \cdot N_x^2 \cdot N_z \cdot M}{p} + \frac{5 \cdot t_{comm} \cdot N_x^2 \cdot N_z \cdot M / k}{p} + 4 \cdot N_x \cdot N_z \cdot N \cdot M \cdot t_{comm}} = \\ \frac{p}{1 + \chi \left(\frac{5}{Nmk} + \frac{4p}{N_x m} \right)}, E_p^G = \frac{S_p^G}{p} \approx \frac{1}{1 + \frac{\chi}{m} \left(\frac{5}{Nk} + \frac{4p}{N_x} \right)}. \tag{14}$$

The contribution to time consumption for inter-processes data transfer is most substantial as expected from the above theoretical estimations of effectiveness. Thus, it is expected that application of geometrical decomposition is less promising than the use of functional decomposition.

Numerical experiment showed validity of theoretical estimation (Table 3).

Table 3. The time of the computer run in the parallel program for the geometrical decomposition

The number of processes	Time, sec
$p = 5$ ($N = 5$)	627
$p = 10$ ($N = 10$)	643
$p = 20$ ($N = 20$)	743

Comparing the results given in Tables 1-3 one can see that the performance of geometrical decomposition is over two times less when compared with the functional decomposition. The reason of that is communication consumption for inter-processors data transfer required for correct solution of the adjoint problem (3)-(4) for entire computational domain divided between active processes [6,7].

In general, this method of the algorithm parallelization for solution of the inverse problem for the conditions considered here provides efficiency of 50%, which indicates good scalability of parallel implementation $p \le 20$.

5.3 Combined Method of Parallelizing

The above approaches to develop the parallel versions of the algorithms to solve the inverse problem of the impurity transport showed rather good results as to the efficiency, but the number of the active processors there is limited: in the case of the functional decomposition – by the number of the N measurements carried out, and in the case of the geometric decomposition – by the size of the network N as well as by the selected network pattern.

It is suggested the use of composite approach with combined functional and geometrical decompositions to increase the number of active processes. That is all p active processes divided into N groups, each of which solves one adjoint problem (3) – (4) using one-dimensional geometrical decomposition. The processes of particular group exchange of computed values near borders of sub-domains. Intermediate computed data are transferred to one of p processes from the rest ones for search of functional minima (6).

Let is make preliminary estimations of time consumption of such parallelization technique.

$$T_p^C \approx \frac{m \cdot t_a \cdot N_x^2 \cdot N_z \cdot M}{p/N} + \frac{5 \cdot t_{comm} \cdot N_x^2 \cdot N_z \cdot M/k}{p/N} + 4 \cdot N_x \cdot N_z \cdot M \cdot t_{comm} +$$
$$+ M \cdot t_{comm} + t_a\left(M + 3N\right) \approx \frac{N \cdot m \cdot t_a \cdot N_x^2 \cdot N_z \cdot M}{p} + \qquad (15)$$
$$+ \frac{5 \cdot N_x^2 \cdot N_z \cdot M/k \cdot t_{comm} \cdot N}{p} + 4N_x \cdot N_z \cdot t_{comm} \cdot M.$$

Then

$$S_p^C = \frac{T_1}{T_p^C} \approx \frac{N \cdot m \cdot t_a \cdot N_x^2 \cdot N_z \cdot M}{\dfrac{N \cdot m \cdot t_a \cdot N_x^2 \cdot N_z \cdot M}{p} + \dfrac{5 \cdot N \cdot t_{comm} \cdot N_x^2 \cdot N_z \cdot M/k}{p} + 4 \cdot N_x \cdot N_z \cdot M \cdot}$$
$$= \frac{p}{1 + \dfrac{\chi}{m}\left(\dfrac{5}{k} + \dfrac{4p}{N \cdot N_x}\right)}, \qquad (16)$$
$$E_p^C = \frac{S_p^C}{p} \approx \frac{1}{1 + \dfrac{\chi}{m}\left(\dfrac{5}{k} + \dfrac{4p}{N \cdot N_x}\right)}, \quad \frac{p}{N} = 1, 2, 3, \ldots$$

Comparing the estimation of the parallel realizations (14) and (16) one may point out that the efficiency of the combined approach is not worse than in the case of the geometric decomposition. However, the number of the applied active processors may be increased by one order of magnitude, which, in case of scaling of the parallel algorithm version may result in a considerable reduction of the computational run time.

The results of numerical calculations with composite approach show decrease of both run time and computational laboriousness of the problem. There is acceleration slowdown with increasing of active processors and with decreasing of the number N of the adjoint problems. The effectiveness of parallel program is 50 per cent, and it is good indicator for such type of problems.

6 Conclusion

In this paper the mathematical statement of solving the inverse problems in order to determine the parameters of an impurity source of the atmospheric air by applying the measured values of the impurity concentrations have been formulated. The mathematical statements are based on the apparatus of adjoint equations and the dual representation of the functional of the concentration of an impurity. The finite-difference methods, the explicit difference schemes, the method of a finite volume are used for the numerical solution problems. To speed up a numerical solution of a problem the following methods of the parallel realization of the problems on the impurity transport have been substantiated and applied: the functional decomposition, the geometrical decomposition and a combined method. The evaluations of acceleration and effectiveness of all the approaches of the parallel realization have been carried out, according to the evaluations of acceleration and effectiveness of the parallel realization methods which have been carried out it was proved that, on the whole, the functional decomposition has an essential advantage as compared with the other approaches.

References

1. Marchuk, G.I.: Mathematical Modeling in Environmental Studies. M. Science, 315 (1982)
2. Panasenko, E.A.: Definition of urban air pollutants, according to observations. Atmospheric and Ocean Optics 22(3), 279–283 (2009)
3. van Leer, B.: Towards the ultimate conservative difference scheme. II. Monotonicity and conservation combined in a second order scheme. J. Comput. Phys. 14, 361–370 (1974)
4. Voevodin, V.V.: Parallel Computing, p. 608. BHV-Peterburg, SPb (2002)
5. Message Passing Interface Forum. MPI: A message - Passing Interface Standart. International Journal of Super Computer Applications 8(314), 165–414 (1994)
6. Malyshkin, V.E.: Concurrent programming multicomputer, p. 296. Izd NSTU, Novosibirsk (2006)
7. Starchenko, A.V.: Parallel computing in problems of environmental protection. In: Second Siberian Workshop on Parallel Computing, pp. S17–S22 (2004)

Parallel Algorithm for Calculation of the Nanodot Magnetization

Konstantin V. Nefedev, Yury P. Ivanov, and Alexey A. Peretyatko

Far Eastern National University, Institution of Physics and Informational Technologies, 8-43, Sukhanova street, Vladivostok, Russia
knefedev@phys.dvgu.ru

Abstract. The task of the calculation of the dipole-dipole interaction between cobalt nanodot and magnetic tip was formalized and the computing algorithm was implemented in parallel C++ code with using of the MPI standard. The parallelization is fulfilled by means of the dividing total cycle and the passing the part of cycle in one process. Our parallel program allows obtain Magnetic-Force-Microscopy (MFM) images and to establish the magnetic configuration which one correspond given experimental MFM picture.

Keywords: simulation of dipole-dipole interaction, MPI standard, parallelization, nanodot, MFM images.

1 Introduction

The decreasing of memory cell size today is one of general trend of the information technologies development. The perspective direction of this field in present time is the researching of the magnetic nanostructure ordered arrays. Magnetic nanoarchitectures are interesting from fundamental point of view and as the possible application of it in the high capacity recording medium, in spintronics, in devises on the base magnetic logic, devices for quantum informatics [1,2]. The detail understanding of interaction nature, character of magnetic state and fundamental objective laws is necessary for the elaboration of magnetic logical nanoelements. However analytical investigation of the phenomena, which ones are on the nanoscales, is joined with difficulties. The model where each particle interacts with each other can have relative simple solution by numerical method. Therefore the numerical simulation of nanomagnetic superstructures and comparison of numerical results with experimental dates is imperative. Note that the numerical simulation is the quite resource-intensive even for single nanoelement a fortiori for ordered arrays, but parallel calculations give wonderful possibilities to essential extend existing class of solvable scientific tasks. In this paper we show how parallel C++ code, which one increases of productivity improvement, can be used for solution of applied task.

2 The Samples and Experimental Data

Cobalt films with the thickness 10 nm are deposited in an ultrahigh vacuum 10^{-10} Torr on naturally oxidized monocrystals (100) Si from effusion cell at a room temperature.

C.H. Hsu and V. Malyshkin (Eds.): MTPP 2010, LNCS 6083, pp. 260–267, 2010.

From above the film was covered 3 nm Cu layer for oxidizing prevention. Investigations have shown that Co films are polycrystal with the uni-axial induced magnetic anisotropy. The radius of the ferromagnetic correlation obtained from MFM images of the films magnetic structure, has made the order of 600 nm. Then the round and square nanodots with the diameter d=600 nm and the period 3d and 2d are patterned by the focused ion Ga+ beams, in each array was 10x10 nanodots.

The structure of films and arrays was explored by high energy electrons diffraction methods (RHEED), a scanning electronic microscopy (SEM) and atomic-force microscopy. The magnetic hysteresis loops of the nanodots array are gained the longitudinal magnitooptical Kerr effect (NanoMOKE-2). To obtain the magnetic structure image and hysteresis loops of the single nanodot the method of magnetic-force microscopy (MFM) and micromagnetic simulation in software package OOMMF [3] are used. Computer simulation of MFM images taking into account interacting of a tip and nanodot has been led.

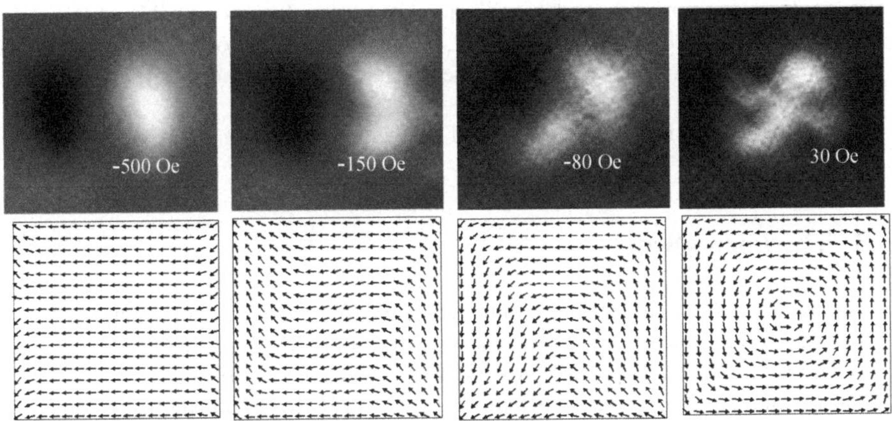

Fig. 1. MFM images of domen micromagnetic structure and possible distribution of magnetization, which one obtained by micromagnetic simulation of square dots in external magnetic field

On fig. 1 for example are shown the MFM images and the results of the micromagnetic simulation of square nanodot from a array with l=3d. It is visible, that at decreasing of an external magnetic field from the subzero saturation to plus there is a transition from the one-domain state (H =-500Oe) to the S-state (H =-150Oe), then to a C-like state (H =-80Oe) which transfers in «four-domain » state (H=30Oe). We will note, that depending on orientation of spins in C- state observed two types of four-domain states with opposite magnetization directions.

The hysteresis loops of a square nanodots array, gained by micromagnetic simulation (fig. 2b), qualitatively also will quantitatively be compounded with the hysteresis loops gained by a magnitooptical Kerr effect (fig. 2a). Magnetization springs on hysteresis loops match to transitions in C-,S- and a 4 four-domain state and back. However magnetic fields in which there is a transition from one magnetic state

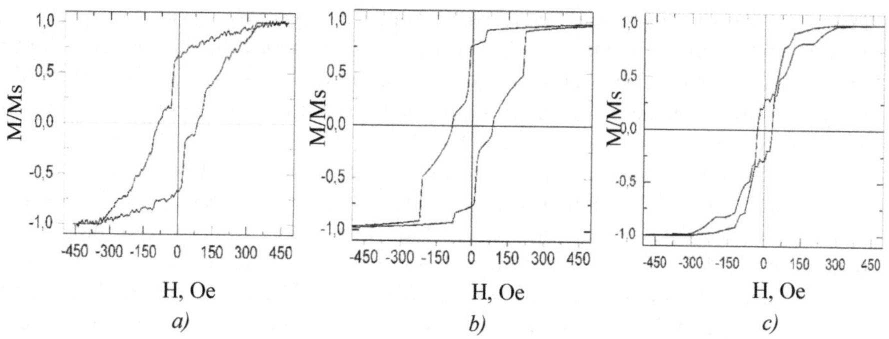

Fig. 2. The hysteresis loops of array square noninteracting nanoparticles, which ones obtained by means of: Kerr effect (a), micromagnetic simulation (b), MFM experimental data (c)

in another, for single nanodot and nanodots array. It was possibly is caused by insufficiently synchronous processes of reversal magnetization of different nanodots in array, and also presence of magnetic states with the opposite magnetization directions.

But the question in that how would we rightly decoding of experimental MFM data.

3 The Formalism

The model of the classical macrospin was used. We made micromagnetic simulation with authoring program which one allows to decoding MFM pictures.

The form of polycrystal is nanodisk with following size 10 nm in height and 600 nm in diameter, for square nanodots linear sizes were 600x600x10 (nm^3). The microstructure of the nanodot is the random distribution of grains, which ones grow out on total height of nanodot. The mean diameter of crystalline grain is 5 nm. In this model the nanosquare consists from 100x100 subdots. The partitions are in the nodes of the simple square lattice. Thus each macrospin has four nearest neighbors in the lattice. For detailed description of formalism see [4].

The interaction energy of tip with the magnetic nanodot (Zeeman energy) can be estimated by well-known rule

$$E = -\int_{V_p} \mathbf{M}(\mathbf{r})\,\mathbf{H}(\mathbf{r})dV\,, \tag{1}$$

where $\mathbf{M}(\mathbf{r})$ – summary magnetic moment of probe in field $\mathbf{H}(\mathbf{r})$. This field was created by the whole nanodot, notably by summary magnetic field of the each partition. In this case in common view the force, which act to probe, can be calculated as

$$\mathbf{F} = -\nabla E = -\int_{V} \nabla \mathbf{M}(\mathbf{r})\,\mathbf{H}(\mathbf{r})dV. \tag{2}$$

Magnetic-force-microscopy can recode only z-component of tip deviation, or z-component of force

$$F_z = \int_V \left(M_x \frac{\partial H_x(r)}{\partial z} + M_y \frac{\partial H_y(r)}{\partial z} + M_z \frac{\partial H_z(r)}{\partial z} \right) dV. \tag{3}$$

Thus, the intensity of the picture pixel correspond to scalar value

$$\frac{\partial F_z}{\partial z} = \int_V \left(M_x \frac{\partial^2 H_x(r)}{\partial z^2} + M_y \frac{\partial^2 H_y(r)}{\partial z^2} + M_z \frac{\partial^2 H_z(r)}{\partial z^2} \right) dV. \tag{4}$$

In case of small number of partitions it is possible to change integration by summation. The form and size of modeling tip have strong value on the result of simulation. In our model we use approach of the magneto-hard tip of the pyramid form, fig. 3.

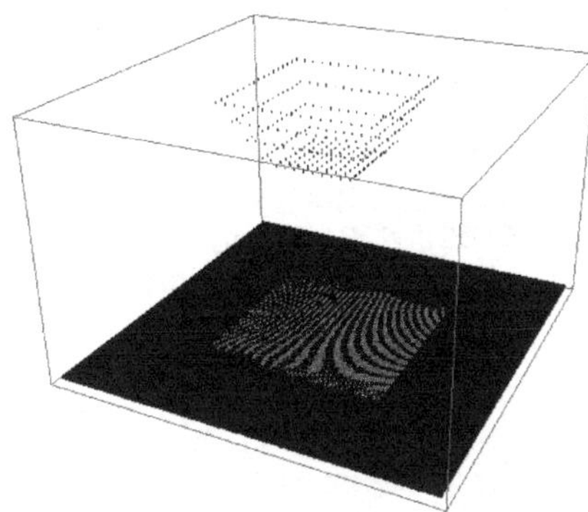

Fig. 3. Simulation of the square nanodot and tip

The magnetic material of the real probe is distributed on surface of pyramid, the same distribution we had in our model. The magneto-hard approach means that the interaction between the tip and nanodot does not influence on magnetic structure it magnetic states.

4 Input and Output Data

Input data were prepared in OOMMF package. This array of double numbers which ones correspond to components vectors magnetization with given coordinates. By means of file stream data redirect into buffer of RAM which one is the dynamical vector. Using of dynamical memory is specified by large number of elements array. Magnetic tip was simulated similar to real form of tip which one used in physical experiment. So tip had form truncated pyramid. Next the in 6 inserted cycles over coordinates of nanodot (x,y,z) and coordinates of tip (x,y,z) the values gradients of force were calculated using formula (4) and store in array. Outcomes of calculations were written in files with formats "nb" and "bmp". The most time demanding calculation is execution of cycles. The tip has to interact with all partition of nanodot and has visit more points than the number of partition for calculation of dispersion fields. Therefore the parallelization was need.

5 The Parallelization and Execution

Total size of source code of program is not too large – 21 KB, till 1000 lines. We used compilation by mpiCC with –O3 optimization. Firstly program was wrote on C++ code and include usual C libraries such as for example "math.h" and "mpi.h".

Program takes file from software package OOMMF with magnetic configuration (it is possible to specify magnetic state by hand), then it calculates MFM image in format of Matematica file (*.nb), additionally the components of nanodot magnetization is calculating.

Explanation of parallelization type in our programm

```
//declaration of datas and functions
. . .
MPI_Init(&argc,&argv);
MPI_Comm_size(MPI_COMM_WORLD,&numprocs);
MPI_Comm_rank(MPI_COMM_WORLD,&myid);
MPI_Get_processor_name(processor_name,&namelen);
fflush(stdout);
while (!done) {
MPI_Bcast(&kna, 1, MPI_DOUBLE, 0,  MPI_COMM_WORLD);
done = 1;
for(int x=myid*sizex/numprocs;
          x <(myid+1)*sizex/numprocs;
                                     x++) {
    for(int y=0; y<sizey; y++){
          for(int z=0; z<sizez; z++){
. . .
      //calculations of force gradient
      }
MPI_Reduce(&kna,&knar,sizex*sizey*sizez*4,      MPI_DOUBLE,
MPI_SUM, 0, MPI_COMM_WORLD);
fflush( stdout );
MPI_Finalize();
. . .
//end of programm
```

Variables "sizex", "sizey" and "sizez" mean number of nodes over direction x, y and z correspondingly. In the result of program code executing we have so many processes so much we set in the "mpirun" command specification, and each process take part from linear size over "x". Of course, used here the type of parallelization is quit trivial, more over the broadcast exchange by messages assumes that variable "myid*sizex/numprocs" must be integer, i.e. assumes exact knowledge the number of processes and processors and the generation of processes is extremely depending from number of free cores. The simplicity of parallel algorithm reduces productiveness. The one MFM image need about ~10^5 seconds (about one day of calculations) in case using of 8 computer cores with 2.2 GHz each, for the resolution of picture 300x300 pixels and about 100 point of tip, and 100x100 double numbers in the input array.

In first version of our program for the simple access to partitions (to value of gradient of force and to components magnetization vector) we have used Cartesian coordinates "x", "y" and "z" which ones were indexes of the double type four-dimensional array with name "`kna`". Interesting to note that if number of free cores were not in corresponding with required number of processes all processes were in "slipping" (wait the generation of other process). After generation of given number of process the calculations were started, and results of calculations summed also in four-dimensional vector "knar";

Next step was full possible optimization of code for calculation of large arrays of nanoparticles – huge number of partition (for example 500x500 numbers in the input array+10% extension). In frame of optimization we change the four-dimensional vector to 1D array, exclude mathematical functions from "math.h" library, such for example as "pow()", change "double" to "float". In addition in this step we used Local Area Multicomputer (LAM) as well as keys for "mpirun" (machinefile) for total loading of nodes of cluster.

These advancements allow enhance the efficiency and possibility of our code. In this way 32 cores (4 nodes with 8 cores each) were used for calculation of 500x500-array (nanoarchitecture 3x3 magnetic nanodots) and total time of calculation was about 6 hours.

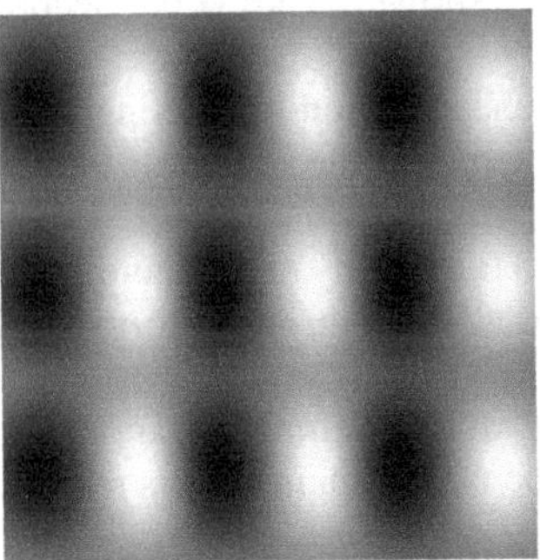

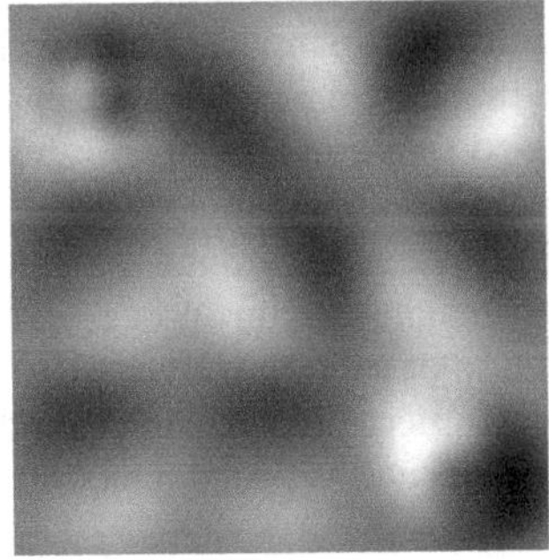

Fig. 4. Simulation of the circle nanodots (total input array 550x550). Fine particles (left side) and zero field (right side).

6 The Comparison with Experiment

The one of main aim of MFM data simulation is interpretation of the experimental MFM images. On fig. 5 for example MFM domain structure images of noninteracting round cobalt nanodots in array with period = 3d (fig. 5c) and l=2d (fig. 5f) in the

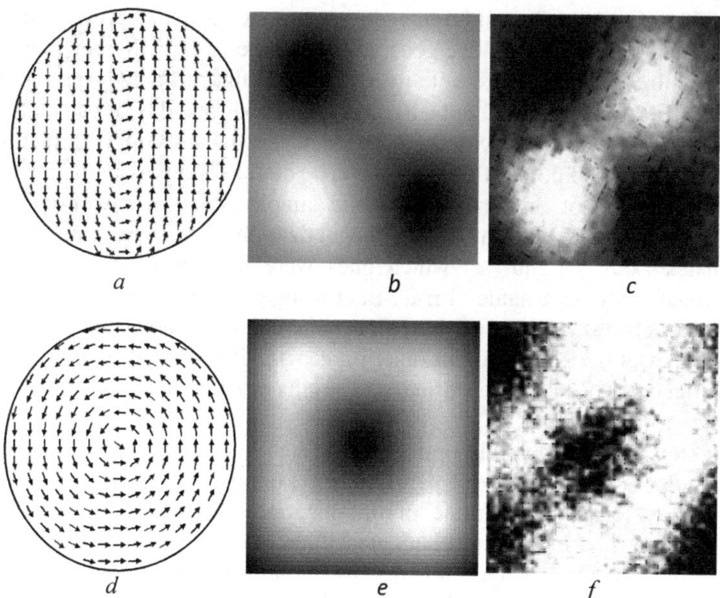

Fig. 5. Magnetic configuration (a,d), simulation MFM images (b,e), experimental MFM images (c, f)

demagnetization state are resulted. The results of micromagnetic simulation of spins allocation in these nanodots, fig. 5a, 5d, and the results of MFM images simulations matching to them, fig. 5b, 5e nearby are shown. The good according of the experimental and simulated images is visible.

The theoretical and experimental researches of non-interacting and interacting square and round nanodots array with a size equal to the region of the ferromagneric correlation has demonstrated:

- In square and round nanodot arrays, related by a dipole-dipole interaction reversal magnetization processes are carried out through formation, bias and annihilation of a magnetic vortex;
- In no interacting square nanodots array in the demagnetization state the four-domain state, and in round nanodots array - the two-domain state is realized.

7 Conclusions

Thus we showed the possibility using of parallel programming technologies for numerical simulation of MFM images. The parallelization by means of the mpi-library allows essential increase the production of calculations and that important to enlarge the class of solvable physical tasks. The analytical calculation of the space distribution of function $\partial F/\partial z$ value at given array magnetic moments demands serious theoretical efforts. More over exact solution of this task not found yet, whereas numerical simulation relatively fast gives solution.

References

1. Chen, Y.J., Huang, T.L., Leong, S.H., et al.: A study of multirow-per-track bit patterned media by spinstand testing and magnetic force microscopy. Appl. Phys. Lett. 93, 102–501 (2008)
2. Imre, A., Csaba, G., Ji, L., Orlov, A., Bernstein, G.H., Porod, W.: Majority Logic Gate for Magnetic Quantum-Dot Cellular Automata. Science 311, 205–208 (2006)
3. `http://math.nist.gov/oommf/`
4. Ovchinnikov, D.V., Bukharev, A.A.: Computer simulation of magnetic-force microscopy images in frame of static model of magnetization distribution and dipole-dipole interaction. Journal of technical physics 71(8), 85–91 (2001)

Advanced Computing Method for Solving of the Polarized-Radiation Transfer Equation*

Andrey Kovtanyuk[1], Konstantin Nefedev[2], and Igor Prokhorov[1]

[1] Institute of Applied Mathematics FEB RAS, Vladivostok, Russia
[2] Far Eastern National University, Vladivostok, Russia

Abstract. The boundary-value problem for polarized-radiation transfer equation in layered medium with Fresnel matching conditions at the boundaries of the medium partition is considered. Parallel numerical algorithm in the MPI environment based on recursive modification of Monte-Carlo method for solving the boundary-value problem is proposed and proved.

1 Introduction

In this work, boundary problem for polarized-radiation transfer equation in layered medium is considered. The description of effects, which appear at the boundaries between different materials, is important for simulation of radiation transfer within a substance. In radiation transfer theory these effects are taken into account by different matching conditions at the boundaries of the medium partition. The most popular is continuous matching conditions of solution at the boundaries of the medium partition.

Boundary problems with more general matching conditions of solution, which describe the reflection-refraction at the boundaries of the medium division, were less examined. Though, from the fiftieth of the last century specialists [1-3] repeatedly payed attention to it and different approaches for finding the boundary-value problem solution were proposed [2,4]. In the papers [5,6] the adequacy of the radiation transfer equation to realistic process was examined, its diffusion approximation was proved. In works [7,8], properties of the scalar radiation transfer equation solution with general matching conditions for 1-D and 3-D were examined.

At the present work, main results of paper [8] are generalized for vector case. The theorem of boundary-value problem solubility was formulated. The numerical recursive algorithm on base of Monte-Carlo method for solving of the boundary-value problem is proposed. Because algorithm of calculation vector-function is quite intricate and nontrivial the using of the environment with automatical parallelization can lead to complication of the debug of the executing

* This work was supported by the Russian Foundation for Basic Research (Grant No. 09- 01-98521), the grant of the competition of integration projects of the Far Eastern Branch, Siberian Branch, and Ural Branch of the Russian Academy of Sciences (Grant No. 09-II-SU-001, 09-II-SO-004).

C.H. Hsu and V. Malyshkin (Eds.): MTPP 2010, LNCS 6083, pp. 268–276, 2010.

single tasks. In connection with it the MPI standard parallelization for considered problem is used. The operation of vector-function calculation at the each point of the layer (or each direction of radiation propagation) is transmitted to single process. Notice that the other possible way of parallelization for this problem is calculation of each Monte Carlo method trajectory in single process.

In conclusion the results of numerical calculations, which demonstrate the influence of refraction, reflection and scattering on polarization and depolarization of radiation, are demonstrated.

2 Problem Formulation. Main Contingencies

Let the set G_0 be some partition of the set $G = (z_0, z_p)$, within which the radiation transfer process is examined $G_0 = \bigcup\limits_{i=1}^{p} G_i$, $G_i = (z_{i-1}, z_i)$.

The planes $z = z_i$ be the interface between the layers G_i. Let us consider the equation of radiation transfer in layered medium for azimuthal symmetry case

$$\nu f_z(z,\nu) + \mu(z) f(z,\nu) = \mu_s(z) \int\limits_{-1}^{1} P(z,\nu,\nu') f(z,\nu') d\nu' + J(z,\nu). \quad (1)$$

Here, $f(z,\nu) = (f_1(z,\nu), f_2(z,\nu))$ is a two-component vector of polarized radiation at the point $z \in G$ in the direction which angle cosine with the positive direction of the axis z is $\nu \in [-1,1]$. $f(z,\nu)$ is associated with the vector of Stokes parameter $(I_{\|}, I_{\perp})$ by the following relations

$$f_1(z,\nu) = \frac{I_{\|}(z,\nu)}{n^2(z)}, \quad f_2(z,\nu) = \frac{I_{\perp}(z,\nu)}{n^2(z)}.$$

Here, $n(z)$ is piecewise-constant refractive index of the medium ($n(z) = n_i$ for $z \in G_i$). The sum $f_1 + f_2$ describes density of radiation flow, $f_1 \geq 0$, $f_2 \geq 0$. The functions μ, μ_s are called respectively the attenuation factor and the scattering coefficient. Two-component vector J describes internal radiation sources, and P is the 2×2 scattering matrix.

As for the coefficients in (1), we assume the following. Functions μ, μ_s, J_i are nonnegative, $\mu \geq \mu_{min} > 0$, and $\mu, \mu_s \in C_b(G_0)$, where $C_b(G_0)$ is the Banach space of functions, bounded and continuous on G_0, with the norm $\|\varphi\|_{C_b(G_0)} = \sup\limits_{x \in G_0} |\varphi(x)|$.

We denote $X = G \times \{[-1,0) \cup (0,1]\}, X_0 = G_0 \times \{[-1,0) \cup (0,1]\}$. We assume, that all the components in the matrix P belong to $C_b(X_0 \times [-1,1] \backslash \{0\})$, $(Pf)_{1,2} \geq 0$ for $f_{1,2} \geq 0$, and

$$\int\limits_{-1}^{1} (p_{i1}(z,\nu,\nu') + p_{i2}(z,\nu,\nu')) d\nu' = 1, \quad i = 1, 2.$$

We define the space $V(X_0)$ formed by the two-component vector-functions $\varphi = (\varphi_1, \varphi_2)$, $\varphi_i \in C_b(X_0)$, with the norm $\|\varphi\|_{V(X_0)} = \max\limits_{i=1,2} \|\varphi_i\|_{C_b(X_0)}$, and let $J \in V(X_0)$.

We introduce the following boundary sets $\Gamma^{\pm} = \Gamma_{int} \cup \Gamma^{\pm}_{ext}$, $\Gamma = \Gamma^+ \cup \Gamma^-$, where $\Gamma_{int} = \bigcup\limits_{i=1}^{p-1} \{z_i \times \{[-1,0) \cup (0,1]\}\}$, $\Gamma^{\pm}_{ext} = \{\{z_0 \times [\mp 1, 0)\} \cup \{z_n \times [\pm 1, 0)\}\}$.

We supplement equation (1) with the boundary condition

$$f^-(z,\nu) = (Bf^+)(z,\nu) + h(z,\nu), \quad (z,\nu) \in \Gamma^-, \tag{2}$$

where $f^{\pm}(z,\nu) = \begin{cases} f(z \pm 0, \nu), & \nu < 0, \\ f(z \mp 0, \nu), & \nu > 0, \end{cases}$ and $f(z \pm 0, \nu) = \lim\limits_{\varepsilon \to +0} f(z \pm \varepsilon, \nu)$. The function h describes the radiation flux that enters the medium G. Let $h \in V(\Gamma^-)$, and $h = 0$ at the set Γ_{int}. The operator B defines matching conditions at the set Γ_{int}, and at the set Γ_{ext} we assume $B = 0$. Thus, equality (2) defines boundary conditions both at the external part of the set G_0 and at its internal boundaries.

We introduce the functions

$$\widetilde{n}_i(\nu) = \begin{cases} n_i/n_{i-1}, \text{ for } & 0 < \nu \le 1, \\ n_{i-1}/n_i, \text{ for } & -1 \le \nu < 0, \end{cases}$$

$$\psi_i(\nu) = \begin{cases} \operatorname{sgn}(\nu)\sqrt{1 - \widetilde{n}_i^2(\nu)(1-\nu^2)}, & \text{for } 1 - \widetilde{n}_i^2(\nu)(1-\nu^2) \ge 0, \\ 0, & \text{for } 1 - \widetilde{n}_i^2(\nu)(1-\nu^2) < 0, \end{cases}$$

We define a matching operator B that will be used to model Fresnel reflection and refraction at the contact boundaries z_i, $\quad i = \overline{1, p-1}$. Let

$$(Bf^+)(z_i, \nu) = R_i(\nu) f^+(z_i, \nu_R) + T_i(\nu) f^+(z_i, \nu_T),$$

where $\nu_R = -\nu$, and $\nu_T = \psi_i(\nu)$ are the directions along which the incident radiation falls onto surface $z = z_i$ and the direction in which the mirror-reflected or Snell-refracted radiation [9], having changed the direction to ν, propagates. Matrix coefficients of reflection and transmission R_i and T_i are defined by the formulas:

$$R_i(\nu) = \begin{pmatrix} R^2_{i,\|} & 0 \\ 0 & R^2_{i,\perp} \end{pmatrix}, \quad T_i(\nu) = \begin{pmatrix} T^2_{i,\|} & 0 \\ 0 & T^2_{i,\perp} \end{pmatrix} \frac{\widetilde{n}_i(\nu)\nu}{\psi_i(\nu)},$$

where

$$R_{i,\|}(\nu) = \frac{\widetilde{n}_i(\nu)\psi_i(\nu) - \nu}{\widetilde{n}_i(\nu)\psi_i(\nu) + \nu}, \quad R_{i,\perp}(\nu) = \frac{\psi_i(\nu) - \widetilde{n}_i(\nu)\nu}{\psi_i(\nu) + \widetilde{n}_i(\nu)\nu},$$

$$T_{i,\|}(\nu) = \frac{2\psi_i(\nu)}{\widetilde{n}_i(\nu)\psi_i(\nu) + \nu}, \quad T_{i,\perp}(\nu) = \frac{2\psi_i(\nu)}{\psi_i(\nu) + \widetilde{n}_i(\nu)\nu}.$$

From the definition of R and T, it follows that:
a) for all ν such that $1 - \widetilde{n}_i^2(\nu)(1-\nu^2) \le 0$, the equalities $T_i(\nu) = 0$, $R_i(\nu) = E$ hold, and for $\nu = \operatorname{sgn}(\nu)\widetilde{n}_i(\nu)/\sqrt{1 + \widetilde{n}_i^2(\nu)}$, the equality $R_{i,\|}(\nu) = 0$ holds.

In optics these cases are named as total internal reflection and reflection at Brewster's angle [9];
b) the elements of matrix coefficients R_i and T_i are continuous over $\nu \in [-1,0) \cup (0,1]$, hence, the operator B maps $V(\Gamma^+)$ into $V(\Gamma^-)$;
c) for the matrix coefficients T_i and R_i, the relation $T_i + R_i = E$ is valid.
Hence, operator $B : V(\Gamma^+) \rightarrow V(\Gamma^-)$ is linear, bounded, nonnegative and $\|B\| \leq 1$.

We define the class $D(X_0)$ in which the solution f of problem (1),(2) is sought.

Definition 1. *We say that vector-function $f = (f_1, f_2)$ belongs to $D(X_0)$ if the following properties hold:*
1) $f_j(z,\nu)$ is a function absolutely continuous over $z \in (z_i, z_{i+1}]$, for all $\nu > 0$, and absolutely continuous over $z \in [z_i, z_{i+1})$, for all $\nu < 0$, $i = \overline{0, p-1}$, $j = 1,2$;
2) $\nu f_z'(z,\nu) + \mu(z) f(z,\nu) \in V(X_0)$;
3) $f^-(z,\nu) \in V(\Gamma^-)$.

3 Some Statements

For further reasoning we introduce the following function

$$\xi(z,\nu) = \begin{cases} z_i, & (z,\nu) \in (z_{i-1}, z_i] \times [-1,0), \\ z_{i-1}, & (z,\nu) \in [z_{i-1}, z_i) \times (0,1], \end{cases} \tag{3}$$

where z_i are boundaries of the layers. Function $\xi(z,\nu)$ is bounded and continuous over X_0, hence, for all $\varphi \in V(\Gamma^-)$ we have $\widetilde{\varphi}(z,\nu) = \varphi(\xi(z,\nu),\nu) \in V(X_0)$, and $\|\varphi\|_{V(\Gamma^-)} = \|\varphi(\xi(z,\nu),\nu)\|_{V(X_0)}$. For further convenience we will use the following designations: $\xi(z,\nu) = \xi$, $D(X_0) = D$, $V(X_0) = V$, $\|\varphi\|_{C_b(G_0)} = \|\varphi\|$. Let us denote:

$$K(z,z',\nu) = \exp\left\{ -\frac{1}{\nu} \int_{z'}^{z} \mu(t)dt \right\}, \quad \overline{\lambda} = \left\| \frac{\mu_s(z)}{\mu(z)} \right\|.$$

In what follows we assume that $\overline{\lambda} < 1$. It follows from the latter inequality that there exists a function $\widetilde{\mu}(z) < \mu(z), \widetilde{\mu} \geq \text{const} > 0$, such that

$$\widetilde{\lambda} = \left\| \frac{\widetilde{\mu}(z)}{\mu(z)} \right\| < 1, \quad \lambda^* = \left\| \frac{\mu_s(z)}{\widetilde{\mu}(z)} \right\| < 1. \tag{4}$$

The differential expression $Lf(z,\nu) = \nu f_z'(z,\nu) + \mu(z) f(z,\nu)$ defines the linear operator $L : D(X_0) \rightarrow V(X_0)$. In the space $D(X_0)$, we introduce the norm:

$$\|\varphi\|_D = \max\left\{ \|\varphi^-\|_{V(\Gamma^-)}, \left\| \frac{L\varphi}{\widetilde{\mu}} \right\|_{V(X_0)} \right\} \tag{5}$$

and notice that the inclusion $D(X_0) \subset V(X_0)$ takes place.

The expressions

$$(A\varphi)(z,\nu) = \frac{1}{\nu}\int\limits_{\xi}^{z} K(z,z',\nu)\varphi(z',\nu)dz', \tag{6}$$

$$(S\varphi)(z,\nu) = \mu_s(z)\int\limits_{-1}^{1} P(z,\nu,\nu')\varphi(z,\nu')d\nu', \tag{7}$$

$$(Tf)(z,\nu) = (Bf^+)(\xi(z,\nu),\nu)K(z,\xi(z,\nu),\nu) + (ASf)(z,\nu) \tag{8}$$

define linear operators $A : V(X_0) \to D(X_0)$, $S : V(X_0) \to V(X_0)$, $T : D(X_0) \to D(X_0)$. According [10] the following statement holds:

Theorem 1. *For a function f to be a solution of problem (1),(2), it is necessary and sufficient that the operator equation*

$$f = f_0 + Tf, \tag{9}$$

$$f_0(z,\nu) = K(z,\xi(z,\nu),\nu)h(\xi(z,\nu),\nu) + (AJ)(z,\nu)$$

be satisfied with the function f in the class $D(X_0)$.

Theorem 2. *Let the inequalities $\|B\| \le 1$, $\overline{\lambda} < 1$, hold, there exists the unique solution of problem (1),(2), which can be found in the form of the Neumann series*

$$f(z,\nu) = \sum_{k=0}^{\infty}(T^k f_0)(z,\nu), \tag{10}$$

and

$$\|T\|_D \le \max\{w,\lambda^*\},$$

where

$$w = q + \widetilde{\lambda}(1-q), \quad q = \max_{1\le i\le p} \exp\left\{-\int\limits_{z_{i-1}}^{z_i} \mu(t)dt\right\}.$$

4 Numerical Experiment

At this section let's consider some numerical experiment of solving problem (1),(2) with Fresnel matching conditions by Monte Carlo method. We suppose that conditions of Theorem 2 hold true. Hence, there exists the unique solution of problems (1),(2), which can be found in form of Neumann series (10). By using Monte Carlo method we can calculate the finite sum

$$\sum_{n=0}^{N}(T^n f_0)(z,\nu). \tag{11}$$

For that, rewrite this sum in form of the following recurrence relation

$$f_n(z,\nu) = Tf_{n-1}(z,\nu) + f_0(z,\nu), \quad n = 1,2,...,N. \tag{12}$$

Let us consider a structure of the operator T. It contains two summands: the first describes reflection-refraction effects and may be calculated exactly, the second describes the contribution of scattering effects. Consider the second summand in more detail for the case of the piecewise-constant coefficients $\mu(z)$, $\mu_s(z)$. Let $z \in G_i$, accordingly we denote $\mu(z) = \mu_i$, $\mu_s(z)/\mu(z) = \lambda_i$. By force of simple transformations we rewrite the integral summand in the right-hand side of (8) in the form

$$I(z,\nu) = \frac{\lambda_i}{\nu}\left(1 - \exp\left(-\frac{\mu_i}{\nu}(z-\xi)\right)\right) \times \\ \times \int_{\xi}^{z}\int_{-1}^{1} \frac{\mu_i \exp\left(-\mu_i(z'-z)/\nu\right)}{1-\exp\left(-\mu_i(z-\xi)/\nu\right)} P(\nu,\nu')f(z',\nu')d\nu' dz'. \tag{13}$$

We can calculate integral in this expression as average of distribution of function f of random variables z', ν' distributed over corresponding intervals (ξ, z) and $(-1,1)$ with corresponding densities

$$\frac{\mu_i \exp\{-\mu_i(z'-z)/\nu\}}{1-\exp\{-\mu_i(z-\xi)/\nu\}}, \quad \frac{1}{2}. \tag{14}$$

In accordance with the Monte Carlo method, we can approximate the integral (13) by the following finite sum

$$\overline{I}(z,\nu) = \frac{2\lambda_i}{M}\left(1 - \exp\left(-\frac{\mu_i}{\nu}(z-\xi)\right)\right)\sum_{k=1}^{M} P(\nu,\nu_k)f(z_k,\nu_k).$$

Here ν_k, z_k, $k = 1,2,...,M$ are independent realizations of the random variables z', ν' distributed with corresponding densities (14). Hence, we can estimate the value of the functions $f_n(z,\nu)$, $n = 1,2,...,N$ by the following expressions

$$f_n(z,\nu) \approx \overline{f}_n(z,\nu) = \frac{1}{M}\sum_{k=1}^{M} s_k(z,\nu), \quad \overline{f}_0(z,\nu) = f_0(z,\nu), \\ s_k(z,\nu) = \left(B\overline{f}_{n-1}|_{\Gamma^+}\right)(\xi,\nu)\exp\left(-\frac{\mu_i}{\nu}(z-\xi)\right) + \\ + \lambda_i\left(1-\exp\left(-\frac{\mu_i}{\nu}(z-\xi)\right)\right)P(\nu,\nu_k)\overline{f}_{n-1}(z_k,\nu_k) + f_0(z,\nu). \tag{15}$$

Then we can calculate recurrence relation (15) by using the standard recursive procedure.

For construction of numerical experiment we took five-layers medium, so that $z_0 = 0$, $z_1 = 1$, $z_2 = 2$, $z_3 = 3$, $z_4 = 4$, $z_5 = 5$, and piecewise-constant functions

$\mu(z)$, $\mu_s(z)$, that are, $\mu(z) = \mu_i$, $\mu_s(z) = \mu_{si}$ for $z \in G_i$. The coefficients values were taken as follows: $\mu_1 = 6$, $\mu_2 = 5$, $\mu_3 = 4$, $\mu_4 = 3$, $\mu_5 = 2$, $\mu_{si} = 0.5\mu_i$, $n_1 = 1.6$, $n_2 = 1.4$, $n_3 = 1.3$, $n_4 = 1.2$, $n_5 = 1.1$. We assumed that matrix P describes Rayleigh scattering (see [11]). The following kind of incoming radiation flow was taken: $h(z,\nu) = (0,0)$ at the boundary $z = 5$ and $h(z,\nu) = (1,0)$ at the boundary $z = 0$.

For this case of the direct problem parameters, all conditions of Theorems 2 are fulfilled and solution of the problems (1), (2) can be calculated by use of the recursive relation (15).

In numerical experiment, the vector function was approximated by a sum of 12 terms of the Neumann series (10), that corresponds to 12 steps in the recursive relation (15), i.e. $N = 12$. In this case guarantee truncation error for the truncated series (11) (or recursive relation (12)) is less than 2.3%, so as it follows from Theorem 2 the series (10) converges as geometric series with the base less then $\max\{w, \lambda^*\} \approx 0.73$

At all calculations, 5000 trajectories (recursive trees) were taken ($M = 5000$). To estimate the accuracy of the method of finding a solution to direct problem (1), (2), the root-mean-square error [12]

$$\delta(z,\nu) = \frac{\sigma(z,\nu)}{\sqrt{M}\,\overline{f}_N(z,\nu)} \tag{16}$$

was found for each value $\overline{f}_N(z,\nu)$ calculated by formula (15). Here $\sigma^2(z,\nu)$ is an estimate of the sample variance

$$\sigma^2(z,\nu) = \frac{1}{M-1}\left(\sum_{k=1}^{M} s_k^2(z,\nu) - M\overline{f}_N^2(z,\nu)\right).$$

For the used value M, the maximum value of the $\delta(z,\nu)$ at all experiments was less then 2.1%.

We calculated values of two characteristics of radiation flow. The first

$$\left|\frac{f_1 - f_2}{f_1 + f_2}\right| \tag{17}$$

describes the polarization state of the radiation flow and the second is density of radiation flow $(f_1 + f_2)$. The calculations are performed at the point $z = 0$ over outgoing directions of radiation propagation.

At the Fig.1 we can see behavior of the quantity (17) at the boundary $z = 0$ over $\nu \in (-1, 0)$. At the values $\nu = 0.48$, $\nu = 0.58$, $\nu = 0.66$ and $\nu = 0.73$ the jumps of the state of polarization take place. This is explained by the effect of the total internal reflection. These jumps can be used for determination of refractive indexes of layers. Note, that jumps of density of radiation flow are ill-defined (see Fig.2).

In proposed recursive algorithm the number of the generated points for each trajectory is estimated as sum of geometric series with the base equal to 3. Hence, these calculations are labor-consuming and the use of parallel programming is justified.

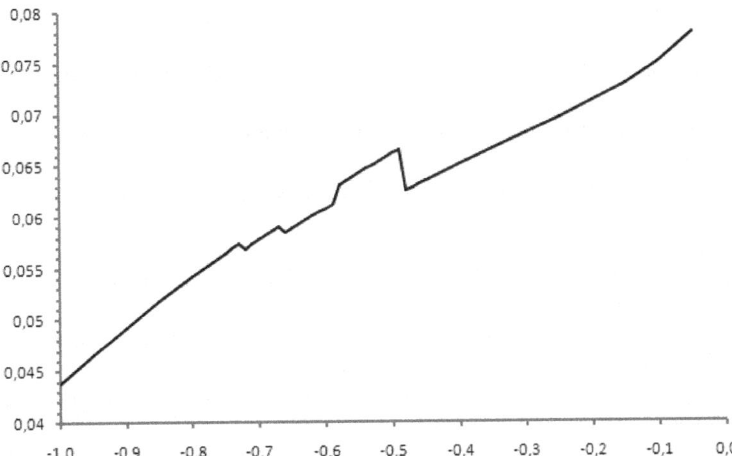

Fig. 1. The polarization state at the boundary $z = 0$ over $\nu \in (-1, 0)$ for the case of linear-polarized incoming flow of radiation

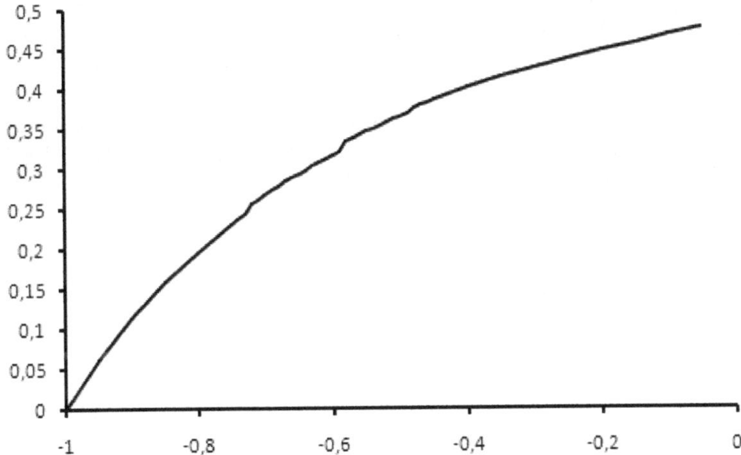

Fig. 2. Density of radiation flow at the boundary $z = 0$ over $\nu \in (-1, 0)$ for the case of linear-polarized incoming flow of radiation

On the other hand, the calculation of characteristics of radiation flow for each direction of radiation propagation is independent process with respect to calculations in other directions. Hence, this algorithm is convenient for parallelization.

The described above numerical experiments were implemented into fortran code with using of the MPI libraries. The parallelization was realized by transfer vector-function calculation at each outgoing direction of radiation propagation into single process. It allowed to decrease computing interval in several times that equals to number of used processes.

References

[1] Rozenberg, G.V.: Stoke's Vector-Parameter. Uspehi Phys. Nauk. 56(1), 77–109 (1955) (in Russian)
[2] Ishimaru, A.: Wave propagation and scattering in random media, vol. 1. Academic Press, London (1978)
[3] Apresyan, L.A., Kravtsov, Y.A.: Radiation Transfer Theory. Nauka, Moscow (1983) (in Russian)
[4] Potapov, V.S.: Method of solution of the radiation transfer equation for an optically thick layer with reflecting boundaries. Theoretical and Mathematical Physics 100(2), 1012–1022 (1994)
[5] Bal, G.: Radiative transfer equations with varying refractive index: a mathematical perspective. J. Opt. Soc. Am. 23(7), 1639–1644 (2006)
[6] Bal, G., Ryzhik, L.: Diffusion Approximation of Radiative Transfer Problems with Interfaces. SIAM J. Appl. Math. 60(6), 1887–1912 (2000)
[7] Prokhorov, I.V.: On the solubility of the boundary-value problem of radiation transport theory with generalized conjugation conditions on the interfaces. Izvestia: Math. 67(6), 1243–1266 (2003)
[8] Prokhorov, I.V., Yarovenko, I.P., Krasnikova, T.V.: An extremum problem for the radiation transfer equation. J. Inv. Ill-Posed Probl. 13(4), 365–382 (2005)
[9] Born, M., Wolf, E.: Principles of Optics. Pergamon, Oxford (1968)
[10] Kovtanyuk, A.E., Prokhorov, I.V.: Boundary-value problem for the polarized-radiation transfer equation in layered medium. Far-Eastren Math. J. 10(1), 50–59 (2010) (in Russian)
[11] Smith, O.J., Siewert, C.E.: The half-space Green's function for an atmosphere with a polarized radiation field. J. Math. Phys. 8(12), 2467–2474 (1967)
[12] Mikhailov, G.A.: Weight Monte-Karlo Methods. SB RAS Publ., Novosibirsk (2000)

Comparative Analysis of Effectiveness of Two Timing-Driven Design Approaches

Iosif Meyerov, Andrey Kamaev, Kirill Kornyakov, and Artem Zhivoderov

N.I. Lobachevsky State University of Nizhni Novgorod, Russia
Computational Mathematics and Cybernetics department
Information Technologies Laboratory (ITLab)
itlab.vlsi@cs.vmk.unn.ru

Abstract. Timing optimization is one of the most significant problems in VLSI design because it determines performance of the designed circuit. Such optimization problems contain hundreds of thousands of variables therefore they may be classified as high performance computing problems. In this paper we present the results of our parallel implementation of two popular approaches to integrated circuit timing optimization. Both approaches are based on the analytical placement computational scheme. The first approach is iterative net-weighting placement. The second executes buffer insertion algorithm after the placement cycle. We analyze effectiveness of both approaches, discuss their drawbacks and suggest ways of improvement.

Keywords: VLSI CAD, timing-driven design, physical design algorithms, placement, buffering, static timing analysis, multicore architecture.

1 Introduction

In deep submicron technology interconnect delay became the bottleneck that determines performance of a circuit. Therefore modern VLSI physical design is often referred to as timing-driven or interconnect-driven design. Timing optimization problem is extremely hard to solve: direct utilization of classical mathematical programming approaches is not suitable because of huge dimension and objective complexity.

A generally accepted way is to use iterative multistage design flow. Among the stages are sizing, placement, routing, buffer insertion and static timing analysis. Several widely used approaches to timing optimization are timing-driven placement (net-weighting and others), buffer insertion, wire and driver sizing.

The original goal of our activity is to develop a joint algorithm for simultaneous placement and buffer insertion. We expect it to be superior to the design flow where placement and buffer insertion stages are interleaved. This paper describes the results of a preliminary stage of our work. We will consider two timing-driven design approaches on the basis of an APlace-like [1, 2] multilevel analytical placement tool implemented by our team. The approaches are net-weighting and buffer insertion. We perform comparative analysis of these approaches and suggest ways of their improvement.

C.H. Hsu and V. Malyshkin (Eds.): MTPP 2010, LNCS 6083, pp. 277–282, 2010.

The rest of the paper is organized as follows. In section 2 we give an overview of our research framework, in section 3 we describe the mentioned above approaches in details. In sections 4 and 5 experimental results and topics for further investigation are discussed. In section 6 a parallel implementation of the design flow is considered, section 7 concludes the paper.

2 Research Framework

All the experiments were performed on the basis of the developed research framework [9]. We use IWLS 2005 suite [7] for benchmarking.

The framework is targeted at physical design stages starting from placement. The following stages are necessary for an adequate research of timing optimization approaches: net-weighted global placement, legalization, detailed placement (iterative local improvement), static timing analysis (STA), buffer insertion and routing (the latter for routability checking). All these components are available in the framework.

Most of the framework components were implemented in our team from scratch. At some stages implementations of third-party algorithms were used [1, 2, 8], some other stages utilize free for academic use third-party source codes and binaries [3, 4, 5, 6]. The third-party modules are used as building blocks for our original design flow. The major features of the design flow will be discussed in section 3.

One of the own components is the static timing analysis tool. It is aimed at rapid timing estimation during the global placement stage. It uses several assumptions to achieve the rapidity: only 2 metal layers, zero length via between layers, slope is not accounted and some other simplifications. Propagation of arrival and required times, calculation of total and worst negative slack (TNS and WNS), critical paths analysis are the major features if this tool.

3 Timing-Driven Approaches under Study

3.1 General Overview

The core of the design algorithm is an analytical APlace-like global placement procedure. This is a multilevel algorithm based on clustering and non-linear programming. Wirelength optimization objective is smoothed using LogSumExp [1] approximation. We use limited-memory variable-metric method from TAO [6] to solve the optimization problem.

Full placement cycle consists of the following stages: global placement, legalization and detailed placement.

3.2 Net-Weighting Approach

Net-weighting is a very popular algorithm for timing optimization because of its simple implementation. Net-weight assignment procedure is performed after full placement cycle and STA. This procedure distributes weights among nets according to their “criticality”. These weights are used as coefficients in wirelength objective during the next placement cycle.

In our framework we use the same net-weighting procedure as in [2].

3.3 Buffer Insertion

Buffer insertion allows reduction of signal delay on nets and as a result satisfaction of general timing constraints. Without buffers signal delay is proportional to the squared wirelength but there is a proven theoretical result which claims that optimal buffering can keep signal delay linear in distance between cells [10].

Within our research work we have implemented the classic Van Ginneken algorithm for RC trees. This algorithm finds optimal (in terms of maximization of required time on net driver) buffers arrangement by solving a dynamic programming problem.

Buffer insertion is executed after full wirelength-driven placement cycle: global placement, legalization and detailed placement. It starts with critical paths analysis. After that critical paths sorted by their negative slack are considered. Every path is buffered and then legalization is again executed.

4 Experimental Results

We used a test set of benchmarks from IWLS 2005 suite. Two types of experiments were performed:

1. Full placement cycle followed by buffer insertion with legalization. Schematically this design flow can be represented as *Global Placement + Legalization + Detailed Placement + STA + Buffering + Legalization + STA*.
2. 9 net-weighting iterations of placement. The design flow for this type of experiments is different. In particular, we do not perform *Detailed Placement* step because it does not consider timing constraints yet (therefore this stage is harmful for timing optimization). Instead we perform 9 iterations of the following cycle: *Global Placement + Legalization + STA + Weights calculation*. The best achieved result among all iterations is then chosen as the output.

It is important to notice that each of the approaches is versatile and adjustable and can include many techniques and stages. The two flows described above are just variants from a variety of possible constructions. Nevertheless they have been chosen as perspective ones and tuned; eventually they produce quite remarkable results.

Experimental results are in tables below. Total wire length (TWL) and total negative slack (TNS) are considered. TWL is calculated on the basis of Steiner trees construction.

5 Analysis and Future Work

As one can see from the tables above our current implementation of netlist buffering reduces TNS by 13% at the average impairing TWL by 22%. Net-weighting approach gives 6% improvement on TNS and only 13% loss of quality in TWL. At the same time analysis of absolute values of TNS and TWL of both optimization flows shows that net-weighting produces worse TWL and better TNS as compared with buffering results. Different initial placements can probably be an explanation here: wirelength-driven detailed placement was executed before buffering which led to better TWL but reduced potential reserve of buffering because of high placement density. As a result there was no free space left for buffers.

Table 1. TWL and TNS improvement after Buffer Insertion and following legalization

Benchmark	Number of elements	TWL (nm)	TNS (ns)	Number of buffers inserted	ΔTWL (%)	ΔTNS (%)
s298	141	3,16E+06	9,10E+00	14	9,02	0,46
pci_spoci_ctrl	1 267	4,58E+07	7,90E+01	254	40,92	6,36
tv80	7 161	3,85E+08	1,31E+03	1392	47,53	-15,39
ac97_ctrl	11 855	6,49E+08	2,67E+03	2284	26,19	-15,26
b21	18 718	1,06E+09	2,02E+03	2055	11,42	-19,22
b22	28 317	1,63E+09	2,98E+03	3225	6,11	-17,54
b17	37 117	2,28E+09	5,32E+03	6510	14,94	-22,83
b18	92 048	5,21E+09	1,54E+04	16697	25,03	-20,01
des_perf	98 341	5,82E+09	8,94E+03	15396	19,31	-14,00
				Average Δ	22,27	-13,05

Table 2. Best TWL and TNS improvement achieved after 9 Net-weighting iterations

Benchmark	Number of elements	TWL (nm)	TNS (ns)	ΔTWL (%)	ΔTNS (%)
s298	141	3,72E+06	9,04E+00	4,58	-1,56
pci_spoci_ctrl	1 267	6,05E+07	7,15 E+01	25,14	-9,49
tv80	7 161	4,38E+08	1,34E+03	0,00	0,00
ac97_ctrl	11 855	9,18E+08	2,44E+03	23,15	-8,52
b21	18 718	1,38E+09	1,86E+03	17,66	-9,21
b22	28 317	2,04E+09	2,85E+03	12,84	-5,99
b17	37 117	2,67E+09	5,32E+03	5,68	-2,41
b18	92 048	5,77E+09	1,55E+04	0,00	0,00
des_perf	98 341	8,08E+09	6,81E+03	24,59	-24,25
			Average Δ	12,63	-6,83

Thereby the experimental results show workability of both implementations of timing-driven algorithms. Further activity will be closely connected with tuning of both design flows. Also we plan to develop joint algorithm that should provide effective reservation of free space for buffers in order to increase effectiveness of placement+buffering flow.

6 Parallel Implementation

Solving optimization problems with hundreds of thousands variables in physical design algorithms is extremely time-consuming. This makes maximal utilization of computational resources critical. In our work we focus on multicore systems with shared memory. Nonlinear mathematical programming problem is solved by a set of threads. The idea of parallel implementation is based on execution of independent calculations of various components of objective function and gradient on different computing cores. The results of our work are given in the table below. Experiments were carried out on a system based on Intel Core2 Duo T6600 2.20GHz, 4Gb RAM and Microsoft Windows XP SP3 operating system.

Table 3. Parallel implementation speedup

Benchmark	Number of elements	Global placement runtime improvement
s298	141	27,45%
pci_spoci_ctrl	1 267	14,73%
tv80	7 161	13,29%
ac97_ctrl	11 855	18,80%
b21	18 718	15,05%
b22	28 317	14,52%
b17	37 117	24,67%
b18	92 048	18,30%
des_perf	98 341	16,21%

At present we have *Global Placement* stage parallelized. As one can see from the table parallelization does not give great advantage. The cause of this behavior is large portion of serial code (legalization and STA) in the examined section of the algorithm. However, a 15% – 25% reduction of running time is significant and vital for regular massive experiments. One of the further directions of our research work is to increase efficiency of parallelization.

7 Conclusion

In this paper we considered two approaches for timing optimization of very large integrated circuits. We implemented research infrastructure in which the major attention is paid to placement and buffer insertion algorithms. We investigated effectiveness of net-weighting and buffer insertion algorithms using multilevel APlace computational scheme and static timing analysis tool. Our further work lies in the field of combining placement and buffer insertion algorithms into one joint approach. We expect that this approach will make the timing-driven flow more effective compared to the approaches described in this paper and will let produce placements with better quality. Also we plan to improve parallel version of the application to minimize runtime and use computational resources effectively.

Acknowledgements

The work is done with Intel support. We are grateful to colleagues from Intel Strategic CAD Laboratory in Moscow Andrey Zhmurin, Leonid Kraginskiy and Oleg Venger for their guidance and support. We also would like to thank our colleagues Alexander Belyakov, Dmitry Gribanov, Ilya Lebedev and Nina Kurina for their help.

References

1. Kahng, A.B., Wang, Q.: Implementation and Extensibility of an Analytic Placer. In: Proc. ACM/IEEE International Symposium on Physical Design, pp. 18–25 (2004)
2. Kahng, A.B., Wang, Q.: An Analytic Placer for Mixed-Size Placement and Timing-Driven Placement. In: Proc. Intl. Conf. Computer-Aided Design, pp. 565–572 (2004)
3. BoxRouter: http://www.cerc.utexas.edu/~thyeros/boxrouter/boxrouter.htm
4. FLUTE, http://home.eng.iastate.edu/~cnchu/flute.html
5. Fast Buffer Insertion (FBI), http://dropzone.tamu.edu/~zhuoli/GSRC/fast_buffer_insertion.html
6. Toolkit for Advanced Optimization, http://www.mcs.anl.gov/research/projects/tao/
7. IWLS 2005 Benchmarks, http://www.iwls.org/iwls2005/benchmarks.html
8. Spindler, P., Schlichtmann, U., Johannes, F.M.: Abacus: fast legalization of standard cell circuits with minimal movement. In: Proc. International Symposium on Physical Design, pp. 47–53 (2008)
9. Kamaev, A., Kornyakov, K., Meyerov, I., Sidnev, A., Zhivoderov, A.: Building a Research Framework for Integrated Circuit Physical Design. In: Proc. East-West Design & Test Symposium (2008)
10. Otten, R.: Global wires harmful? In: Proc. Int. Symp. on Physical Design (1998)

Multithreaded Integrated Design of Aiframe Panel Manufacture Processes*

Mikhail Guzev[1], Alexandr Oleinikov[2], Konstantin Bormotin[2], and Oleg Dolgopolik[3]

[1] Institute of applied mathematics, 7 Radio Street, Vladivostok, 690042,
[2] State educational institutional of higher professional educational "Komsomolsk-on-Amur state technical university", 27, Lenina prospect, 681013 Komsomolsk-on-Amur
cvmi@knastu.ru
[3] Joint Stock Company "Komsomolsk-on-Amur Aircraft Production Association" named after Yury Gagarin, 1st, Sovetskaya St., Komsomolsk-on-Amur, 681018, Russia

Abstract. The paper describes the basic features of the developed computer-aided system of design, modeling, and electronic simulation of integral panel manufacturing. Results of system application for three-dimensional stress analysis computations and simulations of panel shaping under various thermomechanical and speed conditions are demonstrated. As is seen from computations, obtaining these solutions without parallelization can last for weeks; therefore, urgent problems can be hardly solved. These computations can be accelerated by using parallelization algorithms. In modern operational systems, the execution thread capability of generating another thread allows building multithreaded programs of recursively called subroutines, recursive calla being substituted by creation of a thread. Based on these activities, a computer-aided design system was designed for manufacturing structural panel elements with complex curvature and engraving. Application of the system for intricate manufacturing processes (blank and die tooling for shaping purposes based on material creeping) assists in eliminating the reject symptoms for wing panels of modern aircraft, considerably reduces the production costs, and improves the product quality.

1 Introduction

High-accuracy numerical simulations are used to design and optimize manufacturing of component parts of a complex structural form with high dimensional accuracy and long operating life. Until recently, it has been a good practice to use simple semiempirical mathematical models for computing the design and

* This work was partly supported by the Russian Foundation for Basic Research (project No. 07-01-00747), by the Russian Ministry for Education and Science (project No. 2.1.1/1686), and by the Komsomolsk-na-Amure Aircraft-Production Association named after Yu.. Gagarin (JSC KNAAPO of JSC Sukhoi Design Bureau), which has approved the publication of these results.

C.H. Hsu and V. Malyshkin (Eds.): MTPP 2010, LNCS 6083, pp. 283–292, 2010.

process parameters. Establishing the basic relations between these models, however, requires accumulation and analysis of vast amounts of experimental and full-size production data, which is material- and time-consuming, leading, under modern business conditions, to manufacturing unprofitability. Therefore, modern design organizations involve the analysis of a priori estimates of strength and also operational and process characteristics of the product being developed. These estimates are based on computer simulations of the most complete theoretical models. Eventually, the capabilities of computer-aided simulation and design systems have a growing impact on the solutions obtained.

Large integral panels are widely used to extend the operating life of the airframe. The stress-strain state (SSS) resulting from panel shaping determines the quality, the geometrical, physical, and mechanical performance of the panel, as well as the geometry of dies and blanks.

Aluminum-alloy wing panels used in components of modern medium-range aircraft have large dimensions, sign-variable bicurvature, irregular internal engraving, difference in height, and variable thickness of stiffening areas (stiffeners). These salient features drastically hamper the application of the thin-walled shell structure deformation theory for describing the deformation of panel blanks and necessitate spatial SSS solutions.

The current SSS simulation of structural elements must well fit within the parallel engineering technology and the CALS technology of computer support of the product life cycle (PLC). The initial reference configuration to for SSS computations (blank geometry) is unknown in general and has to be defined using a model of a prefabricated part determined in the computer-aided design (CAD) system. Existing methods and software solutions used to determine blank configurations would normally utilize geometrical panel data only and ignore panel deformation due to shaping, which may lead to impermissible dimensional errors.

One of the key salient features of high-tensile light alloys is a large difference in (heterogeneous) resistances to strain, compression, and torsion. Material models based of modern computer-aided SSS engineering software (CAE), however, give no models for materials with different resistances. The alloys have limited ductility; that is why the panels considered often are inappropriately classified in terms of conventional shaping technologies using "instant" plastic deformations. In shaping under plasticity conditions, the material life does not extend beyond the panel manufacturing stage, causing the panel to suffer intolerable damage and cracks, which results in rejection of the entire panel.

Slow shaping under steady creep conditions seems promising for extending the material life, minimizing damage and residual stresses, and providing panel manufacturing quality where stresses are restricted by the elastic limit vicinity, while the shape is generated by gradually accumulating creep deformations. Despite the above-mentioned advantages, however, creep conditions have found few applications so far because of poor development of methods for determining a non-stationary 3-D SSS for elastoviscoplastic bodies with different resistances, which have a topologically complex aerohy-drodynamic shape.

As is seen from computations, obtaining these solutions without parallelization can last for weeks; therefore, urgent problems can be hardly solved. These computations can be accelerated by using parallelization algorithms. In modern operational systems, the execution thread capability of generating another thread allows building multithreaded programs of recursively called subroutines, recursive calla being substituted by creation of a thread.

Based on these activities, a computer-aided design system was designed for manufacturing structural panel elements with complex curvature and engraving. Application of the system for intricate manufacturing processes (blank and die tooling for shaping purposes based on material creeping) assists in eliminating the reject symptoms for wing panels of modern aircraft, considerably reduces the production costs, and improves the product quality [1-5].

2 Formulation of the Shaping Simulation Problem

An integral panel is a reinforced shell which, together with stiffeners, is made of a single plate or sheet. Panel shaping normally includes two stages. The first one consists in creating the internal engraving of the panel blank, i.e. ribbing. Ribbing is normally obtained by milling the original panel blank as a plate. During the second stage, the ribbed blank is shaped to provide the aerodynamic shape of the panel.

The most versatile method of spatial discretization of equations of deformable body mechanics (DBM) for random applications used in 3-D deformation solutions is the finite element method (FEM). Originally, the major problems in computer simulations of shaping processes consist in drastic and combined nonlinearity: physical, geometrical, and contact. Appropriate control of all nonlinearities may require significant computer powers to investigate the impact of the finite elements in use on the accuracy and resources of computation for various combinations of these nonlinearities.

The definition of the DBM equations exclusive of the geometrical deformation nonlinearity, i.e., with minor deformations, turns, and displacements, yet, with possible large translational displacements of the body as a rigid unit, is referred to as MNO (Material Nonlinear Only). Two equation definitions are considered for the geometrical nonlinearity, TL (Total Lagrangian) and UL (Updated Lagrangian) [6,7], which have different reference configurations of the deformed body. The reference configuration in the TL definition is the initial body configuration, while that in the UL definition is the current (deformed) configuration.

The TL definition of equations is further considered.

The weak equilibrium equations (virtual work balance equations) are

$$\int_V \mathbf{S} : \delta\mathbf{E} dV = \int_V \rho\mathbf{f} \cdot \delta\mathbf{u} dV + \int_{S_T} \tilde{\mathbf{T}}^* \cdot \delta\mathbf{u} dS + \hat{R}_c, \quad \forall\delta\mathbf{u} \quad (\delta\mathbf{u} = 0 \quad on \quad S_u). \tag{1}$$

Hereinafter: $\mathbf{S}$ is the second tensor of the Piola-Kirchhoff equations, $\mathbf{E}$ is the Green-Lagrange strain tensor, and the vector of surface forces $\tilde{\mathbf{T}}^* \equiv \mathbf{N} \cdot (\mathbf{S} +$

$\mathbf{S} \cdot \nabla \mathbf{u}$) is determined on the part of the surface S_T, on S_u: $\mathbf{u} = \mathbf{u}^*$. Here, $\mathbf{S} \equiv J\mathbf{F}^{-1} \cdot \sigma \cdot \mathbf{F}^{-T}$, $\mathbf{F} \equiv \mathbf{g} + \nabla \mathbf{u}^T$, $J \equiv \det \mathbf{F}$, and $\mathbf{g}$ is the unit tensor.

The kinematic relations (between the Green-Lagrange strain tensor and the displacement gradient tensor) are

$$\mathbf{E} = \frac{1}{2}(\nabla \mathbf{u} + \nabla \mathbf{u}^T + \nabla \mathbf{u} \cdot \nabla \mathbf{u}^T). \tag{2}$$

The constitutive relations are formulated as follows:

$$\dot{\mathbf{S}} = \mathbf{C} : \dot{\mathbf{E}} + \mathbf{\Phi}. \tag{3}$$

Here, $\mathbf{\Phi}$ is the second-rank tensor whose constituents in general may depend on the constituents of the tensor of the Piola-Kirchhoff stresses $\mathbf{S}$ and the Green-Lagrange strain tensor $\mathbf{E}$. For elastic and elastoplastic materials, the tensor $\mathbf{\Phi}$ has to be set equal to zero.

To apply a stepwise procedure of integrating Eqs. (1)-(3) with respect to the deformation parameter t , it is necessary to derive equations of solid body deformation given in increments from these equations. The step $\triangle t$ is assumed to be fairly small. All sought quantities are assumed to be determined by the time t, i.e. Eqs. (1) are satisfied identically. The function dependence on t is further presented by the left superscript: say, ${}^t\mathbf{S}$ and ${}^{t+\triangle t}\mathbf{S}$ are the second tensors of the Piola-Kirchhoff stresses determined at the deformation times t and $t + \triangle t$, respectively. Hereinafter, the symbol $\triangle$ before the quantity refers to its increment from the time t to the time $t + \triangle t$, say, $\triangle \mathbf{S} \equiv^{t+\triangle t} \mathbf{S} -^t \mathbf{S}$ and $\triangle \mathbf{E} \equiv^{t+\triangle t} \mathbf{E} -^t \mathbf{E}$.

Considering Eq. (1) at the time $t + \triangle t$ and linearizing the constitutive relations (3) with respect to the time t, we derive a linearized equation of possible displacements given in increments ($\delta \mathbf{u} = 0$ on S_u):

$$\int_V \delta \mathbf{e} :^t \mathbf{C} : \mathbf{e} dV + \frac{1}{2} \int_V {}^t\mathbf{S} : \delta[\nabla \tilde{\mathbf{u}} \cdot \nabla \tilde{\mathbf{u}}^T] dV = \int_V \rho^{t+\triangle t}\mathbf{f} \cdot \delta \tilde{\mathbf{u}} dV +$$

$$+ \int_{S_T} {}^{t+\triangle t}\tilde{\mathbf{T}}^* \cdot \delta \tilde{\mathbf{u}} dS + {}^{t+\triangle t}\hat{R}_c - \int_V ({}^t\mathbf{S} + {}^t\mathbf{\Phi} \triangle t) : \delta \mathbf{e} dV. \tag{4}$$

Here, $\tilde{\mathbf{u}} \equiv \triangle \mathbf{u}$, $\quad \mathbf{e} \equiv \frac{1}{2}(\nabla \tilde{\mathbf{u}} + \nabla \tilde{\mathbf{u}}^T + \nabla^t \mathbf{u} \cdot \nabla \tilde{\mathbf{u}}^T + \nabla \tilde{\mathbf{u}} \cdot \nabla^t \mathbf{u}^T)$.

A discrete analog of the scalar equations (4) is obtained by the finite element method. In both cases, we obtain an equation of the type [7-9]:

$$\delta \mathbf{U}^{T t}\mathbf{K} \triangle \mathbf{U} = \delta \mathbf{U}^T ({}^{t+\triangle t}\mathbf{R} - {}^t\mathbf{F}), \quad \forall \delta \mathbf{U} \in \mathbf{O}^{N_{eq}}. \tag{5}$$

Where, $\triangle \mathbf{U}$ is the increment vector of nodal displacements; ${}^{t+\triangle t}\mathbf{R}$ and ${}^t\mathbf{F}$ are the vectors of external and internal forces determined at the times $t + \triangle t$ and t, respectively; ${}^t\mathbf{K}$ is the symmetrical stiffness matrix determined at the time t; Neq is the number of the master degrees of freedom.

Owing to the arbitrariness of the vector $\delta \mathbf{U}$, the scalar equation (5) is equivalent to the vector equation:

$${}^t\mathbf{K} \triangle \mathbf{U} =^{t+\triangle t} \mathbf{R} - {}^t\mathbf{F}. \tag{6}$$

Once the increment vector of the nodal displacements $\triangle\mathbf{U}$ has been derived from the system of linear equations (6), the solution ${}^{t+\triangle t}\mathbf{U}$ for the nodal displacement vector at the time instant $t+\triangle t$ is determined by the formula ${}^{t+\triangle t}\mathbf{U} = {}^{t}\mathbf{U}+\triangle\mathbf{U}$. This solution is updated using the Newton-Raphson method until the residual vector at the i-th iteration ${}^{t+\triangle t}\mathbf{R} - {}^{t+\triangle t}\mathbf{F}^{(i)}$ becomes close to the zero vector (in the Euclidean norm sense) within a given relative solution error.

To determine the matrix ${}^{t}\mathbf{K}$ and the vectors ${}^{t+\triangle t}\mathbf{R}$ and ${}^{t}\mathbf{F}$ in the system of algebraic equations (6) and to solve this system, we applied the MSC.Marc 2007 package [6] which takes into account all types of nonlinearities of the DBM equations. The geometric simulation and display of computation results were provided in the MSC.Patran 2007 package [10].

Creeping laws satisfying these requirements may be referred to as the generalized Norton's laws for steady creeping of isotropic media with different strain, compression, and shear characteristics. We consider two types of creep laws.

Type 1. Associated creep law. For such a creep law, a potential relation is assumed to exist and to be formulated as:

$$\dot{\epsilon}^{c} = \frac{\partial\mathbf{\Phi}(\sigma_e,\theta)}{\partial\sigma},$$

where the potential $\mathbf{\Phi}$ is a scalar function generally dependent on both the effective stress and the stress state angle type. We then form a function $\mathbf{\Phi}$ which, for a fixed value of the angle type θ of the stress state θ, is a linear combination of three scalar functions of the effective stress σ_e:

$$\mathbf{\Phi}(\sigma_e,\theta) \equiv \frac{1}{2}[\mathbf{\Phi}_1+\mathbf{\Phi}_2+(\mathbf{\Phi}_2-\mathbf{\Phi}_1)\sin 3\theta+(2\mathbf{\Phi}_0-\mathbf{\Phi}_1-\mathbf{\Phi}_2)(1-|\sin 3\theta|)],$$

where:

$$\mathbf{\Phi}_1(\sigma_e) \equiv \frac{B_1}{n_1+1}\sigma_e^{n_1+1}, \quad \mathbf{\Phi}_2(\sigma_e) \equiv \frac{B_2}{n_2+1}\sigma_e^{n_2+1}, \quad \mathbf{\Phi}_0(\sigma_e) \equiv \frac{B_0}{n_0+1}\sigma_e^{n_0+1}$$

In these equations, B_1, n_1, B_2, n_2, B_0 and n_0 are the creep constants defined through monoaxial compression, strain, and shear (torsion) experiments, respectively, as:

$$\mathbf{\Phi}(\sigma_e,-\frac{\pi}{6}) = \mathbf{\Phi}_1(\sigma_e), \quad \mathbf{\Phi}(\sigma_e,\frac{\pi}{6}) = \mathbf{\Phi}_2(\sigma_e), \quad \mathbf{\Phi}(\sigma_e,0) = \mathbf{\Phi}_0(\sigma_e).$$

Type 2. Non-associated creep law. This law is given by

$$\gamma(\sigma_e,\theta) \equiv \frac{1}{2}[\gamma_1+\gamma_2+(\gamma_2-\gamma_1)\sin 3\theta],$$

where:

$$\gamma_1(\sigma_e) \equiv \frac{3}{2}B_1\sigma_e^{n_1-1}, \quad \gamma_2(\sigma_e) \equiv \frac{3}{2}B_2\sigma_e^{n_2-1}.$$

Here, the material constants B_1, n_1 and B_2, n_2 have a mechanistic interpretation similar to the associated creep law.

3 Composition of Design and Process System

Structurally, the IPTCGEM system (Involutes, Procedure, Tools, Computation Grid, Electronic Model) consists of dedicated software and common packages of engineering (MSC.Marc) and geometrical analysis (Unigraphics UG) (Fig. 1). The T-Reollaw Fortran program determines the thermoreological law for a blank that is not available in the model base for materials available in MSC.Marc package, e.g., to control the difference in alloy resistances and non-uniform anisotropy of elastic and inelastic properties. This program is launched in MSC.Marc via MSC.Marc Mentat and (or) MSC.Patran pre/postprocessors. The input data are characteristics of deformation diagrams for the reference material witness sample under predetermined conditions and deformation temperature. The output data are fields of displacement constituents and also stress and strain tensors of the panel blank. The UnformTheory program implements an iterative

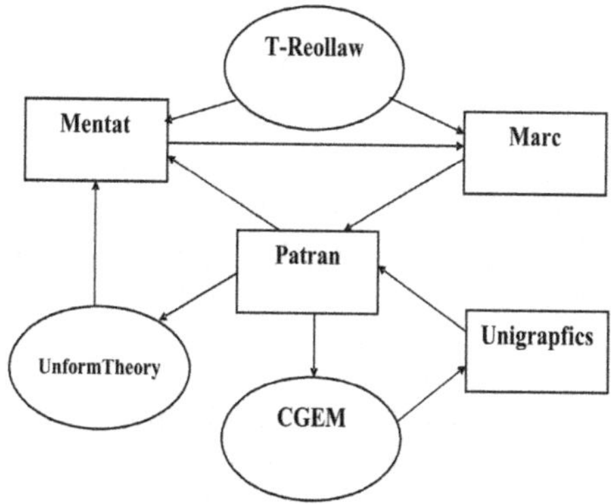

Fig. 1. Functional diagram of the IPTCGEM system

algorithm of successive refinement of the 3-D finite-element sweep of the CAD model of the part with an anticipated blank shape and working outline of die tooling. The algorithm implementation makes the parameters of the computed sweep approach the parameters of the sought blank, and the configuration derived from the computed sweep approach the theoretical contour.

The program implementing the algorithm is encoded using MSC.Patran interface procedures - PCL command language. The input data include the geometrical model of the part, the process parameters and die forming modes, and the tolerance imported from Unigraphics to MSC.Patran. The output data are the surface computation grids for the blank sweep and the working contour of the tooling, which eliminate any reject symptoms of the process.

The CGEM program converts the surface computation grids of the blank and the working contour of the tooling to the UG electronic modules suitable for defining efficient machining programs for part blanks and dies in the CAM UG module. The program consists of two sets of subprograms: structure library and procedures encoded in the Component Pascal language and subroutines to be executed under Unigraphics encoded in the C++ language using the interface procedures of Unigraphics - UG Open:

1) PointLinks - a library containing data structures for presenting necessary objects and procedure of bypassing and searching for objects;

2) Geom - a library providing geometrical procedures and analysis procedures for the computation grid and its fields;

3) ReadEdges - a subprogram for creating a file containing numbers of the rib points of the computation grid; it is launched under Unigraphics. The input data for the subprogram are the 3-D part model and the computation grid for the part; the output data are the file with rib points;

4) BuildEdges - a subprogram for building an electronic 3-D model of the part, based on the computation grid using the file with rib points. The input data for the subprogram are the computation grid of the FEM sweep and the file with rib points for the sweep. The subprogram is launched under Unigraphics. The program operation is aimed at facilitating viewing and evaluating and generates a facet model of the part, facet rib segments lying on the body ribs, body rib splines (if possible to be defined by splines) and planes (if the surface areas are close to the plane) or a set of points on the working layer, which may give a reference cloud to build surfaces using the surface building operation;

5) SearchEdges - a subprogram for building an electronic 3-D model of the part, based on the computation grid through searching for the ribs. The input data are similar to those for Build-Edges. The subprogram is launched under Unigraphics. The output data are the facet and elec-tronic 3-D models of the blank sweep and die tooling with highlighted flat, smooth, cylindrical, and developing surface areas. Identification of these areas by means of the subprogram is used in the CAM UG module to streamline the programs for blank milling at NPC machines.

4 Results of Calculations and Parallel Realization of the IPTCGEM System for a Given Number of Processors

The operation of the IPTCGEM system can be accelerated by using parallel algorithms in performing multithreaded computations on computers with the SMP-architecture.

Serial and multithreaded algorithms of computations with implementation of the IPTCGEM system were performed, in the first variant, on a Pentium 4 3GHz, 2Gb and, in the second variant, on a server with two Intel Dual-Core Xeon 1.6GHz, 2Gb processors under Microsoft Windows XP Professional and Microsoft Windows XP Professional x64 Edition, respectively.

The test computations were performed for the problem of development of a panel with elastoplastic properties. Figure 2 shows the time of problem solving as a function of the number of threads.

The type of parallelism used is based on domain decomposition [6]. The model is decomposed into domains of elements, where each element belongs to one and only one domain. The nodes located on the domain boundaries are duplicated in all domains at the boundary. The total number of elements is, thus, the same as in a serial (nonparallel) run, but the total number of nodes can be greater. The computations in each domain are performed by separate processes (in our case, the kernel) on the computer used. Such an approach appears to be most effective for application on systems with the SMP architecture, because it allows reaching the peak efficiency (the maximum loading of computing kernels) of the multinuclear architecture, since there is no need to synchronize operation of the threads.

The results of testing the distributed computations with matrix decomposition on a server with the SMP-architecture show that the optimal number of threads is equal to the number of processors.

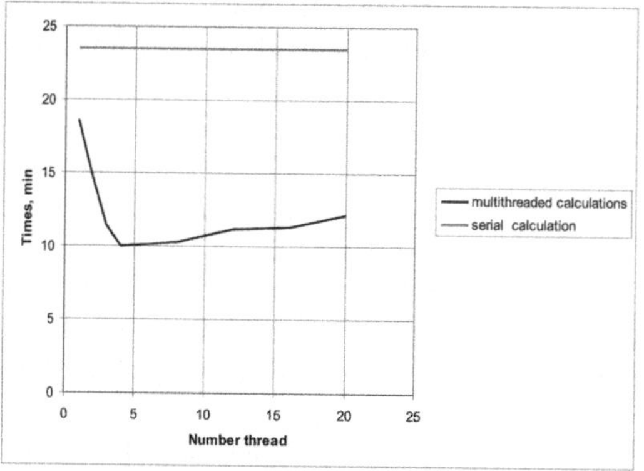

Fig. 2.

If the number of kernels is unknown, we propose a synchronization mechanism built into all advanced operational systems (semaphores). The results of operation of semaphores are an insignificant decline of the computation performance relative to the maximum reachable value and to an increase in the optimal number of threads. The semaphore is initialized by an initial value equal to the number of computation nodes (kernels). Each newly generated thread "expects" this semaphore and then grasps it (reducing the counter associated with a semaphore by one). New streams are not generated as soon as the counter becomes equal to zero. After the end of operation of a thread, it "releases" the semaphore, increasing the counter by one

Figure 3 shows the displacement norm (H) in an iterative process (I) of calculating the panel shape.

Based on the results of industrial tests, we can conclude that the application of the IPTCGEM system ensures that [11]:

- the manufacturing procedure for structural elements and their blanks is carefully developed and can be adapted to manufacturing during the design stage;

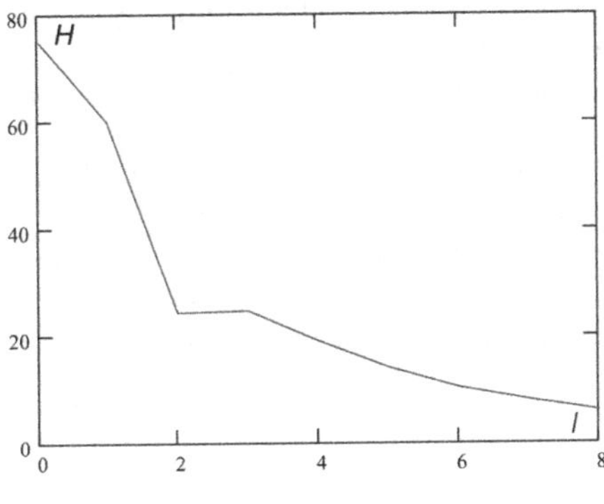

Fig. 3.

- the transition to electronic simulation of procedures is provided; expensive, labor-consuming and long full-size updating of dies, blanks, temperature, and rate performance of die forming are ruled out;
- the part is manufactured during the first run due to in-depth simulation of all manufacturing procedures.

References

1. Oleinikov, A.I.: Simulation of Panel Shaping of the Russian Regional Plane: IX Theoretical and Applied Mechanics All-Russia Congress. In: Novgorod, N. (ed.) Proceedings Annotation, August 22-28, vol. 3: NSU, p. 162 (2006)
2. Guzev, M.A., Oleinikov, A.I., Seryi, S.V., Bormotin, K.S., Pekarsh, A.I.: Virtual Development of Large Sized Parts Shaping Technologies and Tapered Thickness at Cluster Systems. Computational Methods and Programming, vol. 8, pp. 123–129 (2007)
3. Oleinikov, A.I., Dolgopolik, O.D., Bormotin, K.S.: Integrated Multiline Designing System for Air Frame Panels Manufacturing Processes: Concurrent Computational Technologies. In: Work of the International Scientific Conference, pp. 199–206. Pub. YuUr-GU (South Ural State University), Chelyabinsk (2008)
4. Oleinikov, A.I., Korobeynikov, S.N., Bormotin, K.S.: Effect of finite-element type concept at elastoplastic panels shaping simulation. Continuous media mechanics calculation 1(2), 63–73 (2008)
5. Korobeinikov, S.N., Oleynikov, A.I., Gorev, B.V., Bormotin, K.S.: Mathematical simulation of creeping processes for metal products of materials with different properties at strain and compression. Computational methods and programming 9, 346–365 (2008)
6. MARC Users Guide. MSC. Software Corporation, vol. A: Theory and Users Information, p. 770; vol. B: Element Library, p. 837 (2007)
7. Korobeinikov, S.N.: Solids nonlinear deformation, p. 262. Siberian Department of the RAS, Novosibirsk (2000)
8. Bathe, K.-J.: Finite Element Procedures. Prentice Hall, New Jersey (1996)

9. Zeinkiewicz, O.C., Taylor, R.L.: The Finite Element Method. Butterworth-Heinemann, Oxford (2000)
10. MSC.Patran Reference Manuals. MSC. Software Corporation: Geometry Modeling, p. 734 (2007)
11. Oleinikov, A.I., Pekarsh, A.I., Bakaev, V.V., Sarikov, S.A., Dolgopolik, O.D.: Preparation of manufacture of difficult details of double signvariable curvature by a method of the is finite-element analysis of geometrical model with complex working out of form-building equipment, plain of a detail and recommendations about technological process. SAPR i graphica (2), 88–96 (2009) (in Russia)

Efficient Biorthogonal Lanczos Algorithm on Message Passing Parallel Computer

Sun Kyung Kim

School of Computer and Information Technology, Daegu University, Korea
skkim@daegu.ac.kr

Abstract. Many important scientific and engineering problems require the computation of a small number of eigenvalues of large nonsymmetric matrices. The biorthogonal Lanczos method is one of the methods to solve that problem. In this paper, we introduce the *s*-step biorthogonal Lanczos method generating reduction matrices which are similar to reduction matrices generated by the standard biorthogonal Lanczos method. The *s*-step generalization of biorthogonal Lanczos method enhances parallel properties by forming *s* simultaneous search direction vectors. The *s*-step biorthogonal Lanczos method has the minimized synchronization points, which resulted in the minimized global communication compared to the standard biorthogonal Lanczos method.

Keywords: biorthogonal Lanczos algorithm, *s*-step method, message passing parallel computer.

1 Introduction

Memory contention on shared memory machines constitutes a severe bottleneck for achieving their maximum performances. The same is true for communication costs on a message passing system. It would be desirable to have method for specific problem, which has low communication cost compared to the computation cost. This is interpreted as a small number of main memory access for the shared memory systems and a small number of global communications for the message passing systems. It also reduces the need for frequent synchronizations of the processors. Linear algebra algorithms, which are implemented efficiently on parallel computers, have also been studied [2,3,7]. Many important scientific and engineering problems require the computation of a small number of eigenvalues of large nonsymmetric matrices. The biorthogonal Lanczos method is one of the methods to solve that problem [5,6]. The biorthogonal Lanczos method generates two sets of vectors that are biorthogonal, and a sequence of nonsymmetric but tridiagonal Lanczos matrices [1]. We want to reduce the need for frequent synchronizations of the processors on the parallel computer. In this paper, the *s*-step biorthogonal Lanczos method forms independent direction vectors using repeated matrix vector multiplication of the coefficient matrix with a single direction or residual vector. This provides coarser granularity and increases parallelism by computing simultaneously all inner products involved in the evaluation of parameters used in advancing the iterations. In the *s*-step method, *s*

C.H. Hsu and V. Malyshkin (Eds.): MTPP 2010, LNCS 6083, pp. 293–299, 2010.

consecutive steps of the standard method are performed simultaneously. This means, for example, that the inner products needed during s steps of the standard method can be performed simultaneously.

2 The Biorthogonal Lanczos Method

Let A be a $n\times n$ nonsymmetric matrix. There are many tridiagonal matrices similar to A and T_n is one of them. Then for some matrix $Q_n = (q_1, \ldots, q_n)$, we have

$$Q_n^{-1} A Q_n = T_n \quad (2.1)$$

Let $P_n = (p_1, \ldots, p_n)$ and replace (2.1) by two separate relations

$$P_n^T Q_n = I$$
$$P_n^T A Q_n = T_n$$

By equating columns on each side of $AQ_n = Q_n T_n$ and $P_n^T A = T_n P_n^T$ in the natural increasing order, we obtain the following equations: For each $j < n$

$$AQ_j = Q_j T_j + r_j e_j^T$$
$$P_j^T A = T_j P_j^T + s_j^T e_j$$

Where r_j and s_j are residual vectors. A and two initial vectors p_1 and q_1 essentially determine all the other elements of P_j, Q_j and T_j.

In this method, the right space Q_j is a Krylov subspace $Q_j = \text{span}[q_1, Aq_1, \ldots, A^{j-1}q_1]$ and the left space P_j is a Krylov subspace $P_j = \text{span}[p_1, A^T q_1, \ldots, (A^T)^{j-1} p_1]$ and $P_j^T Q_j = I_j$. The eigenvalues of the biorthogonal Lanczos matrices T_j are called Ritz values of A.

For many matrices and for relatively small j, compared to n, several of the extreme eigenvalues of A are well approximated by the corresponding Ritz values. The right Ritz vector $Q_j y(=z)$ obtained from a right eigenvector y of a given T_j is an approximation to a corresponding right eigenvector of A, and the left Ritz vector $P_j \hat{y}(=\hat{z})$ obtained from a left eigenvector $\hat{y}$ of a given T_j is an approximation to a corresponding left eigenvectors of A. A simple version of biorthogonal algorithm can be formulated as follows:

Algorithm 1. The biorthogonal Lanczos Algorithm

$q_0 = 0, p_0 = 0$ and $\beta_0 = 0, \gamma_0 = 0$ Choose q_1 and p_1 with $(p_1, q_1) = 1$.
For j=1 **until** Convergence **Do**

1. Compute and store Aq_j, $p_j^T A$
2. $\alpha_j = (Aq_j, p_j)$
3. $r_j = Aq_j - \gamma_{j-1} q_{j-1} - \alpha_j q_j$
   $s_j^T = p_j^T A - \beta_{j-1} p_{j-1}^T - \alpha_j q_j^T$
4. $\beta_j \gamma_j = (r_j, s_j)$
5. $q_{j+1} = r_j / \beta_j$
   $p_{j+1} = s_j / \gamma_j$

EndFor

Note that $p_j^TA = A^Tp_j$ requires multiplication by the transpose of A. Let T_j be the tridiagonal matrix at step j

$$T_j = \begin{bmatrix} \alpha_1 & \gamma_1 & & & \\ \beta_1 & \alpha_2 & \gamma_2 & & \\ & \cdot & \cdot & \cdot & \cdot \\ & & & & \gamma_{j-1} \\ & & & \beta_{j-1} & \alpha_j \end{bmatrix}$$

For each j, the nonsymmetric Lanczos matrix T_j is the biorthogonal projection of A onto the Krylov subspaces spanned by the Q_j and P_j. The coefficients β_j and γ_j in T_j are not uniquely defined by step 4 of algorithm 1. Q_j and P_j are biorthogonal with any choices of β_j, γ_j which satisfy equation $\beta_j\gamma_j = (r_j, s_j)$. One possible choice is $\beta_j = \sqrt{|(r_j, s_j)|}$ and $\gamma_j = \text{sign}(r_j, s_j)\beta_j$ [4]. Observe that the continuation of this recursion requires that $(r_j, s_j) \neq 0$ for any j. $(r_j, s_j) = 0$ causes the algorithm to break down.

3 *s*-Step Biorthogonal Lanczos Method

Let us denote k by the iteration number in the s-step biorthogonal Lanczos method. We will denote by V and W the s-step biorthogonal Lanczos vectors instead of p and q of the standard biorthogonal Lanczos method. Given the vectors $v_k^1, v_k^2, \ldots, v_k^s$ we will use $\overline{V}_k$ (each of dimension n) to denote matrix of $\{v_k^1, v_k^2, \ldots, v_k^s\}$. Given the vectors $w_k^1, w_k^2, \ldots, w_k^s$ we will use $\overline{W}_k$ (each of dimension n) to denote matrix of $\{w_k^1, w_k^2, \ldots, w_k^s\}$. The subspace $\overline{V}_k$ is spanned by $\{v_k^1, Av_k^1, \ldots, A^{s-1}v_k^1\}$ so that $\overline{V}_k$ is made of orthogonal to the subspaces $\overline{W}_{k-1}, \overline{W}_{k-2}, \ldots, \overline{W}_1$. Also the subspace $\overline{W}_k$ is spanned by $\{w_k^1, A^Tw_k^1, \ldots, (A^T)^{s-1}w_k^1\}$ so that $\overline{W}_k$ is made orthogonal to the subspaces $\overline{V}_{k-1}, \overline{V}_{k-2}, \ldots, \overline{V}_1$.

Let $s \times s$ matrix $\overline{W}_k^T\overline{V}_k$ be nonsingular. Then LU decomposition with row exchanging can be applied to the matrix $\overline{W}_k^T\overline{V}_k$ as follows:

$$\overline{P}_k\left(\overline{W}_k^T\overline{V}_k\right) = \overline{L}_k * \overline{U}_k$$

where $\overline{P}_k$ is a permutation matrix, $\overline{L}_k$ is a $s \times s$ lower triangular matrix and $\overline{U}_k$ is a $s \times s$ upper triangular matrix.

Remark 1

If $\hat{T}_j$ is a tridiagonal matrix generated by the standard block Lanczos algorithm and $\overline{T}_k = U_k^{-1}\hat{T}_jU_k$ where $j = s*k$ and $U_k = \text{diag}(\overline{U}_1, \overline{U}_2, \ldots, \overline{U}_k)$, then $\overline{T}_k$ becomes a nonsymmetric matrix similar to $\hat{T}_j$ as follows:

$$\overline{T}_k = \begin{bmatrix} G_1 & E_1 & & & \\ F_1 & G_2 & E_2 & & \\ & \cdot & \cdot & \cdot & \cdot \\ & & & & E_{k-1} \\ & & & F_{k-1} & G_k \end{bmatrix}$$

where G_j and E_j are $s \times s$ matrices. Here is a $s \times s$ matrix F_j whose only nonzero element is at location $(1, s)$. Also the matrix $L_k^{-1}\hat{T}_j L_k$, where $L_k = \mathrm{diag}(\overline{L}_1, \overline{L}_2, \ldots, \overline{L}_k)$ and $j = s*k$ becomes the same type of a nonsymmetric matrix $\overline{T}_k$. We now give the defining equations of the s-step biorthogonal Lanczos method in the form of an algorithm.

Algorithm 2. The s-step biorthogonal Lanczos algorithm

$\overline{V}_0 = 0, \overline{W}_0 = 0,$

$$\overline{V}_1 = [v_1^1, Av_1^1, \ldots, A^{s-1}v_1^1]$$
$$\overline{W}_1 = [w_1^1, A^T w_1^1, \ldots, (A^T)^{s-1}w_1^1]$$
$$[b_0^j] = 0, [\overline{b}_0^j] = 0, 1 \le j \le s$$

For $k = 1$ **until** Convergence **Do**

Select $[a_k^j], [b_{k-1}^j]$, $1 \le j \le s$ to orthogonalize $\overline{V}_k$ against $\overline{W}_{k-1}$.
Also select $[\overline{a}_k^j], [\overline{b}_{k-1}^j]$, $1 \le j \le s$ to orthogonalize $\overline{W}_k$ against $\overline{V}_{k-1}$.
These give $v_{k+1}^1 = Av_k^s - \overline{V}_{k-1}b_{k-1}^s - \overline{V}_k a_k^s$
$w_{k+1}^1 = Aw_k^s - \overline{W}_{k-1}\overline{b}_{k-1}^s - \overline{W}_k \overline{a}_k^s$.
Select $[t_k^j]$, $2 \le j \le s$ to orthogonalize $\{Av_{k+1}^1, \ldots, A^{s-1}v_{k+1}^1\}$against $\overline{W}_k$
which gives $v_{k+1}^j = A^{j-1}v_{k+1}^1 - \overline{V}_k t_k^j$.
Select $[\overline{t}_k^j]$, $2 \le j \le s$ to orthogonalize $\{Aw_{k+1}^1, \ldots, A^{s-1}w_{k+1}^1\}$ against $\overline{V}_k$
which gives $w_{k+1}^j = (A^T)^{j-1}w_{k+1}^1 - \overline{W}_k \overline{t}_k^j$ for $j = 2, \ldots, s$.

EndFor

Let A be a $n \times n$ nonsymmetric matrix and $V_k = \{\overline{V}_1, \overline{V}_2, \ldots, \overline{V}_k\}$, $W_k = \{\overline{W}_1, \overline{W}_2, \ldots, \overline{W}_k\}$. In the s-step biorthogonal Lanczos method, the Ritz values of A in V_k are the eigenvalues λ_k of $\overline{T}_k$.

4 Analysis of the *s*-Step Biorthogonal Lanczos Method

The test problem was derived from the five-point discretization of the following partial differential equation:

$$-(bu_x)_x - (cu_x)_x + (du)_x + (eu)_y + fu = g$$

on the unit square, where

$$b(x,y) = e^{-xy}, \qquad c(x,y) = e^{xy}$$

$$d(x,y) = \beta(x+y), \qquad e(x,y) = \gamma(x+y)$$

$$f(x,y) = \frac{1}{1+x+y}$$

subject to the Dirichlet boundary conditions u =0 on the boundary. In the test we took γ =50.0, β = 1.0, which yielded a nonsymmetric matrix. Table 1 shows that matrices generated by the standard and the s-step biorthogonal Lanczos method have the same largest eigenvalues, but $s > 5$ in the s-step method loss of accuracy for eigenvalues has been observed. We tested the methods on the problem of size N=4096. We also compared to the largest eigenvalues computed by the standard, the s-step method with those by Saad's program code* using the Arnoldi method. Saad's code [6] uses reorthogonalization and deflated iteration to compute eigenvalues. The stopping criterion is $\varepsilon = 10^{-6}$ in Saad's program. In the standard and the s-step biorthogonal Lanczos methods, we find the largest eigenvalues after a reduced matrix of a certain size is generated, so these biorthogonal Lanczos methods require minimal storage and time.

Table 1. Largest eigenvalues using the biorthogonal Lanczos methods

T_j	Code*	standard	2-step
10×10	0.10204000×10^2	0.98652673×10^2	0.98652673×10^2
20×20	0.10204000×10^2	0.10202484×10^2	0.10202484×10^2
30×30	0.10203900×10^2	0.10204000×10^2	0.10204000×10^2
40×40	0.10203872×10^2	0.10204000×10^2	0.10204000×10^2

T_j	3-step	4-step	5-step	6-step
10×10	-	-	0.98652673×10^2	-
20×20	-	0.10202484×10^2	0.10202491×10^2	-
30×30	0.10204000×10^2	-	0.10202016×10^2	0.10197283×10^2
40×40	-	0.10204000×10^2	0.10202015×10^2	-

During s iterations of the standard biorthogonal Lanczos algorithm, $2s$ inner products should be separately repeated at $2s$ times. Thus on message passing system, lots of times are needed for global communication, which in turn reduces the efficiency of the parallel system. However, the s-step parallel algorithm in this paper can perform all the needed inner products at once during s iterations of the corresponding standard method, the communication time necessary for inner products is reduced by a factor of $1/(2s)$ compared to the standard method. All the inner products needed for s iterations of the standard method are performed at the same time in the s-step method, so that only one global communication is required and the communication cost is decreased on a message passing system. The s-step biorthogonal Lanczos method needs $s+1$ times of matrix-vector operations compared to s times in the standard methods. We compare the computational work and the data communications of the s-step biorthogonal Lanczos method to the standard one. Table 2 shows the number of vector operations and the data communications during a single iteration of the s-step biorthogonal Lanczos algorithm and s iterations of the corresponding standard biorthogonal Lanczos method. As shown in Table 2, the communication cost can be greatly decreased by introducing more effective algorithm in parallel process.

Table 2. Comparison of the numbers of the vector operations and the data communications : s iterations of the standard method and 1 iteration of the s-step biorthogonal Lanczos method.

	standard Biorthogonal Lanczos	s-step Biorthogonal Lanczos
Dot products	$2s$	$2s$
Vector updates	$5s$	$2s(s+1)$
Matrix×vector	s	$s+1$
global communication	$2s$	1
local communication	s	1

5 Conclusion

Parallel processing systems equipped with from a few to many processors are now being used in many areas. As the number of the processors involved in the parallel system is increased, the relative importance of the communication cost grows. In this paper, we proposed the s-step biorthogonal Lanczos method suitable to reduce the communication cost. The s-step biorthogonal Lanczos method generates reduction matrices which are similar to reduction matrices generated by the standard biorthogonal Lanczos method, therefore the s-step method has the same eigenvalues as the standard method. The s-step biorthogonal Lanczos method algorithm showed similar convergence properties as the standard method, but s-step method is more effective in the parallel system because a large amount of the inner products can be done at once. This process can reduce the data communication time in a message passing system. The s-step method has the better performance compared to the standard method in Message Passing parallel computer.

Acknowledgments

This work was supported by Daegu University Grant.

References

1. Golub, G.H., Van Loan, C.F.: MATRIX Computations. Johns Hopkins University Press (1996)
2. Kurzak, J., Alvaro, W., Dongarra, J.: Optimizing matrix multiplication for a short-vector SIMD architevture – CELL processor. Parallel Computing 35 (2009)
3. Kim, S.K., Chronopoulos, A.T.: A Class of Lanczos Algorithms Implemented on Parallel Computers. Parallel Computing 17, 763–778 (1991)
4. Parlett, B.N., Taylor, D.R., Liu, Z.A.: A lookahead algorithm for unsymmetric matrices. Math. Comp. 44, 105–124 (1985)
5. Saad, Y.: The Lanczos biorthogonalization algorithm and other oblique projection methods for solving large unsymmetric systems. SIAM J. Numer. Anal. 19, 485–506 (1982)
6. Saad, Y.: Partial eigensolutions of large nonsymmetric matrices, Research Report YALUE/DCS/RR-397, Yale Univ. (1985)
7. Sim, L.C., Leedham, G., Jian, L.C., Schroder, H.: Fast solution of large N × N matrix equations in an MIMD-SIMD Hybrid System. Parallel Computing 29, 1669–1684 (2003)

Author Index

GPSR Compliance

The European Union's (EU) General Product Safety Regulation (GPSR) is a set of rules that requires consumer products to be safe and our obligations to ensure this.

If you have any concerns about our products, you can contact us on ProductSafety@springernature.com

In case Publisher is established outside the EU, the EU authorized representative is:

Springer Nature Customer Service Center GmbH
Europaplatz 3
69115 Heidelberg, Germany

Batch number: 09485837

Printed by Printforce, the Netherlands